Reeds Astro Navigation Tables

Lt Cdr Harry J Baker
RD RNR MRIN

2011 EDITION

ADLARD COLES NAUTICAL
London

ACKNOWLEDGEMENT

This product is derived in part from material obtained from the UK Hydrographic Office with the permission of the UK Hydrographic Office and Her Majesty's Stationery Office.

© British Crown Copyright, 2011. All rights reserved.

NOTICE: The UK Hydrographic Office (UKHO) and its licensors make no warranties or representations, express or implied, with respect to this product. The UKHO and its licensors have not verified the information within this product or quality assured it.

Astronomical data generated by HM Nautical Almanac Office

2011 edition published by Adlard Coles Nautical
an imprint of A & C Black Publishers Ltd
36 Soho Square, London W1D 3QY
www.adlardcoles.com

Astronomical data copyright © Council for the Central Laboratory
of the Research Councils 2010
Explanatory text copyright © Harry Baker 2010

ISBN 978 1408 1 2336 2

Note: While all reasonable care has been taken in the publication of this book, the publisher takes no responsibility for the use of the methods or products described in the book.

CONTENTS

Editor's Notes for the Cruising Yachtsman

GENERAL

You might call this the common sense almanac. It has been devised and produced by practical ocean navigators, and has these main advantages:

- All the information which you need in order to navigate by the heavenly bodies – Sun, Moon, Planets, Stars – is contained herein.
- The slim volume is designed to open flat on your chart table.

We have not attempted to make this a text book on astro-navigation, and we have excluded data which are not relevant to the task in hand, i.e. finding out where you are.

Nowadays nearly all ocean-going yachts are fitted with GPS. If you also carry a sextant (and know how to use it) plus our little volume you will be able to navigate confidently by the heavenly bodies; this will not only add to your safety but will give you great personal satisfaction. Perhaps one day you will achieve the ultimate observed position with simultaneous sights of Sun, Moon and Venus!

Of course there are other almanacs, and there are also dedicated navigational computers which carry ephemerides for some or all of the heavenly bodies. The latter are expensive and, by their nature, vulnerable. We believe that the cruising yachtsman should carry his almanac in indestructible book form and be able to reduce his sights without the need to refer to any data outside his almanac. We also believe that the method of sight reduction must be in simple, easily-explained, steps and must be the same for all bodies.

People who are learning astro-navigation have asked us how they can practise ashore with their sextants and with the tables; so we have included a section (page 6) which tells them how to go about it.

All navigational working is done using Greenwich Mean Time (GMT), which is the same as Universal Time (UT). GMT is therefore used in all examples and no reference is made to other definitions of time.

These descriptions, in accordance with traditional marine practice, are used for sextant values:

Sextant altitude – as read from the sextant

Observed altitude – SA corrected for index error

Apparent altitude – OA corrected for dip

True altitude – AA corrected for all other influences.

LAYOUT

The almanac is in five different parts, although it is thought better to adhere to simple page numbering, as shown on the contents page.

PART ONE is this section, the editor's notes, including a full description of the volume.

PART TWO starts with a comprehensive set of examples for all the normal use of the almanac by the practical navigator. The intention has been to group the examples so that they follow each other naturally. It is recommended that you should work through them in comfort ashore so that you can find your way around the almanac without difficulty at sea. The section concludes with supplementary information which may be useful.

PART THREE is the ephemerides. It consists of thirty-seven monthly pages – three for each month plus one for the Pole Star. The first of the monthly pages is devoted to Greenwich Hour Angle and Declination of the Sun, and Greenwich Hour Angle of the first point of Aries. Be careful not to confuse them – it is a fairly common error! On the next monthly page GHA and DEC of the Moon and the four navigational Planets are tabulated. The third and final monthly page carries all the necessary Star information – Magnitude, Transit, Declination and Sidereal Hour Angle for 60 selected Stars. Alongside the Stars are astronomical and complementary data for the Sun and Moon, thus completing all the necessary ephemeral details. This section ends on page 44 with the Pole Star table; this looks like a perennial correction table but in fact it changes annually.

PART FOUR is the necessary collection of tables to produce increments and corrections for the main tabulations in the ephemerides, and also to correct observed altitudes. Also slipped in are some notes to help you to identify and use the Planets, and some observations about the Moon's phases. There is a refraction table which is not actually used in any of the worked examples; it is there because when you practise ashore you will need this little table.

PART FIVE includes all the tables – Versines, Log Cosines and ABC, which will enable you to reduce your sights, finding calculated altitude and azimuth. These may look rather daunting at first; but after working through a few examples you will find that they come quite easily. We have resisted producing a sight form for workings because most people have their own ideas in this respect. Right at the back, on page 68, is a simple Star chart.

ALMANAC DETAILS

TO FIND THE GEOGRAPHICAL POSITION OF HEAVENLY BODIES

- Greenwich hour angle (GHA) and declination (DEC) of the SUN are tabulated at intervals of two hours.

- GHA and DEC of the MOON are tabulated at intervals of six hours.

- GHA and DEC of the four navigational PLANETS are tabulated at intervals of twenty-four hours.

- GHA of the first point of ARIES is tabulated at intervals of two hours.

- Sidereal hour angle (SHA) and DEC of the principal sixty navigational STARS are tabulated each month.

Incremental tables enable the navigator to obtain values of GHA and DEC for any given time. Note that, by their nature, these tables may not produce a precisely accurate value; but differences are trivial and of no importance for the practical yacht navigator.

TO FIND CALCULATED ALTITUDE AND AZIMUTH

The versine method is used to determine the calculated altitude (CA) of the body from the navigator's estimated position (EP).

An important advantage of the versine method is that your sight is plotted from your EP, rather than from an assumed position, thus eliminating large intercepts and the consequent risk of plotting errors.

ABC tables are used to determine azimuth (AZ).

For those navigators who prefer to use simple scientific calculators formulae are provided for obtaining CA and AZ.

TO FIND TRUE ALTITUDE

Tables are provided to correct observed altitude (OA) into true altitude (TA) for all bodies.

OTHER USEFUL DATA

- Daily pages for sunrise, sunset, twilight, moonrise, moonset.

- Alphabetical index of principal Stars.

- Eclipses in 2011.

- Meridian passage time for all bodies.

- Arc to time table.

- Limited use of the almanac in 2012.

EXAMPLES

Examples are included for all workings, using the 2011 almanac data.

OTHER RECOMMENDED BOOKS

Reeds Sextant Simplified by Dag Pike, published by Adlard Coles Nautical. ISBN 0 7136 6705 2.

This is a comprehensive description of the sextant and how to use it.

Celestial Navigation For Yachtsmen by Mary Blewitt, published by Adlard Coles Nautical. ISBN 0 7136 6422 3.

This is the classic primer, including a simple and elegant description of astro-navigation theory.

ERRORS AND OMISSIONS

We have carefully checked the text and tables of this volume. However, neither the publisher nor the editor can accept any responsibility for the consequences of any errors or omissions which might have occurred.

FINALLY

2011 is the fifteenth year of publication of these tables.

Our thanks are due to Mr George Whitchurch for his help in checking the calculations.

We shall be happy to receive comments and suggestions from teachers, students and users of astro-navigation.

Don t forget your 2012 almanac. It will be published in September, 2011. If you cannot obtain a copy from your normal source, contact:

Adlard Coles Nautical
36 Soho Square
London W1D 3QY

www.adlardcoles.com

Harry Baker, *Editor*

Examples

In some tables interpolation by eye is necessary.

The first four examples guide you through the process of finding GHA and DEC for all the heavenly bodies; these values are required for the sight reduction process. Then there are examples for applying corrections to observed altitude, followed by establishing meridian passage times for all bodies, finding times of sunrise, moonrise etc. and use of Polaris to find latitude. Finally, there are examples on the use of the tables for establishing CA and AZ in the sight reduction process.

To find GHA and DEC of the SUN on Monday, 10th January at 11h 23m 13s

Tabulated values are read off from the first monthly page. Aries is included in the same table so be careful not to confuse. GHA increments are found in the table on page 47. DEC can be interpolated by eye.

Tabulated value at 10h *monthly page*	328° 09.3
Increment for lh 23m *page 47*	20° 45.0
Increment for 13s *page 47*	03.3
Required GHA	348° 57.6
DEC *monthly page, by inspection*	S 21° 58.1

When interpolating by eye for DEC take care to observe if it is increasing or decreasing.

To find GHA and DEC of the MOON on Friday, 14th January at 9h 18m 41s

Tabulated values are read off from the second monthly page. At the same time note mean variation per hour for both GHA and DEC. The mean variation per hour for GHA is always 14 degrees plus a number of minutes of arc; but only the minutes are shown in the column. For GHA enter table on page 51 with arguments incremental whole hours and mean variation per hour; then go to the minutes table on page 52 to complete the increments. For DEC simply multiply whole hours by mean variation per hour and then obtain the increment for minutes from the table on page 53.

GHA mean variation per hour *monthly page*	14° 30.0
GHA tabulated value at 6h *monthly page*	163° 35.7
Increment for 3h *page 51*	43° 30.0
Increment for 18m *page 52*	4° 21.0

Increment for 41s *page 52*	09.9
Required GHA	211° 36.6
DEC mean variation per hour *monthly page*	07.1
DEC tabulated value at 6h *monthly page*	N19° 40.6
Increment for 3h *mental arithmetic*	21.3
Increment for 19m *page 53*	02.2
Required DEC	N20° 04.1

When extracting mean variation per hour for DEC take care to check if DEC is increasing or decreasing.

To find GHA and DEC of VENUS on Friday, 11th March at 14h 28m 19s

Tabulated values are read off from the second monthly page. At the same time note mean variation per hour for both GHA and DEC. For GHA enter table on page 48 with arguments incremental whole hours and mean variation per hour; then go to the minutes table on page 49 to complete the increments.

For DEC enter the single table on page 50 with arguments total incremental time and mean variation per hour.

GHA mean variation per hour *monthly page*	14° 59.4
Tabulated value at 0h *monthly page*	215° 24.1
Increment for 14h *page 48*	209° 51.6
Increment for 28m *page 49*	6° 59.7
Increment for 19s *page 49*	04.8
Required GHA	432° 20.2 −360° 72° 20.2
DEC mean variation per hour *monthly page*	−0.7
Tabulated value at 0h *monthly page*	S17° 20.6
Increment for 14h 28m *page 50*	−10.1
Required DEC	S17° 10.5

When extracting mean variation per hour for DEC take care to check if DEC is increasing or decreasing.

To find GHA and DEC of SIRIUS on Friday, 15th April at 17h 26m 14s

To find GHA of the Star it is necessary first to establish GHA Aries and then add sidereal hour angle (SHA) of the Star. Tabulated value for GHA Aries is read off from the first monthly page; the tabulation is in the same table as the Sun so be careful not to confuse. The table on page 47, right-hand side of page, is entered to establish the increments. SHA of the Star is then read off from the third monthly page, and DEC is read off at the same time. Neither SHA nor DEC change appreciably during the month so they are applied as read.

Tabulated value GHA ARIES at 16h *monthly page*	83°	28.1
Increment for 1h 26m *page 47*	21°	33.6
Increment for 14s *page 47*		03.5
GHA ARIES	105°	05.2
SHA SIRIUS *monthly page*	258°	35.2
Required GHA	363°	40.4
	–360°	
	3°	40.4
Required DEC *monthly page*	S 16°	44.2

To find TRUE ALTITUDE of the SUN

On Monday, 5th December the observed altitude (OA) of the Sun's lower limb was 19° 24'.8. Height of eye (HE) was 8 feet.

Enter table on page 45 (left hand side) with arguments OA and HE. Interpolate by eye. Remember to apply the monthly correction, found at bottom of the table. The total correction is always additive.

OA	19°	24.8
Main correction *page 45*		+10.7
Correction for month *at foot of page*		+00.3
TA	19°	35.8

Note that if the upper limb is observed (this is unusual) you must go to the monthly page to find the semi-diameter. Deduct twice this value from the OA and then use the table as described above.

To find TRUE ALTITUDE of PLANET or STAR

OA of body 37° 28.4'. HE 10 feet.

Enter the table on page 45, right hand side, with arguments OA and HE. Interpolate by eye. The correction is always subtractive.

OA	37°	28.4
Total correction *page 45*		–04.3
TA	37°	24.1

To find TRUE ALTITUDE of MOON

On Tuesday, 16th August the OA of the Moon's upper limb was 51° 14.6'. HE 10 feet.

Table on page 46 is entered with arguments OA and horizontal parallax (HP). HP is read from the Moon data on the monthly page; the value is for 12h and interpolation may be necessary. Note that when using the left hand main table, i.e. upper limb, the sign changes from plus to minus when the OA is greater than 62°. Finally a HE correction, found at the foot of the table, must be made.

HP *monthly page*	54.4
OA	51° 14.6
Main correction *page 46, interpolation by eye*	+08.9
HE correction *at foot of table*	+06.7
TA	51° 30.2

To find the time of MERIDIAN PASSAGE of the SUN on Saturday, 19th November in longitude 54° 15' W

For all practical purposes the GMT of transit of Sun and Planets can be found simply by converting vessel's longitude to time, using the arc to time table. One degree of longitude equals four minutes of time. In West longitudes transit occurs after Greenwich transit, in East longitudes before.

Read off from the monthly page the GMT of transit at Greenwich; then convert arc (your longitude) to time from the table on page 53.

GMT of transit at Greenwich *monthly page*	11h 45m
Arc to time 54° *page 53*	+3 36
Arc to time 15' *page 53*	+01
GMT of transit in 54° 15' W	15h 22m

To find the time of MERIDIAN PASSAGE of JUPITER on Saturday, 8th October in longitude 33° 19'W

The method is similar to that for the Sun. Read off from the second monthly page the GMT of transit at Greenwich then do the arc to time conversion from the table on page 53.

GMT of transit at Greenwich *monthly page*	01h 19m
Arc to time 33° *page 53*	+2 12
Arc to time 19' *page 53, nearest minute*	+01
GMT of transit in 33° 19 'W	03h 32m

To find the time of MERIDIAN PASSAGE of Star ARCTURUS on Thursday, 8th December in longitude 84° 55' E

On the monthly Star pages the time of the meridian passage of the star at Greenwich is given once only, and this is for the first day of the month. The star transits approximately four minutes earlier each day. Thus you must first calculate the Greenwich transit time for the day in question by deducting four minutes for each completed day, and then apply the arc to time adjustment.

Transit at Greenwich on 1st December (monthly page) is at 09h 37m. Deduct 28m for seven completed days; so the time of transit at Greenwich on 8th December is 09h 09m.

GMT of transit at Greenwich *monthly page less correction*	9h 09m
Arc to time 84° *page 53*	–5 36
Arc to time 55' *page 53, nearest minute*	–04
GMT of transit in 84° 55'E	3h 29m

To find the time of MERIDIAN PASSAGE of the MOON on Monday, 25th April in longitude 64° W

The monthly page shows the time of meridian passage at Greenwich. The basic calculation of the GMT of meridian passage in your longitude is the same as for Sun and Planets; but a further correction must be applied because of the Moon's continuous movement around the earth. This correction is derived from a value called Diff. in minutes of time, and is used, together with longitude, to enter the table on page 54. Diff. changes daily and is found on the third monthly page.

GMT of transit at Greenwich *monthly page, also note Diff. 45m*	06h 20m
Arc to time 64 *page 53*	+4 16
Diff./longitude correction *page 54*	+08
GMT of transit	10h 44m

To find times of SUNRISE, SUNSET and CIVIL TWILIGHT

These times are tabulated on the third monthly page for the Greenwich meridian and latitude 52° N. To obtain the times for other latitudes you must enter the correction table which is alongside. Note that the correction table is accurate for the 15th of each month; thus interpolation is sometimes necessary.

To find GMT of SUNRISE on Sunday, 15th May in 30° N 42° E

GMT of sunrise at Greenwich *monthly page*	4h 07m
Latitude correction 30° N *table alongside*	+1 00
Arc to time 42° E *page 53*	–2 48
GMT of sunrise	2h 19m

Supposing you wish to find sunrise on 31st May in the same latitude. You must note the correction for the middle of May, i.e. +1h 00m. Then look up the correction for mid June, which is +1h 19m. Since end May lies midway between these two values the accurate correction, by interpolation, is +1h 10m.

Exactly the same working is used for sunset, morning civil twilight and evening civil twilight.

To find times of MOONRISE and MOONSET

The yachtsman is unlikely to need the precise times of moonrise and moonset. The phase of the Moon will give him a rough idea (see notes on page 51). The monthly pages give the exact times for the Greenwich meridian in latitude 52° N. By applying longitude (arc to time) a rough idea can be obtained in moderate latitudes, although this can be an hour or more out depending on the position of the observer and the declination of the Moon.

To find approximate GMT of MOONRISE on Monday, 23rd May in position 40° N 20° E.

GMT of moonrise at Greenwich *monthly page*	00h 03m
Arc to time 20° E *Page 53*	–1 20
Approx. GMT of moonrise	22h 43m*

*On Sunday 22nd May

To find LATITUDE from sextant observation of POLARIS

Obtain LHA ARIES and enter table, page 44, to obtain value Q. Apply this value to your TA to establish your latitude.

To find latitude by POLARIS on Saturday, 22nd October at 18h 49m. Estimated position 24° 30'N 22° 15''W. OA 24° 44.0''. HE 7 feet.

GHA ARIES at 18h *monthly page*	300°	49.3
Increment for 49m *page 47*	12°	17.0
Longitude W	–22°	15.0
LHA ARIES at 18h 49m	290°	51.3
OA	24°	44.0
Altitude total correction *page 45*		–04.5
TA	24°	39.5
Q correction *page 44*		+15.0
LAT	24°	54.5

Note that the Pole Star table changes annually. It is therefore part of the ephemerides rather than the correction tables.

Using the VERSINE formula to find CALCULATED ALTITUDE

The formula is used for all bodies.

Vers CZD = Vers LHA x Cos LAT x Cos DEC + Vers (LAT ± DEC)

In the brackets, LAT and DEC are added if different names, i.e. one is N and the other is S, and subtracted if same names.

CZD is calculated zenith distance. Calculated altitude (CA) is 90° – CZD.

Solution requires tables of natural versines and log versines. These are combined in the same table; so take care to differentiate. A table of log cosines is also provided; so the arithmetic required is only addition.

The following example shows the process. You will probably wish to produce your own blank pro formas in order to facilitate the work.

LAT 25° 30.0'N. DEC of body S 15° 25.0'. LHA of body 335° 24.0'.

1. Log vers LHA   8.9579
   When degrees are printed at the bottom minutes are read (upwards) on the right. Take care to enter the units, i.e. 8.9579 not 9579. Units are not printed in every column (for space considerations) so be careful when units are changing, e.g. from 8 to 9.

2. Log cos LAT   9.9555
   Again, units are not printed in every column. Ignore the integrated table of log sines, with degrees on the bottom and minutes on the right.

3. Log cos DEC   9.9841

4. Add 1 + 2 + 3   28.8975

5. Discard ten's digit   8.8975

6. Natural versine of 5   0.0789
   By inspection of versine table. Interpolation may be necessary. Again, units are not printed in every column.

7. LAT + DEC   40° 55.0'
   Adding LAT and DEC because different names.

8. Natural versine of 7   0.2443

9. Add 6 + 8   0.3232

10. CZD   47° 24.5
    By inspection of versine table.

11. CA is 90° – CZD =   42° 35.5

Using the ABC tables to find AZIMUTH

The tables are arranged so that A + B = C. C is the azimuth, and is in quadrantal notation. Table A is entered using arguments LHA and LAT. Table B is entered using arguments LHA and DEC. Table C is entered using arguments A + B and LAT.

This example shows the process, and is for the same sight as the previous (versine) example.

Precision in interpolation is not necessary because plus or minus one degree of bearing in plotting is not critical.

LAT 25° 30.0'N. DEC of body S 15° 25.0'. LHA of body 335° 24.0'.

1. From table A   +1.040
   LHA in column, LAT in row. Labelled + because of note at top of table. Interpolation necessary.

2. From table B   +0.665
   LHA in column, DEC in row. Labelled + because of note at top of table. Interpolation necessary.

3. Add A+ B   +1.705
   In cases where this is a minus quantity ignore sign; but the sign is used in step 5.

4. From table C, AZ is   33°
   A + B in row, LAT in column.

The AZ is in quadrantal notation, which is expressed from North to East or West or from South to East or West. The method of applying the AZ in order to obtain the true bearing is explained at the bottom of table C.

5. The value of A + B was positive; so the AZ is named South.

6. LHA is greater than 180° so AZ is named South towards East.

7. AZ is 33° so true bearing of body is 180° – 33° = 147°.

In practice you know that the observed body was in the South-East so you can apply the AZ without resorting to this working. It is only really necessary if the body bears almost East or almost West.

PRACTISING ASHORE

It can be very useful to practise ashore. If you are new to astro-navigation it gives you the opportunity to become familiar with your sextant and with the routine of recording altitudes and times. It will also help you to become adept at using the tables and perhaps to develop your own pro forma for the sight reduction process.

What you have to do is create an artificial horizon. This is easy. Choose a calm day or night and put a large bucket or bowl of water on the ground. The surface of the water provides the artificial horizon because it is at a tangent to the curvature of the earth. Windy weather is no good because the water has to be completely smooth so that it can act as a mirror.

To make a start, the Moon when full or nearly full is the best object. There is usually less wind at night, and the Moon can be picked up quite easily without dazzle. The Sun is also an easy object on a calm day; but you must remember to use your sextant shades, as you would at sea, in order to avoid risk of eye damage. It is a good deal more difficult, but still feasible, to isolate the reflections of bright Planets and Stars.

To take the altitude, stand so that you can see the reflection of the body on the water. With your sextant bring the body down (as you would at sea) and superimpose it on the water reflection. You are using the sextant in the normal way except that instead of bringing the body down until it kisses the horizon you must bring it down so that it *covers* the water reflection of the body.

Just as you would at sea, take several observations and then calculate average times and sextant altitudes. To obtain observed altitude we first apply your sextant's index error and then halve the result. Here is an example:

Sextant altitude	104° 23.8
Index error	−6.4
	104° 17.4
Divide by 2 =	52° 8.7

which is observed altitude.

You will not be able to observe altitudes much above 55° because your sextant arc covers only about 110°. It is difficult to handle low altitudes so the best range is about 15° to 55°.

The sight reduction process is standard; but in finding true altitudes we have to vary the method because the only corrections required are for refraction (all bodies) and, for the Moon, horizontal parallax.

For Sun, Planet and Star go to the Mean Refraction table on page 46 and enter it with apparent altitude which, for our purposes here, is observed altitude. The correction is always subtractive, and this is the only correction required.

The Moon is a little more complicated. From the daily page make a note of horizontal parallax; the value is for 1200 GMT so you will probably need to interpolate. Then go to the Moon Altitude Total Correction table on page 46. There is a section for the upper limb and another for the lower limb. Enter each table with arguments observed altitude and horizontal parallax. Take the mean of these two values and add 9.8 minutes. The result is the total and only correction. Ignore the height of eye correction at the bottom of the table.

Here's wishing you very small intercepts!

Compass Checking by AMPLITUDE

The integrated table of LOG COSINES and LOG SINES makes it easy to calculate amplitudes by using the formula: SIN AMPLITUDE = SIN DEC (of the observed body) ÷ COS LAT (of the observer).

Note that SINES are read at the bottom of the table, with minutes read upwards from the right.

This example shows the method: LAT 46° 20'N, Sunset, Sun's DEC N 9° 30'.

1.	Log sin 9° 30'	9.2176
2.	Add 10	19.2176
3.	Log cosine 46° 20'	9.8391
4.	2–3	9.3785

This is the log sin of the amplitude. Inspect the table and read off a value of 13° 50. The amplitude is converted to a true bearing like this: the bearing is named E for a rising body and W for a setting body; the bearing is named N or S according to the body's DEC. The true bearing of the body is therefore W 13° 50'N or, more familiarly, 283° 50'.

This is the bearing your compass would show if there were no compass error. You can now apply known variation and therefore identify the deviation for your present heading.

Formulae for SCIENTIFIC CALCULATORS

Simple formulae exist for finding CA and AZ using a scientific calculator.

Sin CA = Cos LAT x Cos DEC x Cos LHA ± (Sin LAT x Sin DEC)

The final bracketed value is added if LAT and DEC are same names, subtracted if LAT and DEC are different names.

Sin AZ = (Sin LHA x Cos DEC) ÷ Cos CA

Azimuth is in quadrantal notation.

JANUARY 2011

SUN AND ARIES

The page contains four large astronomical data tables ("SUN AND ARIES") giving GHA (Greenwich Hour Angle) and Dec. (Declination) of the Sun, and GHA of Aries, tabulated against GMT (in hours: 00–22, even hours) for each day in January 2011.

Left panel — Saturday, 1st January to Wednesday, 5th January

GMT	SUN GHA ° ´	Dec. ° ´	ARIES GHA ° ´
	Saturday, 1st January		
00	179 12.1	S23 02.7	100 18.2
02	209 11.5	23 02.3	130 23.1
04	239 10.9	23 01.9	160 28.1
06	269 10.3	23 01.5	190 33.0
08	299 09.8	23 01.1	220 37.9
10	329 09.2	23 00.7	250 42.8
12	359 08.6	23 00.3	280 47.8
14	29 08.0	22 59.9	310 52.7
16	59 07.4	22 59.5	340 57.6
18	89 06.8	22 59.1	11 02.6
20	119 06.2	22 58.7	41 07.5
22	149 05.6	S22 58.2	71 12.4
	Sunday, 2nd January		
00	179 05.0	S22 57.8	101 17.3
02	209 04.4	22 57.4	131 22.3
04	239 03.9	22 57.0	161 27.2
06	269 03.3	22 56.5	191 32.1
08	299 02.7	22 56.1	221 37.1
10	329 02.1	22 55.6	251 42.0
12	359 01.5	22 55.2	281 46.9
14	29 01.0	22 54.7	311 51.8
16	59 00.4	22 54.3	341 56.8
18	88 59.8	22 53.8	12 01.7
20	118 59.2	22 53.4	42 06.6
22	148 58.6	S22 52.9	72 11.6
	Monday, 3rd January		
00	178 58.0	S22 52.4	102 16.5
02	208 57.4	22 52.0	132 21.4
04	238 56.8	22 51.5	162 26.3
06	268 56.3	22 51.0	192 31.3
08	298 55.7	22 50.6	222 36.2
10	328 55.1	22 50.1	252 41.1
12	358 54.5	22 49.6	282 46.1
14	28 54.0	22 49.1	312 51.0
16	58 53.4	22 48.6	342 55.9
18	88 52.8	22 48.1	13 00.8
20	118 52.2	22 47.6	43 05.8
22	148 51.6	S22 47.1	73 10.7
	Tuesday, 4th January		
00	178 51.1	S22 46.6	103 15.6
02	208 50.5	22 46.1	133 20.6
04	238 49.9	22 45.6	163 25.5
06	268 49.4	22 45.1	193 30.4
08	298 48.8	22 44.6	223 35.3
10	328 48.2	22 44.1	253 40.3
12	358 47.7	22 43.5	283 45.2
14	28 47.1	22 43.0	313 50.1
16	58 46.5	22 42.5	343 55.1
18	88 45.9	22 41.9	14 00.0
20	118 45.4	22 41.4	44 04.9
22	148 44.8	S22 40.9	74 09.8
	Wednesday, 5th January		
00	178 44.2	S22 40.3	104 14.8
02	208 43.7	22 39.8	134 19.7
04	238 43.1	22 39.2	164 24.6
06	268 42.5	22 38.7	194 29.6
08	298 42.0	22 38.1	224 34.5
10	328 41.4	22 37.6	254 39.4
12	358 40.9	22 37.0	284 44.3
14	28 40.3	22 36.5	314 49.3
16	58 39.7	22 35.9	344 54.2
18	88 39.2	22 35.3	14 59.1
20	118 38.6	22 34.8	45 04.0
22	148 38.1	S22 34.2	75 09.0

Second panel — Thursday, 6th January to Monday, 10th January

GMT	SUN GHA ° ´	Dec. ° ´	ARIES GHA ° ´
	Thursday, 6th January		
00	178 37.5	S22 33.6	105 13.9
02	208 37.0	22 33.1	135 18.8
04	238 36.4	22 32.4	165 23.8
06	268 35.8	22 31.8	195 28.7
08	298 35.3	22 31.3	225 33.6
10	328 34.7	22 30.7	255 38.5
12	358 34.2	22 30.1	285 43.5
14	28 33.6	22 29.5	315 48.4
16	58 33.1	22 28.9	345 53.3
18	88 32.5	22 28.3	15 58.3
20	118 32.0	22 27.7	46 03.2
22	148 31.4	S22 27.0	76 08.1
	Friday, 7th January		
00	178 30.9	S22 26.4	106 13.0
02	208 30.3	22 25.8	136 18.0
04	238 29.8	22 25.2	166 22.9
06	268 29.3	22 24.6	196 27.8
08	298 28.7	22 23.9	226 32.8
10	328 28.2	22 23.3	256 37.7
12	358 27.6	22 22.7	286 42.6
14	28 27.1	22 22.0	316 47.5
16	58 26.6	22 21.4	346 52.5
18	88 26.0	22 20.8	16 57.4
20	118 25.5	22 20.1	47 02.3
22	148 24.9	S22 19.5	77 07.3
	Saturday, 8th January		
00	178 24.4	S22 18.8	107 12.2
02	208 23.9	22 18.2	137 17.1
04	238 23.3	22 17.5	167 22.0
06	268 22.8	22 16.8	197 27.0
08	298 22.3	22 16.2	227 31.9
10	328 21.7	22 15.5	257 36.8
12	358 21.2	22 14.8	287 41.8
14	28 20.7	22 14.2	317 46.7
16	58 20.1	22 13.5	347 51.6
18	88 19.6	22 12.8	17 56.5
20	118 19.1	22 12.1	48 01.5
22	148 18.6	S22 11.4	78 06.4
	Sunday, 9th January		
00	178 18.0	S22 10.7	108 11.3
02	208 17.5	22 10.1	138 16.3
04	238 17.0	22 09.4	168 21.2
06	268 16.5	22 08.7	198 26.1
08	298 15.9	22 08.0	228 31.0
10	328 15.4	22 07.3	258 36.0
12	358 14.9	22 06.6	288 40.9
14	28 14.4	22 05.8	318 45.8
16	58 13.9	22 05.1	348 50.7
18	88 13.4	22 04.4	18 55.7
20	118 12.8	22 03.7	49 00.6
22	148 12.3	S22 03.0	79 05.5
	Monday, 10th January		
00	178 11.8	S22 02.3	109 10.5
02	208 11.3	22 01.5	139 15.4
04	238 10.8	22 00.8	169 20.3
06	268 10.3	22 00.1	199 25.2
08	298 09.8	21 59.3	229 30.2
10	328 09.3	21 58.6	259 35.1
12	358 08.7	21 57.8	289 40.0
14	28 08.2	21 57.1	319 45.0
16	58 07.7	21 56.4	349 49.9
18	88 07.2	21 55.6	19 54.8
20	118 06.7	21 54.9	49 59.7
22	148 06.2	S21 54.1	80 04.7

Third panel — Tuesday, 11th January to Saturday, 15th January

GMT	SUN GHA ° ´	Dec. ° ´	ARIES GHA ° ´
	Tuesday, 11th January		
00	178 05.7	S21 53.3	110 09.6
02	208 05.2	21 52.6	140 14.5
04	238 04.7	21 51.8	170 19.5
06	268 04.2	21 51.0	200 24.4
08	298 03.7	21 50.3	230 29.3
10	328 03.2	21 49.5	260 34.2
12	358 02.7	21 48.7	290 39.2
14	28 02.2	21 47.9	320 44.1
16	58 01.7	21 47.2	350 49.0
18	88 01.3	21 46.4	20 54.0
20	118 00.8	21 45.6	50 58.9
22	148 00.3	S21 44.8	81 03.8
	Wednesday, 12th January		
00	177 59.8	S21 44.0	111 08.7
02	207 59.3	21 43.2	141 13.7
04	237 58.8	21 42.4	171 18.6
06	267 58.3	21 41.6	201 23.5
08	297 57.8	21 40.8	231 28.4
10	327 57.4	21 40.0	261 33.4
12	357 56.9	21 39.2	291 38.3
14	27 56.4	21 38.3	321 43.2
16	57 55.9	21 37.5	351 48.2
18	87 55.4	21 36.7	21 53.1
20	117 54.9	21 35.9	51 58.0
22	147 54.5	S21 35.0	82 02.9
	Thursday, 13th January		
00	177 54.0	S21 34.2	112 07.9
02	207 53.5	21 33.4	142 12.8
04	237 53.0	21 32.6	172 17.7
06	267 52.6	21 31.7	202 22.7
08	297 52.1	21 30.9	232 27.6
10	327 51.6	21 30.0	262 32.5
12	357 51.2	21 29.2	292 37.4
14	27 50.7	21 28.3	322 42.4
16	57 50.2	21 27.5	352 47.3
18	87 49.8	21 26.6	22 52.2
20	117 49.3	21 25.8	52 57.2
22	147 48.8	S21 24.9	83 02.1
	Friday, 14th January		
00	177 48.4	S21 24.0	113 07.0
02	207 47.9	21 23.2	143 11.9
04	237 47.4	21 22.3	173 16.9
06	267 47.0	21 21.4	203 21.8
08	297 46.5	21 20.5	233 26.7
10	327 46.1	21 19.7	263 31.7
12	357 45.6	21 18.8	293 36.6
14	27 45.2	21 17.9	323 41.5
16	57 44.7	21 17.0	353 46.4
18	87 44.3	21 16.1	23 51.4
20	117 43.8	21 15.2	53 56.3
22	147 43.4	S21 14.3	84 01.2
	Saturday, 15th January		
00	177 42.9	S21 13.4	114 06.2
02	207 42.5	21 12.5	144 11.1
04	237 42.0	21 11.6	174 16.0
06	267 41.6	21 10.7	204 20.9
08	297 41.1	21 09.8	234 25.9
10	327 40.7	21 08.9	264 30.8
12	357 40.2	21 08.0	294 35.7
14	27 39.8	21 07.1	324 40.7
16	57 39.4	21 06.2	354 45.6
18	87 38.9	21 05.2	24 50.5
20	117 38.5	21 04.3	54 55.4
22	147 38.1	S21 03.4	85 00.4

Fourth panel — Sunday, 16th January to Monday, 31st January

GMT	SUN GHA ° ´	Dec. ° ´	ARIES GHA ° ´
	Sunday, 16th January		
00	177 37.6	S21 02.4	115 05.3
02	207 37.2	21 01.5	145 10.2
04	237 36.8	21 00.6	175 15.2
06	267 36.3	20 59.6	205 20.1
08	297 35.9	20 58.7	235 25.0
10	327 35.5	20 57.7	265 29.9
12	357 35.0	20 56.8	295 34.9
14	27 34.6	20 55.8	325 39.8
16	57 34.2	20 54.9	355 44.7
18	87 33.8	20 53.9	25 49.6
20	117 33.3	20 53.0	55 54.6
22	147 32.9	S20 52.0	85 59.5
	Monday, 17th January		
00	177 32.5	S20 51.0	116 04.4
02	207 32.1	20 50.1	146 09.4
04	237 31.7	20 49.1	176 14.3
06	267 31.2	20 48.1	206 19.2
08	297 30.8	20 47.1	236 24.1
10	327 30.4	20 46.2	266 29.1
12	357 30.0	20 45.2	296 34.0
14	27 29.6	20 44.2	326 38.9
16	57 29.2	20 43.2	356 43.9
18	87 28.8	20 42.2	26 48.8
20	117 28.4	20 41.2	56 53.7
22	147 28.0	S20 40.2	86 58.6
	Tuesday, 18th January		
00	177 27.6	S20 39.2	117 03.6
02	207 27.2	20 38.2	147 08.5
04	237 26.8	20 37.2	177 13.4
06	267 26.4	20 36.2	207 18.4
08	297 26.0	20 35.2	237 23.3
10	327 25.6	20 34.2	267 28.2
12	357 25.2	20 33.2	297 33.1
14	27 24.8	20 32.2	327 38.1
16	57 24.4	20 31.2	357 43.0
18	87 24.0	20 30.1	27 47.9
20	117 23.6	20 29.1	57 52.9
22	147 23.2	S20 28.1	87 57.8
	Wednesday, 19th January		
00	177 22.8	S20 27.0	118 02.7
02	207 22.4	20 26.0	148 07.6
04	237 22.0	20 25.0	178 12.6
06	267 21.6	20 23.9	208 17.5
08	297 21.3	20 22.9	238 22.4
10	327 20.9	20 21.9	268 27.4
12	357 20.5	20 20.8	298 32.3
14	27 20.1	20 19.8	328 37.2
16	57 19.7	20 18.7	358 42.1
18	87 19.3	20 17.6	28 47.1
20	117 19.0	20 16.6	58 52.0
22	147 18.6	S20 15.5	88 56.9
	Thursday, 20th January		
00	177 18.2	S20 14.5	119 01.9
02	207 17.8	20 13.4	149 06.8
04	237 17.5	20 12.3	179 11.7
06	267 17.1	20 11.3	209 16.6
08	297 16.7	20 10.2	239 21.6
10	327 16.4	20 09.1	269 26.5
12	357 16.0	20 08.0	299 31.4
14	27 15.6	20 06.9	329 36.4
16	57 15.3	20 05.9	359 41.3
18	87 14.9	20 04.8	29 46.2
20	117 14.5	20 03.7	59 51.1
22	147 14.2	S20 02.6	89 56.1
	Friday, 21st January		
00	177 13.8	S20 01.5	120 01.0
02	207 13.5	20 00.4	150 05.9
04	237 13.1	19 59.3	180 10.9
06	267 12.8	19 58.2	210 15.8
08	297 12.4	19 57.1	240 20.7
10	327 12.0	19 56.0	270 25.6
12	357 11.7	19 54.9	300 30.6
14	27 11.3	19 53.8	330 35.5
16	57 11.0	19 52.7	0 40.4
18	87 10.7	19 51.5	30 45.4
20	117 10.3	19 50.4	60 50.3
22	147 10.0	S19 49.3	90 55.2
	Saturday, 22nd January		
00	177 09.6	S19 48.2	121 00.1
02	207 09.3	19 47.1	151 05.1
04	237 08.9	19 45.9	181 10.0
06	267 08.6	19 44.8	211 14.9
08	297 08.3	19 43.7	241 19.8
10	327 07.9	19 42.5	271 24.8
12	357 07.6	19 41.4	301 29.7
14	27 07.2	19 40.2	331 34.6
16	57 06.9	19 39.1	1 39.6
18	87 06.6	19 37.9	31 44.5
20	117 06.3	19 36.8	61 49.4
22	147 05.9	S19 35.6	91 54.3
	Sunday, 23rd January		
00	177 05.6	S19 34.5	121 59.3
02	207 05.3	19 33.3	152 04.2
04	237 04.9	19 32.2	182 09.1
06	267 04.6	19 31.0	212 14.1
08	297 04.3	19 29.8	242 19.0
10	327 04.0	19 28.7	272 23.9
12	357 03.7	19 27.5	302 28.8
14	27 03.3	19 26.3	332 33.8
16	57 03.0	19 25.1	2 38.7
18	87 02.7	19 24.0	32 43.6
20	117 02.4	19 22.8	62 48.6
22	147 02.1	S19 21.6	92 53.5
	Monday, 24th January		
00	177 01.8	S19 20.4	122 58.4
02	207 01.5	19 19.2	153 03.3
04	237 01.1	19 18.0	183 08.3
06	267 00.8	19 16.8	213 13.2
08	297 00.5	19 15.6	243 18.1
10	327 00.2	19 14.5	273 23.1
12	356 59.9	19 13.2	303 28.0
14	26 59.6	19 12.0	333 32.9
16	56 59.3	19 10.8	3 37.8
18	86 59.0	19 09.6	33 42.8
20	116 58.7	19 08.4	63 47.7
22	146 58.4	S19 07.2	93 52.6
	Tuesday, 25th January		
00	176 58.1	S19 06.0	123 57.5
02	206 57.8	19 04.8	154 02.5
04	236 57.5	19 03.6	184 07.4
06	266 57.2	19 02.3	214 12.3
08	296 57.0	19 01.1	244 17.3
10	326 56.7	18 59.9	274 22.2
12	356 56.4	18 58.6	304 27.1
14	26 56.1	18 57.4	334 32.0
16	56 55.8	18 56.2	4 37.0
18	86 55.5	18 54.9	34 41.9
20	116 55.2	18 53.7	64 46.8
22	146 55.0	S18 52.5	94 51.8
	Wednesday, 26th January		
00	176 54.7	S18 51.2	124 56.7
02	206 54.4	18 50.0	155 01.6
04	236 54.1	18 48.7	185 06.5
06	266 53.9	18 47.5	215 11.5
08	296 53.6	18 46.2	245 16.4
10	326 53.3	18 45.0	275 21.3
12	356 53.0	18 43.7	305 26.3
14	26 52.8	18 42.4	335 31.2
16	56 52.5	18 41.2	5 36.1
18	86 52.2	18 39.9	35 41.0
20	116 52.0	18 38.6	65 46.0
22	146 51.7	S18 37.4	95 50.9
	Thursday, 27th January		
00	176 51.4	S18 36.1	125 55.8
02	206 51.2	18 34.8	156 00.8
04	236 50.9	18 33.5	186 05.7
06	266 50.6	18 32.3	216 10.6
08	296 50.4	18 31.0	246 15.5
10	326 50.1	18 29.7	276 20.5
12	356 49.9	18 28.4	306 25.4
14	26 49.6	18 27.1	336 30.3
16	56 49.4	18 25.8	6 35.3
18	86 49.1	18 24.5	36 40.2
20	116 48.9	18 23.2	66 45.1
22	146 48.6	S18 21.9	96 50.0
	Friday, 28th January		
00	176 48.4	S18 20.6	126 55.0
02	206 48.1	18 19.3	156 59.9
04	236 47.9	18 18.0	187 04.8
06	266 47.6	18 16.7	217 09.8
08	296 47.4	18 15.4	247 14.7
10	326 47.2	18 14.1	277 19.6
12	356 46.9	18 12.8	307 24.5
14	26 46.7	18 11.5	337 29.5
16	56 46.4	18 10.1	7 34.4
18	86 46.2	18 08.8	37 39.3
20	116 46.0	18 07.5	67 44.2
22	146 45.7	S18 06.2	97 49.2
	Saturday, 29th January		
00	176 45.5	S18 04.8	127 54.1
02	206 45.3	18 03.5	157 59.0
04	236 45.1	18 02.2	188 04.0
06	266 44.8	18 00.8	218 08.9
08	296 44.6	17 59.5	248 13.8
10	326 44.4	17 58.2	278 18.7
12	356 44.2	17 56.8	308 23.7
14	26 43.9	17 55.5	338 28.6
16	56 43.7	17 54.1	8 33.5
18	86 43.5	17 52.8	38 38.5
20	116 43.3	17 51.4	68 43.4
22	146 43.1	S17 50.1	98 48.3
	Sunday, 30th January		
00	176 42.9	S17 48.7	128 53.2
02	206 42.6	17 47.4	158 58.2
04	236 42.4	17 46.0	189 03.1
06	266 42.2	17 44.6	219 08.0
08	296 42.0	17 43.3	249 13.0
10	326 41.8	17 41.9	279 17.9
12	356 41.6	17 40.5	309 22.8
14	26 41.4	17 39.2	339 27.7
16	56 41.2	17 37.8	9 32.7
18	86 41.0	17 36.4	39 37.6
20	116 40.8	17 35.0	69 42.5
22	146 40.6	S17 33.7	99 47.5
	Monday, 31st January		
00	176 40.4	S17 32.3	10 31.8
02	206 40.2	17 30.9	40 36.7
04	236 40.0	17 29.5	70 41.7
06	266 39.8	17 28.1	100 46.6
08	296 39.6	S17 26.7	250 12.1
10	326 39.4	17 25.3	280 17.0
12	356 39.2	17 23.9	310 22.0
14	26 39.1	S17 22.5	340 26.9

MOON

Day	GMT hr	GHA ° '	Mean Var/hr 14°+	Dec ° '	Mean Var/hr '
1 Sat	0	223 57.6	26.0	S22 50.4	4.4
	6	310 34.1	26.0	S23 17.3	3.5
	12	37 10.0	25.9	S23 39.1	2.7
	18	123 45.5	25.9	S23 55.7	1.8
2 Sun	0	210 21.1	26.0	S24 07.1	0.9
	6	296 56.9	26.0	S24 13.3	0.1
	12	23 33.4	26.2	S24 14.2	0.7
	18	110 10.8	26.4	S24 10.0	1.7
3 Mon	0	196 49.5	26.7	S24 00.7	2.5
	6	283 29.7	27.0	S23 46.4	3.3
	12	10 11.8	27.3	S23 27.4	4.1
	18	96 55.9	27.8	S23 03.6	4.7
4 Tu	0	183 42.3	28.2	S22 35.4	5.4
	6	Eclipse of the Sun occurs today			
	12	357 22.6	29.0	S21 26.3	6.8
	18	84 16.8	29.6	S20 45.8	7.4
5 Wed	0	171 13.8	30.0	S20 01.8	7.9
	6	258 13.6	30.5	S19 14.3	8.5
	12	345 16.2	31.0	S18 23.7	9.0
	18	72 21.7	31.4	S17 30.2	9.4
6 Th	0	159 29.9	31.8	S16 34.0	9.8
	6	246 40.9	32.3	S15 35.3	10.2
	12	333 54.5	32.7	S14 34.4	10.5
	18	61 10.5	33.1	S13 31.5	10.8
7 Fri	0	148 28.9	33.4	S12 26.8	11.1
	6	235 49.5	33.8	S11 20.4	11.4
	12	323 12.2	34.1	S10 12.7	11.5
	18	50 36.8	34.4	S 9 03.8	11.6
8 Sat	0	138 03.1	34.7	S 7 53.8	11.8
	6	225 30.9	34.9	S 6 43.0	11.9
	12	313 00.1	35.0	S 5 31.4	12.0
	18	40 30.4	35.2	S 4 19.3	12.1
9 Sun	0	128 01.6	35.4	S 3 06.8	12.1
	6	215 33.6	35.4	S 1 54.1	12.2
	12	303 06.1	35.5	S 0 41.2	12.1
	18	30 39.0	35.5	N 0 31.6	12.2
10 Mon	0	118 12.0	35.5	N 1 44.3	12.0
	6	205 44.9	35.4	N 2 56.6	12.0
	12	293 17.6	35.3	N 4 08.6	11.9
	18	20 49.7	35.2	N 5 20.0	11.8
11 Tu	0	108 21.1	35.1	N 6 30.7	11.6
	6	195 51.7	34.9	N 7 40.6	11.5
	12	283 21.1	34.6	N 8 49.6	11.3
	18	10 49.1	34.4	N 9 57.4	11.1
12 Wed	0	98 15.7	34.1	N11 04.0	10.9
	6	185 40.5	33.7	N12 09.2	10.6
	12	273 03.5	33.5	N13 12.9	10.3
	18	0 24.3	33.1	N14 14.9	10.0
13 Th	0	87 42.8	32.6	N15 15.0	9.6
	6	174 58.9	32.3	N16 13.0	9.3
	12	262 12.3	31.7	N17 08.8	8.8
	18	349 23.1	31.3	N18 02.1	8.4
14 Fri	0	76 30.9	30.8	N18 52.8	7.9
	6	163 35.7	30.3	N19 40.6	7.4
	12	250 37.5	29.8	N20 25.3	6.8
	18	337 36.2	29.2	N21 06.8	6.2
15 Sat	0	64 31.7	28.7	N21 44.7	5.6
	6	151 24.1	28.2	N22 18.8	5.0
	12	238 13.4	27.7	N22 48.9	4.2
	18	324 59.7	27.2	N23 14.7	3.5
16 Sun	0	51 43.1	26.7	N23 36.2	2.7
	6	138 23.8	26.3	N23 52.9	1.9
	12	225 01.9	25.9	N24 04.8	1.1
	18	311 37.8	25.6	N24 11.6	0.2
17 Mon	0	38 11.7	25.3	N24 13.2	0.7
	6	124 43.9	25.1	N24 09.4	1.7
	12	211 14.6	24.9	N24 00.2	2.6
	18	297 44.4	24.9	N23 45.4	3.4
18 Tu	0	24 13.4	24.8	N23 25.0	4.4
	6	110 42.1	24.8	N22 59.1	5.3
	12	197 10.8	24.9	N22 27.7	6.2
	18	283 39.8	25.0	N21 50.7	7.2
19 Wed	0	10 09.5	25.1	N21 08.5	8.0
	6	96 40.0	25.3	N20 21.0	8.8
	12	183 11.8	25.5	N19 28.5	9.7
	18	269 44.9	25.8	N18 31.2	10.4
20 Th	0	356 19.6	26.0	N17 29.4	11.1
	6	82 56.0	26.4	N16 23.4	11.7
	12	169 34.1	26.7	N15 13.3	12.3
	18	256 14.0	27.0	N13 59.6	12.9
21 Fri	0	342 55.7	27.3	N12 42.7	13.4
	6	69 39.2	27.6	N11 22.8	13.8
	12	156 24.5	27.8	N10 00.3	14.1
	18	243 11.3	28.1	N 8 35.5	14.4
22 Sat	0	329 59.7	28.4	N 7 09.0	14.7
	6	56 49.5	28.5	N 5 40.9	14.8
	12	143 40.5	28.7	N 4 11.7	15.0
	18	230 32.6	28.9	N 2 41.8	15.1
23 Sun	0	317 25.5	29.0	N 1 11.5	15.0
	6	44 19.1	29.1	S 0 18.8	15.0
	12	131 13.2	29.1	S 1 48.8	14.9
	18	218 07.6	29.1	S 3 18.2	14.7
24 Mon	0	305 02.1	29.0	S 4 46.6	14.5
	6	31 56.5	29.0	S 6 13.6	14.2
	12	118 50.7	28.9	S 7 39.1	13.9
	18	205 44.3	28.8	S 9 02.6	13.6
25 Tu	0	292 37.4	28.7	S10 23.9	13.1
	6	19 29.7	28.5	S11 42.7	12.7
	12	106 21.0	28.3	S12 58.8	12.2
	18	193 11.4	28.1	S14 11.9	11.6
26 Wed	0	280 00.6	28.0	S15 21.6	11.0
	6	6 48.7	27.7	S16 27.9	10.3
	12	93 35.5	27.6	S17 30.4	9.8
	18	180 21.1	27.3	S18 29.0	9.0
27 Th	0	267 05.5	27.2	S19 23.5	8.3
	6	353 48.7	27.0	S20 13.6	7.6
	12	80 30.8	26.8	S20 59.3	6.7
	18	167 12.0	26.7	S21 40.3	6.0
28 Fri	0	253 52.2	26.6	S22 16.5	5.2
	6	340 31.8	26.6	S22 47.9	4.4
	12	67 11.0	26.5	S23 14.3	3.5
	18	153 49.8	26.5	S23 35.7	2.7
29 Sat	0	240 28.6	26.5	S23 52.0	1.8
	6	327 07.7	26.5	S24 03.3	1.0
	12	53 47.2	26.7	S24 09.5	0.2
	18	140 27.4	26.9	S24 10.7	0.7
30 Sun	0	227 08.6	27.1	S24 06.9	1.5
	6	313 51.1	27.3	S23 58.2	2.3
	12	40 35.0	27.6	S23 44.8	3.1
	18	127 20.6	27.9	S23 26.7	3.8
31 Mon	0	214 08.2	28.3	S23 04.0	4.6
	6	300 57.7	28.6	S22 37.0	5.3
	12	27 49.5	29.1	S22 05.9	5.9
	18	114 43.7	29.4	S21 30.7	6.5

PLANETS

VENUS						JUPITER				
Mer Pass h m	GHA ° '	Mean Var/hr 15°+	Dec ° '	Mean Var/hr '	Day	GHA ° '	Mean Var/hr 15°+	Dec ° '	Mean Var/hr '	Mer Pass h m
08 47	228 13.1	0.0	S15 16.5	0.6	1 Sat	102 57.9	2.2	S 2 32.0	0.1	17 06
08 47	228 14.0	0.0	S15 30.0	0.6	2 SUN	103 49.6	2.2	S 2 28.6	0.1	17 02
08 47	228 14.3	0.0	S15 43.5	0.6	3 Mon	104 41.3	2.1	S 2 25.1	0.1	16 59
08 47	228 13.9	0.0	S15 57.0	0.6	4 Tu	105 32.8	2.1	S 2 21.5	0.1	16 55
08 47	228 13.0	14	S16 10.4	0.6	5 Wed	106 24.2	2.1	S 2 17.9	0.1	16 52
08 47	228 11.5	59.9	S16 23.7	0.5	6 Th	107 15.4	2.1	S 2 14.3	0.2	16 49
08 47	228 09.5	59.9	S16 36.9	0.5	7 Fri	108 06.5	2.1	S 2 10.5	0.2	16 45
08 48	228 06.9	59.9	S16 50.0	0.5	8 Sat	108 57.5	2.1	S 2 06.8	0.2	16 42
08 48	228 03.8	59.9	S17 02.9	0.5	9 SUN	109 48.3	2.1	S 2 03.0	0.2	16 38
08 48	228 00.2	59.8	S17 15.7	0.5	10 Mon	110 39.0	2.1	S 1 59.1	0.2	16 35
08 48	227 56.0	59.8	S17 28.3	0.5	11 Tu	111 29.6	2.1	S 1 55.2	0.2	16 32
08 49	227 51.3	59.8	S17 40.7	0.5	12 Wed	112 20.1	2.1	S 1 51.2	0.2	16 28
08 49	227 46.2	59.8	S17 52.9	0.5	13 Th	113 10.5	2.1	S 1 47.2	0.2	16 25
08 49	227 40.5	59.7	S18 04.9	0.5	14 Fri	114 00.7	2.1	S 1 43.1	0.2	16 22
08 50	227 34.4	59.7	S18 16.6	0.5	15 Sat	114 50.9	2.1	S 1 39.0	0.2	16 18
08 50	227 27.8	59.7	S18 28.1	0.5	16 SUN	115 40.9	2.1	S 1 34.8	0.2	16 15
08 51	227 20.8	59.7	S18 39.3	0.5	17 Mon	116 30.8	2.1	S 1 30.6	0.2	16 12
08 51	227 13.3	59.7	S18 50.2	0.4	18 Tu	117 20.6	2.1	S 1 26.4	0.2	16 08
08 52	227 05.3	59.7	S19 00.7	0.4	19 Wed	118 10.3	2.1	S 1 22.1	0.2	16 05
08 52	226 57.0	59.6	S19 11.0	0.4	20 Th	118 59.9	2.1	S 1 17.7	0.2	16 02
08 53	226 48.2	59.6	S19 20.9	0.4	21 Fri	119 49.3	2.1	S 1 13.3	0.2	15 59
08 54	226 38.9	59.6	S19 30.5	0.4	22 Sat	120 38.7	2.1	S 1 08.9	0.2	15 55
08 54	226 29.3	59.6	S19 39.7	0.4	23 SUN	121 28.0	2.0	S 1 04.4	0.2	15 52
08 55	226 19.3	59.6	S19 48.5	0.4	24 Mon	122 17.2	2.0	S 0 59.9	0.2	15 49
08 56	226 08.9	59.6	S19 56.9	0.3	25 Tu	123 06.2	2.0	S 0 55.4	0.2	15 45
08 56	225 58.2	59.5	S20 04.9	0.3	26 Wed	123 55.2	2.0	S 0 50.8	0.2	15 42
08 57	225 47.1	59.5	S20 12.4	0.3	27 Th	124 44.1	2.0	S 0 46.1	0.2	15 39
08 58	225 35.6	59.5	S20 19.6	0.3	28 Fri	125 32.9	2.0	S 0 41.5	0.2	15 36
08 59	225 23.8	59.5	S20 26.3	0.3	29 Sat	126 21.6	2.0	S 0 36.8	0.2	15 32
09 00	225 11.7	59.5	S20 32.5	0.2	30 SUN	127 10.2	2.0	S 0 32.0	0.2	15 29
09 00	224 59.3	59.5	S20 38.3	0.2	31 Mon	127 58.7	2.0	S 0 27.3	0.2	15 26

VENUS, Av. Mag. –4.5
SHA January 5 124; 10 119; 15 113; 20 108; 25 102; 30 96

JUPITER, Av. Mag. –2.3
SHA January 5 2; 10 1; 15 1; 20 0; 25 359; 30 358

MARS						SATURN				
Mer Pass h m	GHA ° '	Mean Var/hr 15°+	Dec ° '	Mean Var/hr '	Day	GHA ° '	Mean Var/hr 15°+	Dec ° '	Mean Var/hr '	Mer Pass h m
12 39	170 15.5	0.4	S23 09.6	0.3	1 Sat	264 01.4	2.4	S 4 19.4	0.0	06 23
12 38	170 24.6	0.4	S23 03.3	0.3	2 SUN	264 58.1	2.4	S 4 20.1	0.0	06 19
12 37	170 33.7	0.4	S22 56.9	0.3	3 Mon	265 54.8	2.4	S 4 20.8	0.0	06 15
12 37	170 42.9	0.4	S22 50.2	0.3	4 Tu	266 51.6	2.4	S 4 21.5	0.0	06 12
12 36	170 52.1	0.4	S22 43.2	0.3	5 Wed	267 48.5	2.4	S 4 22.1	0.0	06 08
12 36	171 01.4	0.4	S22 35.9	0.3	6 Th	268 45.5	2.4	S 4 22.7	0.0	06 04
12 35	171 10.7	0.4	S22 28.4	0.3	7 Fri	269 42.6	2.4	S 4 23.2	0.0	06 00
12 34	171 20.1	0.4	S22 20.7	0.3	8 Sat	270 39.7	2.4	S 4 23.8	0.0	05 56
12 34	171 29.5	0.4	S22 12.7	0.3	9 SUN	271 37.0	2.4	S 4 24.2	0.0	05 53
12 33	171 39.0	0.4	S22 04.4	0.4	10 Mon	272 34.4	2.4	S 4 24.6	0.0	05 49
12 32	171 48.6	0.4	S21 55.9	0.4	11 Tu	273 31.9	2.4	S 4 25.0	0.0	05 45
12 32	171 58.2	0.4	S21 47.1	0.4	12 Wed	274 29.4	2.4	S 4 25.4	0.0	05 41
12 31	172 07.9	0.4	S21 38.1	0.4	13 Th	275 27.1	2.4	S 4 25.7	0.0	05 37
12 30	172 17.7	0.4	S21 28.9	0.4	14 Fri	276 24.9	2.4	S 4 25.9	0.0	05 33
12 30	172 27.6	0.4	S21 19.4	0.4	15 Sat	277 22.7	2.4	S 4 26.1	0.0	05 30
12 29	172 37.5	0.4	S21 09.7	0.4	16 SUN	278 20.7	2.4	S 4 26.3	0.0	05 26
12 28	172 47.5	0.4	S20 59.7	0.4	17 Mon	279 18.8	2.4	S 4 26.4	0.0	05 22
12 28	172 57.6	0.4	S20 49.5	0.4	18 Tu	280 16.9	2.4	S 4 26.5	0.0	05 18
12 27	173 07.8	0.4	S20 39.1	0.4	19 Wed	281 15.2	2.4	S 4 26.6	0.0	05 14
12 26	173 18.1	0.4	S20 28.4	0.5	20 Th	282 13.5	2.4	S 4 26.6	0.0	05 10
12 26	173 28.4	0.4	S20 17.5	0.5	21 Fri	283 12.0	2.4	S 4 26.6	0.0	05 06
12 25	173 38.9	0.4	S20 06.4	0.5	22 Sat	284 10.6	2.4	S 4 26.5	0.0	05 02
12 24	173 49.4	0.4	S19 55.1	0.5	23 SUN	285 09.2	2.4	S 4 26.4	0.0	04 59
12 24	174 00.0	0.4	S19 43.5	0.5	24 Mon	286 08.0	2.4	S 4 26.3	0.0	04 55
12 23	174 10.8	0.5	S19 31.7	0.5	25 Tu	287 06.8	2.5	S 4 26.1	0.0	04 51
12 22	174 21.6	0.5	S19 19.7	0.5	26 Wed	288 05.8	2.5	S 4 25.8	0.0	04 47
12 21	174 32.5	0.5	S19 07.5	0.5	27 Th	289 04.8	2.5	S 4 25.6	0.0	04 43
12 21	174 43.5	0.5	S18 55.1	0.5	28 Fri	290 04.0	2.5	S 4 25.3	0.0	04 39
12 20	174 54.6	0.5	S18 42.5	0.5	29 Sat	291 03.2	2.5	S 4 24.9	0.0	04 35
12 19	175 05.8	0.5	S18 29.7	0.5	30 SUN	292 02.6	2.5	S 4 24.5	0.0	04 31
12 18	175 17.1	0.5	S18 16.6	0.5	31 Mon	293 02.1	2.5	S 4 24.1	0.0	04 27

MARS, Av. Mag. +1.1
SHA January 5 67; 10 62; 15 58; 20 54; 25 50; 30 46

SATURN, Av. Mag. +0.7
SHA January 5 164; 10 163; 15 163; 20 163; 25 163; 30 163

JANUARY 2011

STARS

No.	Name	Mag	Transit h m	Dec ° '	SHA ° '
	0h GMT January 1				
ψ	ARIES January 1		17 16	—	—
1	Alpheratz	2.1	17 25	N29 09.3	357 45.5
2	Ankaa	2.4	17 43	S42 14.9	353 17.4
3	Schedar	2.2	17 57	N56 36.2	349 42.7
4	Diphda	2.0	18 00	S17 55.6	348 57.7
5	Achernar	0.5	18 54	S57 11.1	335 27.9
6	POLARIS	2.0	20 02	N89 19.1	318 29.2
7	Hamal	2.0	19 23	N23 31.0	328 02.7
8	Acamar	3.2	20 14	S40 15.8	315 19.4
9	Menkar	2.5	20 18	N 4 08.0	314 16.7
10	Mirfak	1.8	20 41	N49 54.2	308 42.6
11	Aldebaran	0.9	21 52	N16 31.9	290 51.1
12	Rigel	0.1	22 30	S 8 11.4	281 13.4
13	Capella	0.1	22 33	N46 00.6	280 36.6
14	Bellatrix	1.6	22 41	N 6 21.5	278 33.6
15	Elnath	1.7	22 42	N28 37.0	278 14.5
16	Alnilam	1.7	22 52	S 1 11.8	275 47.8
17	Betelgeuse	0.1–1.2	23 11	N 7 24.5	271 02.8
18	Canopus	−0.7	23 39	S52 42.2	263 56.4
19	Sirius	−1.5	0 04	S16 44.0	258 34.9
20	Adhara	1.5	0 18	S28 59.4	255 13.5
21	Castor	1.6	0 54	N31 51.7	246 09.8
22	Procyon	0.4	0 59	N 5 11.6	245 01.2
23	Pollux	1.1	1 05	N27 59.8	243 29.4
24	Avior	1.9	1 41	S59 32.7	234 18.1
25	Suhail	2.2	2 27	S43 28.7	222 53.3
26	Miaplacidus	1.7	2 32	S69 45.7	221 39.3
27	Alphard	2.0	2 46	S 8 42.5	217 57.5
28	Regulus	1.4	3 27	N11 54.6	207 45.1
29	Dubhe	1.8	4 23	N61 41.1	193 53.2
30	Denebola	2.1	5 08	N14 30.4	182 35.2
31	Gienah	2.6	5 34	S17 36.3	175 54.0
32	Acrux	1.3	5 45	S63 09.5	173 11.1
33	Gacrux	1.6	5 50	S57 10.4	172 02.8
34	Mimosa	1.3	6 06	S59 44.8	167 54.0
35	Alioth	1.8	6 12	N55 53.6	166 22.0
36	Spica	1.0	6 43	S11 13.2	158 33.1
37	Alkaid	1.9	7 06	N49 15.1	153 00.2
38	Hadar	0.6	7 22	S60 25.4	148 50.5
39	Menkent	2.1	7 25	S36 25.4	148 09.7
40	Arcturus	0.0	7 34	N19 07.3	145 57.3
41	Rigil Kent	−0.3	7 58	S60 52.6	139 54.3
42	Zuben'ubi	2.8	8 09	S16 05.3	137 07.5
43	Kochab	2.1	8 08	N74 06.2	137 20.2
44	Alphecca	2.2	8 52	N26 40.4	126 12.6
45	Antares	1.0	9 47	S26 27.3	112 28.6
46	Atria	1.9	10 07	S69 02.6	107 32.5
47	Sabik	2.4	10 28	S15 44.3	102 14.8
48	Shaula	1.6	10 51	S37 06.6	96 24.6
49	Rasalhague	2.1	10 52	N12 33.1	96 08.4
50	Eltanin	2.2	11 14	N51 29.1	90 47.4
51	Kaus Aust	1.9	11 42	S34 22.7	83 46.5
52	Vega	0.0	11 54	N38 47.6	80 40.6
53	Nunki	2.0	12 13	S26 16.9	76 00.8
54	Altair	0.8	13 08	N 8 53.9	62 10.3
55	Peacock	1.9	13 43	S56 42.0	53 22.5
56	Deneb	1.3	13 58	N45 19.3	49 33.2
57	Enif	2.4	15 01	N 9 55.6	33 49.1
58	Al Na'ir	1.7	15 25	S46 54.5	27 46.2
59	Fomalhaut	1.2	16 14	S29 33.9	15 26.1
60	Markab	2.5	16 21	N15 16.0	13 40.3

SUN AND MOON

SUN

Yr	Day of Mth	Week	Semi-Diam	Transit h m	Twilight h m	Lat 52°N Sunrise h m	Sunset h m	Twilight h m
1	1	Sat	16.3	12 03	07 28	08 08	15 59	16 39
2	2	Sun	16.3	12 04	07 28	08 08	16 00	16 40
3	3	Mon	16.3	12 04	07 27	08 08	16 01	16 40
4	4	Tu	16.3	12 05	07 27	08 07	16 02	16 42
5	5	Wed	16.3	12 05	07 27	08 07	16 03	16 43
6	6	Th	16.3	12 06	07 27	08 07	16 05	16 44
7	7	Fri	16.3	12 06	07 26	08 07	16 06	16 45
8	8	Sat	16.3	12 07	07 26	08 06	16 07	16 46
9	9	Sun	16.3	12 07	07 25	08 06	16 09	16 47
10	10	Mon	16.3	12 07	07 25	08 05	16 10	16 49
11	11	Tu	16.3	12 08	07 24	08 04	16 12	16 51
12	12	Wed	16.3	12 08	07 24	08 04	16 13	16 53
13	13	Th	16.3	12 09	07 23	08 03	16 15	16 54
14	14	Fri	16.3	12 09	07 22	08 02	16 16	16 55
15	15	Sat	16.3	12 09	07 22	08 01	16 18	16 57
16	16	Sun	16.3	12 10	07 21	08 00	16 19	16 58
17	17	Mon	16.3	12 10	07 20	07 59	16 21	17 00
18	18	Tu	16.3	12 11	07 19	07 58	16 22	17 01
19	19	Wed	16.3	12 11	07 18	07 57	16 24	17 03
20	20	Th	16.3	12 11	07 17	07 55	16 26	17 04
21	21	Fri	16.3	12 12	07 16	07 54	16 27	17 06
22	22	Sat	16.3	12 12	07 14	07 53	16 29	17 07
23	23	Sun	16.3	12 12	07 13	07 52	16 31	17 09
24	24	Mon	16.3	12 12	07 12	07 51	16 33	17 11
25	25	Tu	16.3	12 13	07 10	07 49	16 34	17 12
26	26	Wed	16.3	12 13	07 09	07 48	16 36	17 14
27	27	Th	16.3	12 13	07 08	07 47	16 38	17 15
28	28	Fri	16.3	12 13	07 08	07 45	16 40	17 17
29	29	Sat	16.3	12 13	07 07	07 44	16 42	17 19
30	30	Sun	16.3	12 13	07 05	07 42	16 43	17 20
31	31	Mon	16.3	12 13	07 05	07 42	16 45	17 22

MOON

Yr	Day of Mth	Week	Age days	Transit (Upper) h m	Diff m	Semi-diam	Hor Par	Lat 52°N Moonrise h m	Moonset h m
1	1	Sat	27	09 25	57	15.7	57.6	05 31	13 16
2	2	Sun	28	10 22	56	15.6	57.1	06 35	14 09
3	3	Mon	29	11 18	53	15.4	56.6	07 27	15 12
4	4	Tu	00	12 11	50	15.3	56.1	08 07	16 21
5	5	Wed	01	13 01	47	15.1	55.6	08 37	17 33
6	6	Th	02	13 48	44	15.0	55.1	09 00	18 45
7	7	Fri	03	14 32	41	14.9	54.7	09 19	19 55
8	8	Sat	04	15 13	41	14.8	54.4	09 36	21 03
9	9	Sun	05	15 54	40	14.8	54.2	09 50	22 11
10	10	Mon	06	16 34	41	14.8	54.2	10 05	23 18
11	11	Tu	07	17 15	43	14.8	54.3	10 20	— —
12	12	Wed	08	17 58	46	14.9	54.6	10 37	00 25
13	13	Th	09	18 44	49	15.0	55.1	10 58	01 34
14	14	Fri	10	19 33	52	15.2	55.7	11 24	02 45
15	15	Sat	11	20 25	56	15.4	56.5	11 58	03 54
16	16	Sun	12	21 21	58	15.6	57.3	12 43	05 01
17	17	Mon	13	22 19	59	15.8	58.2	13 42	06 00
18	18	Tu	14	23 18	57	16.1	59.0	14 53	06 49
19	19	Wed	15	24 15	—	16.3	59.7	16 15	07 28
20	20	Th	16	00 15	56	16.4	60.2	17 41	07 58
21	21	Fri	17	01 11	53	16.5	60.4	19 08	08 22
22	22	Sat	18	02 04	52	16.5	60.4	20 34	08 43
23	23	Sun	19	02 56	51	16.4	60.2	21 59	09 02
24	24	Mon	20	03 48	52	16.3	59.8	23 23	09 22
25	25	Tu	21	04 39	53	16.1	59.3	— —	09 43
26	26	Wed	22	05 32	54	16.0	58.7	00 46	10 07
27	27	Th	23	06 26	55	15.8	58.1	02 06	10 37
28	28	Fri	24	07 21	56	15.7	57.5	03 22	11 16
29	29	Sat	25	08 17	55	15.5	56.9	04 28	12 04
30	30	Sun	26	09 12	53	15.4	56.4	05 23	13 02
31	31	Mon	27	10 05	50	15.2	55.9	06 06	14 08

Lat Corr to Sunrise, Sunset etc.

Lat °	Twilight h m	Sunrise h m	Sunset h m	Twilight h m
N70	SBH	SBH	SBH	−1 56
68	+2 34	+1 57	−2 33	−1 32
66	+1 53	+1 32	−1 53	−1 13
64	+1 25	+1 13	−1 25	−0 57
62	+1 04	+0 57	−1 04	−0 44
N60	+0 47	+0 44	−0 47	−0 33
58	+0 33	+0 32	−0 32	−0 23
56	+0 21	+0 22	−0 20	−0 14
54	+0 10	+0 14	−0 09	−0 07
50	−0 08	+0 07	+0 06	+0 06
N45	−0 26	−0 06	+0 27	+0 20
40	−0 41	−0 20	+0 42	+0 32
35	−0 54	−0 33	+0 54	+0 42
30	−1 05	−0 43	+1 05	+0 54
20	−1 24	−0 52	+1 24	+1 05
N10	−1 41	−1 08	+1 41	+1 24
0	−1 56	−1 24	+1 56	+1 41
S10	−2 13	−1 39	+2 11	+1 55
20	−2 30	−1 55	+2 28	+2 14
30	−2 49	−2 13	+2 48	+2 36
35	−3 01	−2 30	+3 00	+2 49
40	−3 15	−2 49	+3 13	+3 06
45	−3 26	−3 01	+3 29	+3 26
S50	−3 51	−3 15	+3 49	+3 52

NOTES

The corrections to sunrise etc. are for middle of January.
SBH means Sun below Horizon.

Phases of the Moon

		d	h	m
●	New Moon	4	09	03
☽	First Quarter	12	11	31
○	Full Moon	19	21	21
☽	Last Quarter	26	12	57

	d	h
Apogee	10	06
Perigee	22	00

10

SUN AND ARIES FEBRUARY 2011 SUN AND ARIES

GMT	SUN GHA ° '	SUN Dec ° '	ARIES GHA ° '
Tuesday, 1st February			
00	176 38.1	S17 15.5	130 51.5
02	206 38.0	17 14.1	160 56.5
04	236 37.8	17 12.7	191 01.4
06	266 37.6	17 11.3	221 06.3
08	296 37.4	17 09.9	251 11.2
10	326 37.3	17 08.5	281 16.2
12	356 37.1	17 07.0	311 21.1
14	26 36.9	17 05.6	341 26.0
16	56 36.7	17 04.2	11 31.0
18	86 36.6	17 02.8	41 35.9
20	116 36.4	17 01.3	71 40.8
22	146 36.2	S16 59.9	101 45.7
Wednesday, 2nd February			
00	176 36.1	S16 58.5	131 50.7
02	206 35.9	16 57.0	161 55.6
04	236 35.8	16 55.6	192 00.5
06	266 35.6	16 54.2	222 05.5
08	296 35.4	16 52.7	252 10.4
10	326 35.3	16 51.3	282 15.3
12	356 35.1	16 49.8	312 20.2
14	26 35.0	16 48.4	342 25.2
16	56 34.8	16 46.9	12 30.1
18	86 34.7	16 45.5	42 35.0
20	116 34.5	16 44.0	72 39.9
22	146 34.4	S16 42.6	102 44.9
Thursday, 3rd February			
00	176 34.2	S16 41.1	132 49.8
02	206 34.1	16 39.7	162 54.7
04	236 33.9	16 38.2	192 59.7
06	266 33.8	16 36.7	223 04.6
08	296 33.7	16 35.3	253 09.5
10	326 33.5	16 33.8	283 14.4
12	356 33.4	16 32.3	313 19.4
14	26 33.2	16 30.9	343 24.3
16	56 33.1	16 29.4	13 29.2
18	86 33.0	16 27.9	43 34.2
20	116 32.8	16 26.4	73 39.1
22	146 32.7	S16 25.0	103 44.0
Friday, 4th February			
00	176 32.6	S16 23.5	133 48.9
02	206 32.5	16 22.0	163 53.9
04	236 32.3	16 20.5	193 58.8
06	266 32.2	16 19.0	224 03.7
08	296 32.1	16 17.5	254 08.7
10	326 32.0	16 16.0	284 13.6
12	356 31.8	16 14.5	314 18.5
14	26 31.7	16 13.0	344 23.4
16	56 31.6	16 11.6	14 28.4
18	86 31.5	16 10.1	44 33.3
20	116 31.4	16 08.6	74 38.2
22	146 31.2	S16 07.0	104 43.2
Saturday, 5th February			
00	176 31.1	S16 05.5	134 48.1
02	206 31.0	16 04.0	164 53.0
04	236 30.9	16 02.5	194 57.9
06	266 30.8	16 01.0	225 02.9
08	296 30.7	15 59.5	255 07.8
10	326 30.6	15 58.0	285 12.7
12	356 30.5	15 56.5	315 17.6
14	26 30.4	15 55.0	345 22.6
16	56 30.3	15 53.4	15 27.5
18	86 30.2	15 51.9	45 32.4
20	116 30.1	15 50.4	75 37.4
22	146 30.0	S15 48.9	105 42.3

GMT	SUN GHA ° '	SUN Dec ° '	ARIES GHA ° '
Sunday, 6th February			
00	176 29.9	S15 47.3	135 47.2
02	206 29.8	15 45.8	165 52.1
04	236 29.7	15 44.3	195 57.1
06	266 29.6	15 42.7	226 02.0
08	296 29.5	15 41.2	256 06.9
10	326 29.4	15 39.7	286 11.9
12	356 29.4	15 38.1	316 16.8
14	26 29.3	15 36.6	346 21.7
16	56 29.2	15 35.0	16 26.6
18	86 29.1	15 33.5	46 31.6
20	116 29.0	15 32.0	76 36.5
22	146 28.9	S15 30.4	106 41.4
Monday, 7th February			
00	176 28.9	S15 28.9	136 46.4
02	206 28.8	15 27.3	166 51.3
04	236 28.7	15 25.8	196 56.2
06	266 28.6	15 24.2	227 01.1
08	296 28.6	15 22.6	257 06.1
10	326 28.5	15 21.1	287 11.0
12	356 28.4	15 19.5	317 15.9
14	26 28.3	15 18.0	347 20.9
16	56 28.3	15 16.4	17 25.8
18	86 28.2	15 14.8	47 30.7
20	116 28.1	15 13.3	77 35.6
22	146 28.1	S15 11.7	107 40.6
Tuesday, 8th February			
00	176 28.0	S15 10.1	137 45.5
02	206 28.0	15 08.5	167 50.4
04	236 27.9	15 07.0	197 55.3
06	266 27.8	15 05.4	228 00.3
08	296 27.8	15 03.8	258 05.2
10	326 27.7	15 02.2	288 10.1
12	356 27.7	15 00.7	318 15.1
14	26 27.6	14 59.1	348 20.0
16	56 27.6	14 57.5	18 24.9
18	86 27.5	14 55.9	48 29.8
20	116 27.5	14 54.3	78 34.8
22	146 27.4	S14 52.7	108 39.7
Wednesday, 9th February			
00	176 27.4	S14 51.1	138 44.6
02	206 27.3	14 49.5	168 49.6
04	236 27.3	14 47.9	198 54.5
06	266 27.3	14 46.3	228 59.4
08	296 27.2	14 44.7	259 04.3
10	326 27.2	14 43.1	289 09.3
12	356 27.1	14 41.5	319 14.2
14	26 27.1	14 39.9	349 19.1
16	56 27.1	14 38.3	19 24.1
18	86 27.0	14 36.7	49 29.0
20	116 27.0	14 35.1	79 33.9
22	146 27.0	S14 33.5	109 38.8
Thursday, 10th February			
00	176 27.0	S14 31.9	139 43.8
02	206 26.9	14 30.3	169 48.7
04	236 26.9	14 28.7	199 53.6
06	266 26.9	14 27.0	229 58.6
08	296 26.9	14 25.4	260 03.5
10	326 26.8	14 23.8	290 08.4
12	356 26.8	14 22.2	320 13.3
14	26 26.8	14 20.5	350 18.3
16	56 26.8	14 18.9	20 23.2
18	86 26.8	14 17.3	50 28.1
20	116 26.7	14 15.7	80 33.1
22	146 26.7	S14 14.0	110 38.0

GMT	SUN GHA ° '	SUN Dec ° '	ARIES GHA ° '
Friday, 11th February			
00	176 26.7	S14 12.4	140 42.9
02	206 26.7	14 10.8	170 47.8
04	236 26.7	14 09.1	200 52.8
06	266 26.7	14 07.5	230 57.7
08	296 26.7	14 05.9	261 02.6
10	326 26.7	14 04.2	291 07.5
12	356 26.7	14 02.6	321 12.5
14	26 26.7	14 00.9	351 17.4
16	56 26.7	13 59.3	21 22.3
18	86 26.7	13 57.6	51 27.3
20	116 26.7	13 56.0	81 32.2
22	146 26.7	S13 54.3	111 37.1
Saturday, 12th February			
00	176 26.7	S13 52.7	141 42.0
02	206 26.7	13 51.0	171 47.0
04	236 26.7	13 49.4	201 51.9
06	266 26.7	13 47.7	231 56.8
08	296 26.7	13 46.1	262 01.8
10	326 26.7	13 44.4	292 06.7
12	356 26.7	13 42.7	322 11.6
14	26 26.8	13 41.1	352 16.5
16	56 26.8	13 39.4	22 21.5
18	86 26.8	13 37.7	52 26.4
20	116 26.8	13 36.1	82 31.3
22	146 26.8	S13 34.4	112 36.3
Sunday, 13th February			
00	176 26.8	S13 32.7	142 41.2
02	206 26.9	13 31.1	172 46.1
04	236 26.9	13 29.4	202 51.0
06	266 26.9	13 27.7	232 56.0
08	296 26.9	13 26.0	263 00.9
10	326 27.0	13 24.4	293 05.8
12	356 27.0	13 22.7	323 10.8
14	26 27.0	13 21.0	353 15.7
16	56 27.1	13 19.3	23 20.6
18	86 27.1	13 17.6	53 25.5
20	116 27.1	13 15.9	83 30.5
22	146 27.2	S13 14.3	113 35.4
Monday, 14th February			
00	176 27.2	S13 12.6	143 40.3
02	206 27.2	13 10.9	173 45.3
04	236 27.3	13 09.2	203 50.2
06	266 27.3	13 07.5	233 55.1
08	296 27.4	13 05.8	264 00.0
10	326 27.4	13 04.1	294 05.0
12	356 27.4	13 02.4	324 09.9
14	26 27.5	13 00.7	354 14.8
16	56 27.5	12 59.0	24 19.8
18	86 27.6	12 57.3	54 24.7
20	116 27.6	12 55.6	84 29.6
22	146 27.7	S12 53.9	114 34.5
Tuesday, 15th February			
00	176 27.7	S12 52.2	144 39.5
02	206 27.8	12 50.5	174 44.4
04	236 27.8	12 48.8	204 49.3
06	266 27.9	12 47.1	234 54.3
08	296 28.0	12 45.3	264 59.2
10	326 28.0	12 43.6	295 04.1
12	356 28.1	12 41.9	325 09.0
14	26 28.1	12 40.2	355 14.0
16	56 28.2	12 38.5	25 18.9
18	86 28.3	12 36.8	55 23.8
20	116 28.3	12 35.0	85 28.8
22	146 28.4	S12 33.3	115 33.7

GMT	SUN GHA ° '	SUN Dec ° '	ARIES GHA ° '
Wednesday, 16th February			
00	176 28.5	S12 31.6	145 38.6
02	206 28.5	12 29.9	175 43.5
04	236 28.6	12 28.1	205 48.5
06	266 28.7	12 26.4	235 53.4
08	296 28.7	12 24.7	265 58.3
10	326 28.8	12 23.0	296 03.2
12	356 28.9	12 21.2	326 08.2
14	26 29.0	12 19.5	356 13.1
16	56 29.0	12 17.8	26 18.0
18	86 29.1	12 16.0	56 23.0
20	116 29.2	12 14.3	86 27.9
22	146 29.3	S12 12.6	116 32.8
Thursday, 17th February			
00	176 29.4	S12 10.8	146 37.7
02	206 29.5	12 09.1	176 42.7
04	236 29.5	12 07.3	206 47.6
06	266 29.6	12 05.6	236 52.5
08	296 29.7	12 03.8	266 57.5
10	326 29.8	12 02.1	297 02.4
12	356 29.9	12 00.3	327 07.3
14	26 30.0	11 58.6	357 12.2
16	56 30.1	11 56.8	27 17.2
18	86 30.2	11 55.1	57 22.1
20	116 30.3	11 53.3	87 27.0
22	146 30.4	S11 51.6	117 32.0
Friday, 18th February			
00	176 30.5	S11 49.8	147 36.9
02	206 30.6	11 48.1	177 41.8
04	236 30.7	11 46.3	207 46.7
06	266 30.8	11 44.6	237 51.7
08	296 30.9	11 42.8	267 56.6
10	326 31.0	11 41.0	298 01.5
12	356 31.1	11 39.3	328 06.5
14	26 31.2	11 37.5	358 11.4
16	56 31.3	11 35.7	28 16.3
18	86 31.4	11 34.0	58 21.2
20	116 31.5	11 32.2	88 26.2
22	146 31.6	S11 30.4	118 31.1
Saturday, 19th February			
00	176 31.7	S11 28.7	148 36.0
02	206 31.8	11 26.9	178 41.0
04	236 31.9	11 25.1	208 45.9
06	266 32.1	11 23.3	238 50.8
08	296 32.2	11 21.6	268 55.7
10	326 32.3	11 19.8	299 00.7
12	356 32.4	11 18.0	329 05.6
14	26 32.5	11 16.2	359 10.5
16	56 32.6	11 14.4	29 15.4
18	86 32.8	11 12.7	59 20.4
20	116 32.9	11 10.9	89 25.3
22	146 33.0	S11 09.1	119 30.2
Sunday, 20th February			
00	176 33.1	S11 07.3	149 35.2
02	206 33.3	11 05.5	179 40.1
04	236 33.4	11 03.7	209 45.0
06	266 33.5	11 01.9	239 49.9
08	296 33.7	11 00.2	269 54.9
10	326 33.8	10 58.4	299 59.8
12	356 33.9	10 56.6	330 04.7
14	26 34.0	10 54.8	0 09.7
16	56 34.2	10 53.0	30 14.6
18	86 34.3	10 51.2	60 19.5
20	116 34.5	10 49.4	90 24.4
22	146 34.6	S10 47.6	120 29.4

GMT	SUN GHA ° '	SUN Dec ° '	ARIES GHA ° '
Monday, 21st February			
00	176 34.7	S10 45.8	150 34.3
02	206 34.9	10 44.0	180 39.2
04	236 35.0	10 42.2	210 44.2
06	266 35.2	10 40.4	240 49.1
08	296 35.3	10 38.6	270 54.0
10	326 35.4	10 36.8	300 58.9
12	356 35.6	10 35.0	331 03.9
14	26 35.7	10 33.2	1 08.8
16	56 35.9	10 31.3	31 13.7
18	86 36.0	10 29.5	61 18.6
20	116 36.2	10 27.7	91 23.6
22	146 36.3	S10 25.9	121 28.5
Tuesday, 22nd February			
00	176 36.5	S10 24.1	151 33.4
02	206 36.6	10 22.3	181 38.4
04	236 36.8	10 20.5	211 43.3
06	266 36.9	10 18.6	241 48.2
08	296 37.1	10 16.8	271 53.1
10	326 37.3	10 15.0	301 58.1
12	356 37.4	10 13.2	332 03.0
14	26 37.6	10 11.4	2 07.9
16	56 37.7	10 09.5	32 12.9
18	86 37.9	10 07.7	62 17.8
20	116 38.1	10 05.9	92 22.7
22	146 38.2	S10 04.1	122 27.6
Wednesday, 23rd February			
00	176 38.4	S10 02.2	152 32.6
02	206 38.6	10 00.4	182 37.5
04	236 38.7	9 58.6	212 42.4
06	266 38.9	9 56.7	242 47.4
08	296 39.1	9 54.9	272 52.3
10	326 39.2	9 53.1	302 57.2
12	356 39.4	9 51.2	333 02.1
14	26 39.6	9 49.4	3 07.1
16	56 39.7	9 47.6	33 12.0
18	86 39.9	9 45.7	63 16.9
20	116 40.1	9 43.9	93 21.9
22	146 40.3	S 9 42.1	123 26.8
Thursday, 24th February			
00	176 40.4	S 9 40.2	153 31.7
02	206 40.6	9 38.4	183 36.6
04	236 40.8	9 36.5	213 41.6
06	266 41.0	9 34.7	243 46.5
08	296 41.2	9 32.9	273 51.4
10	326 41.3	9 31.0	303 56.4
12	356 41.5	9 29.2	334 01.3
14	26 41.7	9 27.3	4 06.2
16	56 41.9	9 25.5	34 11.1
18	86 42.1	9 23.6	64 16.1
20	116 42.3	9 21.8	94 21.0
22	146 42.5	S 9 19.9	124 25.9
Friday, 25th February			
00	176 42.6	S 9 18.1	154 30.9
02	206 42.8	9 16.2	184 35.8
04	236 43.0	9 14.4	214 40.7
06	266 43.2	9 12.5	244 45.6
08	296 43.4	9 10.6	274 50.6
10	326 43.6	9 08.8	304 55.5
12	356 43.8	9 06.9	335 00.4
14	26 44.0	9 05.1	5 05.3
16	56 44.2	9 03.2	35 10.3
18	86 44.4	9 01.3	65 15.2
20	116 44.6	8 59.5	95 20.1
22	146 44.8	S 8 57.6	125 25.1

GMT	SUN GHA ° '	SUN Dec ° '	ARIES GHA ° '
Saturday, 26th February			
00	176 45.0	S 8 55.8	155 30.0
02	206 45.2	8 53.9	185 34.9
04	236 45.4	8 52.0	215 39.8
06	266 45.6	8 50.2	245 44.8
08	296 45.8	8 48.3	275 49.7
10	326 46.0	8 46.4	305 54.6
12	356 46.2	8 44.6	335 59.6
14	26 46.4	8 42.7	6 04.5
16	56 46.6	8 40.8	36 09.4
18	86 46.8	8 38.9	66 14.3
20	116 47.1	8 37.1	96 19.3
22	146 47.3	S 8 35.2	126 24.2
Sunday, 27th February			
00	176 47.5	S 8 33.3	156 29.1
02	206 47.7	8 31.4	186 34.1
04	236 47.9	8 29.6	216 39.0
06	266 48.1	8 27.7	246 43.9
08	296 48.3	8 25.8	276 48.8
10	326 48.5	8 23.9	306 53.8
12	356 48.8	8 22.1	336 58.7
14	26 49.0	8 20.2	7 03.6
16	56 49.2	8 18.3	37 08.6
18	86 49.4	8 16.4	67 13.5
20	116 49.6	8 14.5	97 18.4
22	146 49.9	S 8 12.6	127 23.3
Monday, 28th February			
00	176 50.1	S 8 10.8	157 28.3
02	206 50.3	8 08.9	187 33.2
04	236 50.5	8 07.0	217 38.1
06	266 50.8	8 05.1	247 43.1
08	296 51.0	8 03.2	277 48.0
10	326 51.2	8 01.3	307 52.9
12	356 51.4	7 59.4	337 57.8
14	26 51.7	7 57.5	8 02.8
16	56 51.9	7 55.6	38 07.7
18	86 52.1	7 53.8	68 12.6
20	116 52.4	7 51.9	98 17.6
22	146 52.6	S 7 50.0	128 22.5

FEBRUARY 2011

MOON

Day	GMT hr	GHA	Mean Var/hr 14°+	Dec	Mean Var/hr
1 Tu	0	201 40.3	29.9	S20 51.8	7.1
	6	288 39.3	30.3	S20 09.2	7.7
	12	15 40.9	30.7	S19 23.3	8.2
	18	102 45.1	31.2	S18 34.2	8.7
2 Wed	0	189 51.8	31.5	S17 42.1	9.2
	6	277 01.0	31.9	S16 47.3	9.6
	12	4 12.6	32.3	S15 49.9	10.0
	18	91 26.6	32.8	S14 50.3	10.3
3 Th	0	178 42.8	33.1	S13 48.5	10.7
	6	266 01.2	33.5	S12 44.8	10.9
	12	353 21.7	33.8	S11 39.4	11.2
	18	80 44.1	34.1	S10 32.5	11.4
4 Fri	0	168 08.2	34.3	S 9 24.2	11.6
	6	255 33.9	34.6	S 8 14.8	11.8
	12	343 01.1	34.8	S 7 04.5	11.9
	18	70 29.6	35.1	S 5 53.3	12.0
5 Sat	0	157 59.2	35.1	S 4 41.6	12.1
	6	245 29.7	35.2	S 3 29.3	12.1
	12	333 01.1	35.3	S 2 16.8	12.1
	18	60 33.0	35.4	S 1 04.1	12.1
6 Sun	0	148 05.3	35.4	N 0 08.6	12.1
	6	235 37.9	35.4	N 1 21.2	12.0
	12	323 10.5	35.4	N 2 33.4	12.0
	18	50 43.0	35.3	N 3 45.3	11.9
7 Mon	0	138 15.1	35.3	N 4 56.5	11.8
	6	225 46.8	35.1	N 6 07.1	11.6
	12	313 17.8	35.0	N 7 16.8	11.5
	18	40 47.9	34.8	N 8 25.5	11.3
8 Tu	0	128 17.0	34.6	N 9 33.2	11.0
	6	215 44.8	34.4	N10 39.5	10.8
	12	303 11.3	34.1	N11 44.4	10.5
	18	30 36.1	33.8	N12 48.0	10.2
9 Wed	0	117 59.3	33.5	N13 49.4	9.9
	6	205 20.5	33.2	N14 49.1	9.6
	12	292 39.7	32.8	N15 46.9	9.2
	18	19 56.6	32.4	N16 42.4	8.8
10 Th	0	107 11.3	32.0	N17 35.5	8.4
	6	194 23.5	31.6	N18 26.1	7.9
	12	281 33.1	31.1	N19 13.9	7.4
	18	8 40.1	30.7	N19 58.9	6.9
11 Fri	0	95 44.4	30.2	N20 40.7	6.4
	6	182 45.9	29.7	N21 19.2	5.8
	12	269 44.6	29.3	N21 54.1	5.2
	18	356 40.5	28.8	N22 25.4	4.5
12 Sat	0	83 33.7	28.4	N22 52.7	3.8
	6	170 24.2	28.0	N23 16.0	3.1
	12	257 12.1	27.5	N23 34.9	2.3
	18	343 57.5	27.1	N23 49.3	1.6
13 Sun	0	70 40.5	26.8	N23 59.1	0.8
	6	157 21.4	26.5	N24 04.1	0.1
	12	244 00.4	26.2	N24 04.1	0.9
	18	330 37.7	26.0	N23 58.9	1.8
14 Mon	0	57 13.5	25.7	N23 48.6	2.7
	6	143 48.1	25.6	N23 33.0	3.6
	12	230 21.7	25.5	N23 12.0	4.4
	18	316 54.8	25.4	N22 45.7	5.3
15 Tu	0	43 27.4	25.4	N22 14.0	6.2
	6	129 59.9	25.4	N21 37.0	7.1
	12	216 32.6	25.5	N20 54.8	8.0
	18	303 05.7	25.6	N20 07.4	8.8
16 Wed	0	29 39.3	25.7	N19 15.1	9.6
	6	116 13.7	25.9	N18 17.9	10.3
	12	202 49.0	26.1	N17 16.1	11.1
	18	289 25.3	26.2	N16 09.9	11.7
17 Th	0	16 02.7	26.5	N14 59.6	12.4
	6	102 41.3	26.6	N13 45.5	13.0
	12	189 21.0	26.8	N12 27.8	13.5
	18	276 01.9	27.0	N11 07.8	14.0
18 Fri	0	2 43.9	27.2	N 9 43.3	14.4
	6	89 27.0	27.3	N 8 17.2	14.7
	12	176 11.0	27.5	N 6 49.0	15.0
	18	262 55.9	27.6	N 5 19.1	15.3
19 Sat	0	349 41.5	27.7	N 3 47.9	15.4
	6	76 27.7	27.7	N 2 15.9	15.4
	12	163 14.3	27.8	N 0 43.3	15.4
	18	250 01.3	27.8	S 0 49.3	15.4
20 Sun	0	336 48.3	27.8	S 2 21.6	15.3
	6	63 35.4	27.8	S 3 53.2	15.3
	12	150 22.2	27.7	S 5 23.7	15.1
	18	237 08.7	27.5	S 6 52.7	14.8
21 Mon	0	323 54.8	27.5	S 8 19.8	14.5
	6	50 40.2	27.3	S 9 44.7	14.5
	12	137 24.9	27.2	S11 07.0	13.7
	18	224 08.8	27.0	S12 26.5	13.2
22 Tu	0	310 51.8	26.8	S13 42.8	12.6
	6	37 33.9	26.7	S14 55.6	12.1
	12	124 15.0	26.5	S16 04.6	11.4
	18	210 55.2	26.3	S17 09.6	10.8
23 Wed	0	297 34.4	26.2	S18 10.4	10.0
	6	24 12.7	26.1	S19 06.7	9.4
	12	110 50.3	26.0	S19 58.5	8.5
	18	197 27.2	26.0	S20 45.4	7.8
24 Th	0	284 03.5	26.0	S21 27.4	6.9
	6	10 39.5	26.0	S22 04.4	6.1
	12	97 15.3	26.1	S22 36.2	5.3
	18	183 51.2	26.3	S23 02.9	4.4
25 Fri	0	270 27.3	26.4	S23 24.5	3.6
	6	357 03.9	26.6	S23 40.8	2.6
	12	83 41.3	26.8	S23 51.9	1.8
	18	170 19.7	27.2	S23 58.0	0.9
26 Sat	0	256 59.3	27.5	S23 59.0	0.1
	6	343 40.3	27.8	S23 55.0	0.7
	12	70 23.0	28.2	S23 46.2	1.5
	18	157 07.5	28.5	S23 32.7	2.3
27 Sun	0	243 54.1	28.9	S23 14.6	3.1
	6	330 42.9	29.3	S22 52.1	3.8
	12	57 33.9	29.8	S22 25.3	4.5
	18	144 27.4	30.3	S21 54.5	5.2
28 Mon	0	231 23.4	30.6	S21 19.9	5.8
	6	318 21.8	30.8	S20 41.6	6.5
	12	45 22.9	30.6	S19 59.8	7.0
	18	132 26.5	31.1	S19 14.7	8.1

PLANETS

VENUS

Mer Pass h m	GHA	Mean Var/hr 14°+	Dec	Mean Var/hr
09 01	224 46.6	59.5	S20 43.6	0.2
09 02	224 33.7	59.4	S20 48.4	0.2
09 03	224 20.4	59.4	S20 52.7	0.2
09 04	224 07.0	59.4	S20 56.5	0.1
09 05	223 53.3	59.4	S20 59.8	0.1
09 06	223 39.4	59.4	S21 02.5	0.1
09 07	223 25.3	59.4	S21 04.8	0.1
09 08	223 11.0	59.4	S21 06.5	0.0
09 09	222 56.5	59.4	S21 07.6	0.0
09 10	222 41.8	59.4	S21 08.2	0.0
09 11	222 27.0	59.4	S21 08.3	0.0
09 12	222 12.1	59.4	S21 08.3	0.1
09 13	221 57.1	59.4	S21 06.8	0.1
09 14	221 41.9	59.4	S21 05.1	0.1
09 15	221 26.7	59.4	S21 03.0	0.1
09 16	221 11.3	59.4	S21 00.2	0.2
09 17	220 55.9	59.4	S20 56.9	0.2
09 18	220 40.5	59.4	S20 53.0	0.2
09 19	220 25.0	59.4	S20 48.5	0.2
09 20	220 09.4	59.4	S20 43.4	0.3
09 21	219 53.9	59.3	S20 37.8	0.3
09 22	219 38.3	59.4	S20 31.6	0.3
09 23	219 22.7	59.4	S20 24.8	0.3
09 24	219 07.2	59.4	S20 17.5	0.4
09 25	218 51.7	59.4	S20 09.5	0.4
09 26	218 36.2	59.4	S20 01.0	0.4
09 27	218 20.8	59.4	S19 52.0	0.4
09 28	218 05.5	59.4	S19 42.3	

VENUS, Av. Mag. –4.2
SHA February 5 89; 10 83; 15 77; 20 71; 25 64; 28 61

MARS

Mer Pass h m	GHA	Mean Var/hr 15°+	Dec	Mean Var/hr
12 18	175 28.5	0.5	S18 03.4	0.6
12 17	175 40.0	0.5	S17 50.0	0.6
12 16	175 51.6	0.5	S17 36.4	0.6
12 16	176 03.4	0.5	S17 22.6	0.6
12 15	176 15.2	0.5	S17 08.6	0.6
12 14	176 27.1	0.5	S16 54.5	0.6
12 13	176 39.1	0.5	S16 40.1	0.6
12 12	176 51.3	0.5	S16 25.6	0.6
12 12	177 03.5	0.5	S16 10.9	0.6
12 11	177 15.9	0.5	S15 56.1	0.6
12 11	177 28.4	0.5	S15 41.0	0.6
12 10	177 40.9	0.5	S15 25.9	0.6
12 09	177 53.6	0.5	S15 10.5	0.6
12 08	178 06.4	0.5	S14 55.0	0.7
12 07	178 19.3	0.5	S14 39.4	0.7
12 06	178 32.3	0.6	S14 23.6	0.7
12 05	178 45.4	0.6	S14 07.7	0.7
12 04	178 58.6	0.6	S13 51.6	0.7
12 03	179 11.9	0.6	S13 35.4	0.7
12 02	179 25.3	0.6	S13 19.0	0.7
12 01	179 38.8	0.6	S13 02.6	0.7
12 00	179 52.4	0.6	S12 45.9	0.7
11 59	180 06.2	0.6	S12 29.2	0.7
11 58	180 20.0	0.6	S12 12.3	0.7
11 57	180 33.9	0.6	S11 55.4	0.7
11 56	180 47.9	0.6	S11 38.3	0.7
11 55	181 01.9	0.6	S11 21.1	0.7
11 54	181 16.1	0.6	S11 03.8	

MMARS, Av. Mag. +1.1
SHA February 5 41; 10 38; 15 34; 20 30; 25 26; 28 24

JUPITER

Day	GHA	Mean Var/hr 15°+	Dec	Mean Var/hr	Mer Pass h m
1 Tu	128 47.1	2.0	S 0 22.4	0.2	15 23
2 Wed	129 35.5	2.0	S 0 17.6	0.2	15 20
3 Th	130 23.7	2.0	S 0 12.7	0.2	15 16
4 Fri	131 11.9	2.0	S 0 07.8	0.2	15 13
5 Sat	132 00.0	2.0	S 0 02.9	0.2	15 10
6 SUN	132 48.0	2.0	N 0 02.1	0.2	15 07
7 Mon	133 35.9	2.0	N 0 07.1	0.2	15 04
8 Tu	134 23.8	2.0	N 0 12.2	0.2	15 00
9 Wed	135 11.6	2.0	N 0 17.2	0.2	14 57
10 Th	135 59.3	2.0	N 0 22.3	0.2	14 54
11 Fri	136 46.9	2.0	N 0 27.4	0.2	14 51
12 Sat	137 34.5	2.0	N 0 32.6	0.2	14 48
13 SUN	138 22.0	2.0	N 0 37.7	0.2	14 45
14 Mon	139 09.4	2.0	N 0 42.9	0.2	14 41
15 Tu	139 56.8	2.0	N 0 48.2	0.2	14 38
16 Wed	140 44.1	2.0	N 0 53.4	0.2	14 35
17 Th	141 31.3	2.0	N 0 58.7	0.2	14 32
18 Fri	142 18.5	2.0	N 1 03.9	0.2	14 29
19 Sat	143 05.7	2.0	N 1 09.3	0.2	14 26
20 SUN	143 52.7	2.0	N 1 14.6	0.2	14 23
21 Mon	144 39.7	2.0	N 1 19.9	0.2	14 19
22 Tu	145 26.7	2.0	N 1 25.3	0.2	14 16
23 Wed	146 13.6	1.9	N 1 30.7	0.2	14 13
24 Th	147 00.4	2.0	N 1 36.1	0.2	14 10
25 Fri	147 47.2	1.9	N 1 41.5	0.2	14 07
26 Sat	148 33.9	1.9	N 1 46.9	0.2	14 04
27 SUN	149 20.6	1.9	N 1 52.4	0.2	14 01
28 Mon	150 07.3	1.9	N 1 57.9		13 58

JUPITER, Av. Mag. –2.1
SHA February 5 357; 10 356; 15 355; 20 354; 25 353; 28 353

SATURN

Day	GHA	Mean Var/hr 15°+	Dec	Mean Var/hr	Mer Pass h m
1 Tu	294 01.6	2.5	S 4 23.6	0.0	04 23
2 Wed	295 01.3	2.5	S 4 23.1	0.0	04 19
3 Th	296 01.0	2.5	S 4 22.6	0.0	04 15
4 Fri	297 00.9	2.5	S 4 22.0	0.0	04 11
5 Sat	298 00.8	2.5	S 4 21.3	0.0	04 07
6 SUN	299 00.9	2.5	S 4 20.7	0.0	04 03
7 Mon	300 01.0	2.5	S 4 19.2	0.0	03 59
8 Tu	301 01.3	2.5	S 4 18.4	0.0	03 55
9 Wed	302 01.6	2.5	S 4 17.6	0.0	03 51
10 Th	303 02.0	2.5	S 4 16.8	0.0	03 47
11 Fri	304 02.6	2.5	S 4 15.9	0.0	03 43
12 Sat	305 03.2	2.5	S 4 15.0	0.0	03 39
13 SUN	306 03.9	2.5	S 4 14.0	0.0	03 35
14 Mon	307 04.7	2.5	S 4 13.0	0.0	03 31
15 Tu	308 05.6	2.5	S 4 12.0	0.0	03 27
16 Wed	309 06.6	2.5	S 4 10.9	0.0	03 23
17 Th	310 07.7	2.5	S 4 09.8	0.0	03 19
18 Fri	311 08.8	2.5	S 4 08.7	0.0	03 15
19 Sat	312 10.1	2.6	S 4 07.5	0.1	03 11
20 SUN	313 11.4	2.6	S 4 06.3	0.1	03 07
21 Mon	314 12.8	2.6	S 4 05.1	0.1	03 03
22 Tu	315 14.3	2.6	S 4 03.9	0.1	02 59
23 Wed	316 15.9	2.6	S 4 02.6	0.1	02 54
24 Th	317 17.6	2.6	S 4 01.3	0.1	02 50
25 Fri	318 19.3	2.6	S 3 59.9	0.1	02 46
26 Sat	319 21.1	2.6	S 3 58.5	0.1	02 42
27 SUN	320 23.0	2.6	S 3 57.1	0.1	02 38
28 Mon	321 25.0	2.6	S 3 57.1		02 34

SATURN, Av. Mag. +0.6
SHA February 5 163; 10 163; 15 163; 20 164; 25 164; 28 164

12

FEBRUARY 2011

STARS

No.	Name	Mag	Transit h m	Dec ° '	SHA ° '
	0h GMT February 1				
ψ	ARIES	—		—	—
1	Alpheratz	2.1	15 14	N29 09.2	357 45.5
2	Ankaa	2.4	15 23	S42 14.9	353 17.6
3	Schedar	2.2	15 41	N56 36.2	349 42.9
4	Diphda	2.0	15 58	S17 55.6	348 57.7
5	Achernar	0.5	16 52	S57 11.0	335 28.2
6	POLARIS	2.0	17 59	N89 19.1	318 41.3
7	Hamal	2.0	16 52	N23 31.0	328 02.8
8	Acamar	3.2	18 12	S40 15.8	315 19.6
9	Menkar	2.5	18 16	N 4 08.0	314 16.8
10	Mirfak	1.8	18 39	N49 54.3	308 42.8
11	Aldebaran	0.9	19 50	N16 31.9	290 51.2
12	Rigel	0.1	20 28	S 8 11.5	281 13.5
13	Capella	0.1	20 31	N46 00.6	280 36.7
14	Bellatrix	1.6	20 39	N 6 21.5	278 33.6
15	Elnath	1.7	20 40	N28 37.0	278 14.5
16	Alnilam	1.7	20 50	S 1 11.8	275 47.9
17	Betelgeuse	0.1–1.2	21 09	N 7 24.4	271 02.9
18	Canopus	−0.7	21 37	S52 42.4	263 56.5
19	Sirius	−1.5	21 59	S16 44.1	258 34.9
20	Adhara	1.5	22 12	S28 59.5	255 13.5
21	Castor	1.6	22 48	N31 51.7	246 09.7
22	Procyon	0.4	22 53	N 5 11.6	245 01.2
23	Pollux	1.1	22 59	N27 59.8	243 29.4
24	Avior	1.9	23 35	S59 32.9	234 18.2
25	Suhail	2.2	0 25	S43 28.9	222 53.3
26	Miaplacidus	1.7	0 30	S69 45.9	221 39.2
27	Alphard	2.0	0 45	S 8 42.6	217 57.4
28	Regulus	1.4	1 25	N11 54.5	207 44.9
29	Dubhe	1.8	2 21	N61 41.2	193 52.9
30	Denebola	2.1	3 06	N14 30.3	182 35.0
31	Gienah	2.6	3 32	S17 36.4	175 53.8
32	Acrux	1.3	3 43	S63 09.6	173 10.8
33	Gacrux	1.6	3 48	S57 10.5	172 02.4
34	Mimosa	1.3	4 04	S59 44.9	167 53.6
35	Alioth	1.8	4 10	N55 53.6	166 21.7
36	Spica	1.0	4 42	S11 13.3	158 32.8
37	Alkaid	1.9	5 04	N49 15.1	152 59.9
38	Hadar	0.6	5 20	S60 25.5	148 50.1
39	Menkent	2.1	5 23	S36 25.5	148 09.4
40	Arcturus	0.0	5 32	N19 07.2	145 57.1
41	Rigil Kent	−0.3	5 56	S60 52.7	139 53.9
42	Zuben'ubi	2.8	6 07	S16 05.4	137 07.2
43	Kochab	2.1	6 06	N74 06.2	137 19.6
44	Alphecca	2.2	6 51	N26 40.3	126 12.4
45	Antares	1.0	7 45	S26 27.4	112 28.4
46	Atria	1.9	8 05	S69 02.6	107 32.6
47	Sabik	2.4	8 26	S15 44.3	102 14.6
48	Shaula	1.6	8 49	S37 06.6	96 24.4
49	Rasalhague	2.1	8 51	N12 33.0	96 08.2
50	Eltanin	2.2	9 12	N51 29.0	90 47.2
51	Kaus Aust	1.9	9 40	S34 22.7	83 46.3
52	Vega	0.0	9 52	N38 47.5	80 40.4
53	Nunki	2.0	10 11	S26 16.9	76 00.6
54	Altair	0.8	11 06	N 8 53.8	62 10.2
55	Peacock	1.9	11 41	S56 41.8	53 22.3
56	Enif	2.4	11 56	N 9 55.6	33 49.1
57	Deneb	1.3	12 59	N45 19.2	49 33.1
58	Al Na'ir	1.7	13 23	S46 54.4	27 46.2
59	Fomalhaut	1.2	14 12	S29 33.8	15 26.1
60	Markab	2.5	14 20	N15 16.0	13 40.3

SUN AND MOON

SUN — Lat 52°N

Yr	Day of Mth	Week	Transit h m	Semi-Diam	Twilight h m	Sunrise h m	Sunset h m	Twilight h m
32	1	Tu	12 14	16.3	07 04	07 41	16 47	17 24
33	2	Wed	12 14	16.3	07 02	07 39	16 49	17 26
34	3	Th	12 14	16.3	07 01	07 37	16 51	17 27
35	4	Fri	12 14	16.3	06 59	07 36	16 53	17 29
36	5	Sat	12 14	16.2	06 58	07 34	16 55	17 31
37	6	Sun	12 14	16.2	06 56	07 32	16 56	17 33
38	7	Mon	12 14	16.2	06 55	07 31	16 58	17 34
39	8	Tu	12 14	16.2	06 53	07 29	17 00	17 36
40	9	Wed	12 14	16.2	06 51	07 27	17 02	17 38
41	10	Th	12 14	16.2	06 50	07 25	17 04	17 40
42	11	Fri	12 14	16.2	06 48	07 23	17 06	17 41
43	12	Sat	12 14	16.2	06 46	07 22	17 08	17 43
44	13	Sun	12 14	16.2	06 44	07 20	17 10	17 45
45	14	Mon	12 14	16.2	06 43	07 18	17 11	17 47
46	15	Tu	12 14	16.2	06 41	07 16	17 13	17 48
47	16	Wed	12 14	16.2	06 39	07 14	17 15	17 50
48	17	Th	12 14	16.2	06 37	07 12	17 17	17 52
49	18	Fri	12 14	16.2	06 35	07 10	17 19	17 54
50	19	Sat	12 14	16.2	06 33	07 08	17 21	17 55
51	20	Sun	12 14	16.2	06 31	07 06	17 23	17 57
52	21	Mon	12 14	16.2	06 29	07 04	17 24	17 59
53	22	Tu	12 14	16.2	06 27	07 02	17 26	18 01
54	23	Wed	12 13	16.2	06 25	07 00	17 28	18 02
55	24	Th	12 13	16.2	06 23	06 57	17 30	18 04
56	25	Fri	12 13	16.2	06 21	06 55	17 32	18 06
57	26	Sat	12 13	16.2	06 19	06 53	17 34	18 08
58	27	Sun	12 13	16.2	06 17	06 51	17 35	18 09
59	28	Mon	12 13	16.2	06 15	06 49	17 37	18 11

MOON — Lat 52°N

Yr	Day of Mth	Week	Age days	Transit (Upper) h m	Diff m	Semi-diam	Hor Par	Moonrise h m	Moonset h m
32	1	Tu	28	10 55	48	15.1	55.4	06 39	15 19
33	2	Wed	29	11 43	44	15.0	55.0	07 05	16 30
34	3	Th	00	12 27	43	14.9	54.7	07 25	17 40
35	4	Fri	01	13 10	41	14.8	54.4	07 42	18 49
36	5	Sat	02	13 51	40	14.8	54.1	07 58	19 57
37	6	Sun	03	14 31	41	14.7	54.0	08 12	21 04
38	7	Mon	04	15 12	42	14.7	54.0	08 27	22 11
39	8	Tu	05	15 54	44	14.8	54.2	08 44	23 19
40	9	Wed	06	16 38	46	14.8	54.5	09 03	—
41	10	Th	07	17 24	50	15.0	54.9	09 26	00 28
42	11	Fri	08	18 14	53	15.1	55.6	09 55	01 36
43	12	Sat	09	19 07	55	15.4	56.3	10 34	02 43
44	13	Sun	10	20 02	57	15.6	57.2	11 24	03 44
45	14	Mon	11	20 59	58	15.9	58.2	12 28	04 37
46	15	Tu	12	21 57	56	16.1	59.1	13 43	05 20
47	16	Wed	13	22 53	56	16.4	60.0	15 07	05 54
48	17	Th	14	23 49	54	16.5	60.7	16 34	06 21
49	18	Fri	15	24 43	—	16.6	61.1	18 02	06 45
50	19	Sat	16	00 43	53	16.6	61.2	19 30	07 06
51	20	Sun	17	01 36	54	16.7	61.0	20 58	07 26
52	21	Mon	18	02 30	55	16.6	60.5	22 25	07 47
53	22	Tu	19	03 24	57	16.5	59.8	23 50	08 11
54	23	Wed	20	04 19	56	16.3	59.0	—	08 40
55	24	Th	21	05 16	56	16.1	58.2	01 09	09 16
56	25	Fri	22	06 12	53	15.8	57.4	02 20	10 02
57	26	Sat	23	07 08	51	15.6	56.6	03 19	10 58
58	27	Sun	24	08 01	48	15.4	55.9	04 06	12 01
59	28	Mon	25	08 52		15.1	55.4	04 41	13 10

Lat Corr to Sunrise, Sunset etc.

Lat °	Twilight h m	Sunrise h m	Sunset h m	Twilight h m
N70	+0 48	+1 23	−1 22	−0 49
68	+0 39	+1 06	−1 06	−0 40
66	+0 32	+0 53	−0 53	−0 32
64	+0 26	+0 42	−0 41	−0 26
62	+0 20	+0 32	−0 32	−0 20
N60	+0 15	+0 25	−0 25	−0 16
58	+0 10	+0 17	−0 17	−0 11
56	+0 06	+0 11	−0 11	−0 07
54	+0 03	+0 05	−0 05	−0 04
50	−0 03	−0 05	+0 04	+0 03
N45	−0 10	−0 15	+0 15	+0 10
40	−0 16	−0 23	+0 23	+0 16
35	−0 22	−0 30	+0 31	+0 21
30	−0 27	−0 37	+0 37	+0 26
20	−0 36	−0 48	+0 48	+0 35
N10	−0 45	−0 59	+0 58	+0 44
0	−0 55	−1 08	+1 07	+0 54
S10	−1 05	−1 16	+1 16	+1 04
20	−1 16	−1 28	+1 26	+1 15
30	−1 28	−1 40	+1 38	+1 30
S35	−1 40	−1 48	+1 45	+1 38
40	−1 50	−1 55	+1 53	+1 49
45	−2 02	−2 05	+2 02	+2 00
S50	−2 17	−2 16	+2 13	+2 16

NOTES
The corrections to sunrise etc. are for middle of February.

Phases of the Moon

	d	h	m
● New Moon	3	02	31
☽ First Quarter	11	07	18
○ Full Moon	18	08	36
☾ Last Quarter	24	23	26

	d	h
Apogee	6	23
Perigee	19	07

SUN AND ARIES · MARCH 2011 · SUN AND ARIES

GMT	SUN GHA ° ′	SUN Dec ° ′	ARIES GHA ° ′
Tuesday, 1st March			
00	176 52.8	S 7 48.1	158 27.4
02	206 53.1	7 46.2	188 32.3
04	236 53.3	7 44.3	218 37.3
06	266 53.5	7 42.4	248 42.2
08	296 53.8	7 40.5	278 47.1
10	326 54.0	7 38.6	308 52.1
12	356 54.3	7 36.7	338 57.0
14	26 54.5	7 34.8	9 01.9
16	56 54.7	7 32.9	39 06.8
18	86 55.0	7 31.0	69 11.8
20	116 55.2	7 29.1	99 16.7
22	146 55.5	S 7 27.2	129 21.6
Wednesday, 2nd March			
00	176 55.7	S 7 25.3	159 26.5
02	206 56.0	7 23.4	189 31.5
04	236 56.2	7 21.5	219 36.4
06	266 56.4	7 19.6	249 41.3
08	296 56.7	7 17.6	279 46.3
10	326 56.9	7 15.7	309 51.2
12	356 57.2	7 13.8	339 56.1
14	26 57.4	7 11.9	10 01.0
16	56 57.7	7 10.0	40 06.0
18	86 57.9	7 08.1	70 10.9
20	116 58.2	7 06.2	100 15.8
22	146 58.5	S 7 04.3	130 20.8
Thursday, 3rd March			
00	176 58.7	S 7 02.4	160 25.7
02	206 59.0	7 00.5	190 30.6
04	236 59.2	6 58.5	220 35.5
06	266 59.5	6 56.6	250 40.5
08	296 59.7	6 54.7	280 45.4
10	327 00.0	6 52.8	310 50.3
12	357 00.2	6 50.9	340 55.3
14	27 00.5	6 49.0	11 00.2
16	57 00.8	6 47.0	41 05.1
18	87 01.0	6 45.1	71 10.0
20	117 01.3	6 43.2	101 15.0
22	147 01.6	S 6 41.3	131 19.9
Friday, 4th March			
00	177 01.8	S 6 39.4	161 24.8
02	207 02.1	6 37.4	191 29.8
04	237 02.4	6 35.5	221 34.7
06	267 02.6	6 33.6	251 39.6
08	297 02.9	6 31.7	281 44.5
10	327 03.2	6 29.7	311 49.5
12	357 03.4	6 27.8	341 54.4
14	27 03.7	6 25.9	11 59.3
16	57 04.0	6 24.0	42 04.2
18	87 04.2	6 22.0	72 09.2
20	117 04.5	6 20.1	102 14.1
22	147 04.8	S 6 18.2	132 19.0
Saturday, 5th March			
00	177 05.0	S 6 16.3	162 24.0
02	207 05.3	6 14.3	192 28.9
04	237 05.6	6 12.4	222 33.8
06	267 05.9	6 10.5	252 38.7
08	297 06.2	6 08.5	282 43.7
10	327 06.4	6 06.6	312 48.6
12	357 06.7	6 04.7	342 53.5
14	27 07.0	6 02.7	12 58.5
16	57 07.3	6 00.8	43 03.4
18	87 07.5	5 58.9	73 08.3
20	117 07.8	5 56.9	103 13.2
22	147 08.1	S 5 55.0	133 18.2

GMT	SUN GHA ° ′	SUN Dec ° ′	ARIES GHA ° ′
Sunday, 6th March			
00	177 08.4	S 5 53.1	163 23.1
02	207 08.7	5 51.1	193 28.0
04	237 09.0	5 49.2	223 33.0
06	267 09.2	5 47.3	253 37.9
08	297 09.5	5 45.3	283 42.8
10	327 09.8	5 43.4	313 47.7
12	357 10.1	5 41.4	343 52.7
14	27 10.4	5 39.5	13 57.6
16	57 10.7	5 37.6	44 02.5
18	87 11.0	5 35.6	74 07.4
20	117 11.3	5 33.7	104 12.4
22	147 11.5	S 5 31.7	134 17.3
Monday, 7th March			
00	177 11.8	S 5 29.8	164 22.2
02	207 12.1	5 27.9	194 27.2
04	237 12.4	5 25.9	224 32.1
06	267 12.7	5 24.0	254 37.0
08	297 13.0	5 22.0	284 41.9
10	327 13.3	5 20.1	314 46.9
12	357 13.6	5 18.1	344 51.8
14	27 13.9	5 16.2	14 56.7
16	57 14.2	5 14.3	45 01.7
18	87 14.5	5 12.3	75 06.6
20	117 14.8	5 10.4	105 11.5
22	147 15.1	S 5 08.4	135 16.4
Tuesday, 8th March			
00	177 15.4	S 5 06.5	165 21.4
02	207 15.7	5 04.5	195 26.3
04	237 16.0	5 02.6	225 31.2
06	267 16.3	5 00.6	255 36.2
08	297 16.6	4 58.7	285 41.1
10	327 16.9	4 56.7	315 46.0
12	357 17.2	4 54.8	345 50.9
14	27 17.5	4 52.8	15 55.9
16	57 17.8	4 50.9	46 00.8
18	87 18.1	4 48.9	76 05.7
20	117 18.4	4 47.0	106 10.7
22	147 18.7	S 4 45.0	136 15.6
Wednesday, 9th March			
00	177 19.0	S 4 43.1	166 20.5
02	207 19.3	4 41.1	196 25.4
04	237 19.6	4 39.2	226 30.4
06	267 20.0	4 37.2	256 35.3
08	297 20.3	4 35.2	286 40.2
10	327 20.6	4 33.3	316 45.1
12	357 20.9	4 31.3	346 50.1
14	27 21.2	4 29.4	16 55.0
16	57 21.5	4 27.4	46 59.9
18	87 21.8	4 25.5	77 04.9
20	117 22.1	4 23.5	107 09.8
22	147 22.5	S 4 21.6	137 14.7
Thursday, 10th March			
00	177 22.8	S 4 19.6	167 19.6
02	207 23.1	4 17.6	197 24.6
04	237 23.4	4 15.7	227 29.5
06	267 23.7	4 13.7	257 34.4
08	297 24.0	4 11.8	287 39.4
10	327 24.4	4 09.8	317 44.3
12	357 24.7	4 07.8	347 49.2
14	27 25.0	4 05.9	17 54.1
16	57 25.3	4 03.9	47 59.1
18	87 25.6	4 02.0	78 04.0
20	117 26.0	4 00.0	108 08.9
22	147 26.3	S 3 58.0	138 13.9

GMT	SUN GHA ° ′	SUN Dec ° ′	ARIES GHA ° ′
Friday, 11th March			
00	177 26.6	S 3 56.1	168 18.8
02	207 26.9	3 54.1	198 23.7
04	237 27.3	3 52.2	228 28.6
06	267 27.6	3 50.2	258 33.6
08	297 27.9	3 48.2	288 38.5
10	327 28.2	3 46.3	318 43.4
12	357 28.6	3 44.3	348 48.4
14	27 28.9	3 42.3	18 53.3
16	57 29.2	3 40.4	48 58.2
18	87 29.5	3 38.4	79 03.1
20	117 29.9	3 36.4	109 08.1
22	147 30.2	S 3 34.5	139 13.0
Saturday, 12th March			
00	177 30.5	S 3 32.5	169 17.9
02	207 30.9	3 30.5	199 22.9
04	237 31.2	3 28.6	229 27.8
06	267 31.5	3 26.6	259 32.7
08	297 31.8	3 24.7	289 37.6
10	327 32.2	3 22.7	319 42.6
12	357 32.5	3 20.7	349 47.5
14	27 32.8	3 18.7	19 52.4
16	57 33.2	3 16.8	49 57.3
18	87 33.5	3 14.8	80 02.3
20	117 33.8	3 12.8	110 07.2
22	147 34.2	S 3 10.9	140 12.1
Sunday, 13th March			
00	177 34.5	S 3 08.9	170 17.1
02	207 34.9	3 06.9	200 22.0
04	237 35.2	3 05.0	230 26.9
06	267 35.5	3 03.0	260 31.8
08	297 35.9	3 01.0	290 36.8
10	327 36.2	2 59.1	320 41.7
12	357 36.5	2 57.1	350 46.6
14	27 36.9	2 55.1	20 51.6
16	57 37.2	2 53.2	50 56.5
18	87 37.6	2 51.2	81 01.4
20	117 37.9	2 49.2	111 06.3
22	147 38.2	S 2 47.2	141 11.3
Monday, 14th March			
00	177 38.6	S 2 45.3	171 16.2
02	207 38.9	2 43.3	201 21.1
04	237 39.3	2 41.3	231 26.1
06	267 39.6	2 39.4	261 31.0
08	297 40.0	2 37.4	291 35.9
10	327 40.3	2 35.4	321 40.8
12	357 40.7	2 33.4	351 45.8
14	27 41.0	2 31.5	21 50.7
16	57 41.3	2 29.5	51 55.6
18	87 41.7	2 27.5	82 00.6
20	117 42.0	2 25.6	112 05.5
22	147 42.4	S 2 23.6	142 10.4
Tuesday, 15th March			
00	177 42.7	S 2 21.6	172 15.3
02	207 43.1	2 19.6	202 20.3
04	237 43.4	2 17.7	232 25.2
06	267 43.8	2 15.7	262 30.1
08	297 44.1	2 13.7	292 35.1
10	327 44.5	2 11.7	322 40.0
12	357 44.8	2 09.8	352 44.9
14	27 45.2	2 07.8	22 49.8
16	57 45.5	2 05.8	52 54.8
18	87 45.9	2 03.8	82 59.7
20	117 46.2	2 01.9	113 04.6
22	147 46.6	S 1 59.9	143 09.6

GMT	SUN GHA ° ′	SUN Dec ° ′	ARIES GHA ° ′
Wednesday, 16th March			
00	177 46.9	S 1 57.9	173 14.5
02	207 47.3	1 55.9	203 19.4
04	237 47.6	1 54.0	233 24.3
06	267 48.0	1 52.0	263 29.3
08	297 48.4	1 50.0	293 34.2
10	327 48.7	1 48.0	323 39.1
12	357 49.1	1 46.1	353 44.1
14	27 49.4	1 44.1	23 49.0
16	57 49.8	1 42.1	53 53.9
18	87 50.1	1 40.1	83 58.8
20	117 50.5	1 38.2	114 03.8
22	147 50.9	S 1 36.2	144 08.7
Thursday, 17th March			
00	177 51.2	S 1 34.2	174 13.6
02	207 51.6	1 32.2	204 18.5
04	237 51.9	1 30.3	234 23.5
06	267 52.3	1 28.3	264 28.4
08	297 52.6	1 26.3	294 33.3
10	327 53.0	1 24.3	324 38.3
12	357 53.4	1 22.4	354 43.2
14	27 53.7	1 20.4	24 48.1
16	57 54.1	1 18.4	54 53.0
18	87 54.4	1 16.4	84 58.0
20	117 54.8	1 14.5	115 02.9
22	147 55.2	S 1 12.5	145 07.8
Friday, 18th March			
00	177 55.5	S 1 10.5	175 12.8
02	207 55.9	1 08.5	205 17.7
04	237 56.3	1 06.5	235 22.6
06	267 56.6	1 04.6	265 27.5
08	297 57.0	1 02.6	295 32.5
10	327 57.3	1 00.6	325 37.4
12	357 57.7	0 58.6	355 42.3
14	27 58.1	0 56.7	25 47.3
16	57 58.4	0 54.7	55 52.2
18	87 58.8	0 52.7	85 57.1
20	117 59.2	0 50.7	116 02.0
22	147 59.5	S 0 48.8	146 07.0
Saturday, 19th March			
00	177 59.9	S 0 46.8	176 11.9
02	208 00.3	0 44.8	206 16.8
04	238 00.6	0 42.8	236 21.7
06	268 01.0	0 40.9	266 26.7
08	298 01.4	0 38.9	296 31.6
10	328 01.7	0 36.9	326 36.5
12	358 02.1	0 34.9	356 41.5
14	28 02.5	0 33.0	26 46.4
16	58 02.8	0 31.0	56 51.3
18	88 03.2	0 29.0	86 56.2
20	118 03.6	0 27.0	117 01.2
22	148 03.9	S 0 25.0	147 06.1
Sunday, 20th March			
00	178 04.3	S 0 23.1	177 11.0
02	208 04.7	0 21.1	207 16.0
04	238 05.1	0 19.1	237 20.9
06	268 05.4	0 17.1	267 25.8
08	298 05.8	0 15.2	297 30.7
10	328 06.2	0 13.2	327 35.7
12	358 06.5	0 11.2	357 40.6
14	28 06.9	0 09.2	27 45.5
16	58 07.3	0 07.3	57 50.5
18	88 07.6	0 05.3	87 55.4
20	118 08.0	0 03.3	118 00.3
22	148 08.4	S 0 01.3	148 05.2

GMT	SUN GHA ° ′	SUN Dec ° ′	ARIES GHA ° ′
Monday, 21st March			
00	178 08.8	N 0 00.6	178 10.2
02	208 09.1	0 02.6	208 15.1
04	238 09.5	0 04.6	238 20.0
06	268 09.9	0 06.6	268 25.0
08	298 10.3	0 08.5	298 29.9
10	328 10.6	0 10.5	328 34.8
12	358 11.0	0 12.5	358 39.7
14	28 11.4	0 14.5	28 44.7
16	58 11.7	0 16.4	58 49.6
18	88 12.1	0 18.4	88 54.5
20	118 12.5	0 20.4	118 59.4
22	148 12.9	N 0 22.4	149 04.4
Tuesday, 22nd March			
00	178 13.2	N 0 24.3	179 09.3
02	208 13.6	0 26.3	209 14.2
04	238 14.0	0 28.3	239 19.2
06	268 14.4	0 30.3	269 24.1
08	298 14.7	0 32.2	299 29.0
10	328 15.1	0 34.2	329 33.9
12	358 15.5	0 36.2	359 38.9
14	28 15.9	0 38.1	29 43.8
16	58 16.2	0 40.1	59 48.7
18	88 16.6	0 42.1	89 53.7
20	118 17.0	0 44.1	119 58.6
22	148 17.4	N 0 46.0	150 03.5
Wednesday, 23rd March			
00	178 17.7	N 0 48.0	180 08.4
02	208 18.1	0 50.0	210 13.4
04	238 18.5	0 52.0	240 18.3
06	268 18.9	0 53.9	270 23.2
08	298 19.2	0 55.9	300 28.2
10	328 19.6	0 57.9	330 33.1
12	358 20.0	0 59.8	0 38.0
14	28 20.4	1 01.8	30 42.9
16	58 20.8	1 03.8	60 47.9
18	88 21.1	1 05.8	90 52.8
20	118 21.5	1 07.7	120 57.7
22	148 21.9	N 1 09.7	151 02.7
Thursday, 24th March			
00	178 22.3	N 1 11.7	181 07.6
02	208 22.6	1 13.6	211 12.5
04	238 23.0	1 15.6	241 17.4
06	268 23.4	1 17.6	271 22.4
08	298 23.8	1 19.5	301 27.3
10	328 24.1	1 21.5	331 32.2
12	358 24.5	1 23.5	1 37.2
14	28 24.9	1 25.5	31 42.1
16	58 25.3	1 27.4	61 47.0
18	88 25.7	1 29.4	91 51.9
20	118 26.0	1 31.4	121 56.9
22	148 26.4	N 1 33.3	152 01.8
Friday, 25th March			
00	178 26.8	N 1 35.3	182 06.7
02	208 27.2	1 37.3	212 11.6
04	238 27.5	1 39.2	242 16.6
06	268 27.9	1 41.2	272 21.5
08	298 28.3	1 43.2	302 26.4
10	328 28.7	1 45.1	332 31.4
12	358 29.1	1 47.1	2 36.3
14	28 29.4	1 49.1	32 41.2
16	58 29.8	1 51.0	62 46.1
18	88 30.2	1 53.0	92 51.1
20	118 30.6	1 55.0	122 56.0
22	148 30.9	N 1 56.9	153 00.9

GMT	SUN GHA ° ′	SUN Dec ° ′	ARIES GHA ° ′
Saturday, 26th March			
00	178 31.3	N 1 58.9	183 05.9
02	208 31.7	2 00.8	213 10.8
04	238 32.1	2 02.8	243 15.7
06	268 32.5	2 04.8	273 20.6
08	298 32.8	2 06.7	303 25.6
10	328 33.2	2 08.7	333 30.5
12	358 33.6	2 10.7	3 35.4
14	28 34.0	2 12.6	33 40.4
16	58 34.3	2 14.6	63 45.3
18	88 34.7	2 16.6	93 50.2
20	118 35.1	2 18.5	123 55.1
22	148 35.5	N 2 20.5	154 00.1
Sunday, 27th March			
00	178 35.9	N 2 22.4	184 05.0
02	208 36.2	2 24.4	214 09.9
04	238 36.6	2 26.4	244 14.9
06	268 37.0	2 28.3	274 19.8
08	298 37.4	2 30.3	304 24.7
10	328 37.7	2 32.2	334 29.6
12	358 38.1	2 34.2	4 34.6
14	28 38.5	2 36.1	34 39.5
16	58 38.9	2 38.1	64 44.4
18	88 39.3	2 40.1	94 49.4
20	118 39.6	2 42.0	124 54.3
22	148 40.0	N 2 44.0	154 59.2
Monday, 28th March			
00	178 40.4	N 2 45.9	185 04.1
02	208 40.8	2 47.9	215 09.1
04	238 41.1	2 49.8	245 14.0
06	268 41.5	2 51.8	275 18.9
08	298 41.9	2 53.8	305 23.9
10	328 42.3	2 55.7	335 28.8
12	358 42.6	2 57.7	5 33.7
14	28 43.0	2 59.6	35 38.6
16	58 43.4	3 01.6	65 43.6
18	88 43.8	3 03.5	95 48.5
20	118 44.2	3 05.5	125 53.4
22	148 44.5	N 3 07.4	155 58.4
Tuesday, 29th March			
00	178 44.9	N 3 09.4	186 03.3
02	208 45.3	3 11.3	216 08.2
04	238 45.7	3 13.3	246 13.1
06	268 46.0	3 15.2	276 18.1
08	298 46.4	3 17.2	306 23.0
10	328 46.8	3 19.1	336 27.9
12	358 47.2	3 21.1	6 32.8
14	28 47.5	3 23.0	36 37.8
16	58 47.9	3 25.0	66 42.7
18	88 48.3	3 26.9	96 47.6
20	118 48.7	3 28.9	126 52.6
22	148 49.0	N 3 30.8	156 57.5
Wednesday, 30th March			
00	178 49.4	N 3 32.8	187 02.4
02	208 49.8	3 34.7	217 07.3
04	238 50.2	3 36.7	247 12.3
06	268 50.5	3 38.6	277 17.2
08	298 50.9	3 40.6	307 22.1
10	328 51.3	3 42.5	337 27.1
12	358 51.7	3 44.4	7 32.0
14	28 52.0	3 46.4	37 36.9
16	58 52.4	3 48.3	67 41.8
18	88 52.8	3 50.3	97 46.8
20	118 53.2	3 52.2	127 51.7
22	148 53.5	N 3 54.2	157 56.6

GMT	SUN GHA ° ′	SUN Dec ° ′	ARIES GHA ° ′
Thursday, 31st March			
00	178 53.9	N 3 56.1	188 01.6
02	208 54.3	3 58.0	218 06.5
04	238 54.6	4 00.0	248 11.4
06	268 55.0	N 4 01.9	278 16.3
08	298 55.4	N 4 03.8	308 21.3
10	328 55.8	4 05.8	338 26.2
12	358 56.1	4 07.7	8 31.1
14	28 56.5	N 4 09.7	38 36.1
16	58 56.9	N 4 11.6	68 41.0
18	88 57.3	4 13.5	98 45.9
20	118 57.6	4 15.5	128 50.8
22	148 58.0	N 4 17.4	158 55.8

14 MARCH

MOON — MARCH 2011

Day	GMT hr	GHA ° '	Mean Var/hr 14°+	Dec ° '	Mean Var/hr
16 Wed	0	219 32.6	31.5	S18 26.5	8.6
	6	306 41.2	31.9	S17 35.5	9.0
	12	33 52.2	32.3	S16 41.8	9.4
	18	121 05.6	32.6	S15 45.6	9.8
17 Th	0	208 21.2	32.9	S14 47.2	10.2
	6	295 39.0	33.4	S13 46.6	10.4
	12	22 58.8	33.6	S12 44.2	10.7
	18	110 20.4	33.9	S11 40.0	10.9
18 Fri	0	197 43.8	34.2	S10 34.3	11.2
	6	285 08.8	34.4	S 9 27.2	11.4
	12	12 35.3	34.7	S 8 19.0	11.5
	18	100 03.1	34.8	S 7 09.7	11.7
19 Sat	0	187 32.0	35.0	S 5 59.5	11.8
	6	275 01.9	35.2	S 4 48.7	11.9
	12	2 32.6	35.2	S 3 37.3	12.0
	18	90 03.9	35.3	S 2 25.6	12.0
20 Sun	0	177 35.8	35.4	S 1 13.6	11.9
	6	265 07.9	35.4	S 0 01.5	12.0
	12	352 40.3	35.3	N 1 10.4	12.0
	18	80 12.6	35.3	N 2 22.1	11.8
21 Mon	0	167 44.7	35.2	N 3 33.5	11.7
	6	255 16.5	35.1	N 4 44.3	11.6
	12	342 47.8	35.1	N 5 54.5	11.4
	18	70 18.4	34.9	N 7 03.8	11.2
22 Tu	0	157 48.2	34.8	N 8 12.1	10.8
	6	245 17.0	34.6	N 9 19.4	10.8
	12	332 44.7	34.4	N10 25.6	10.5
	18	60 11.1	34.2	N11 29.8	10.1
23 Wed	0	147 36.0	33.9	N12 32.7	10.1
	6	234 59.0	33.6	N13 33.2	9.9
	12	322 21.1	33.3	N14 33.2	9.5
	18	49 41.0	32.9	N15 30.5	9.1
24 Th	0	136 59.0	32.7	N16 25.5	8.7
	6	224 14.9	32.3	N17 18.1	8.3
	12	311 28.8	32.0	N18 08.2	7.9
	18	38 40.4	31.6	N18 55.5	7.4
25 Fri	0	125 49.7	31.1	N19 40.0	6.9
	6	212 56.8	30.8	N20 21.3	6.3
	12	300 01.5	30.4	N20 59.5	5.8
	18	27 03.9	30.0	N21 34.2	5.1
26 Sat	0	114 04.0	29.6	N22 05.3	4.6
	6	201 01.7	29.2	N22 32.7	3.8
	12	287 57.3	28.9	N22 56.1	3.2
	18	14 50.7	28.5	N23 15.5	2.4
27 Sun	0	101 42.0	28.2	N23 30.6	1.7
	6	188 31.4	27.9	N23 41.4	1.0
	12	275 19.1	27.6	N23 47.7	0.2
	18	2 05.1	27.4	N23 49.3	0.6
28 Mon	0	88 49.6	27.2	N23 46.3	1.4
	6	175 32.9	27.1	N23 38.4	2.2
	12	262 15.1	26.9	N23 25.6	3.0
	18	348 56.5	26.8	N23 07.9	3.8
29 Tu	0	75 37.2	26.7	N22 45.3	4.7
	6	162 17.5	26.7	N22 17.6	5.5
	12	248 57.4	26.6	N21 45.1	6.3
	18	335 37.3	26.6	N21 07.6	7.1
30 Wed	0	62 17.3	26.7	N20 25.2	8.0
	6	148 57.5	26.7	N19 38.1	8.7
	12	235 38.0	26.8	N18 46.4	9.5
	18	322 18.9	26.9	N17 50.1	10.2
31 Th	0	49 00.4	27.0	N16 49.5	10.9
	6	135 42.5	27.1	N15 44.7	11.5
	12	222 25.2	27.2	N14 35.9	12.1
	18	309 08.5	27.4	N13 23.4	12.7

PLANETS — MARCH 2011

VENUS

Day	Mer Pass h m	GHA ° '	Mean Var/hr 14°+	Dec ° '	Mean Var/hr
1 Tu	09 29	217 50.2	59.4	S19 32.1	0.4
2 Wed	09 30	217 35.0	59.4	S19 21.4	0.5
3 Th	09 31	217 21.4	59.4	S19 10.1	0.5
4 Fri	09 32	217 05.0	59.4	S18 58.3	0.5
5 Sat	09 33	216 50.2	59.4	S18 45.9	0.5
6 SUN	09 34	216 35.5	59.4	S18 33.0	0.6
7 Mon	09 35	216 20.9	59.4	S18 19.5	0.6
8 Tu	09 36	216 06.5	59.4	S18 05.5	0.6
9 Wed	09 37	215 52.2	59.4	S17 51.1	0.6
10 Th	09 38	215 38.1	59.4	S17 36.1	0.6
11 Fri	09 39	215 24.1	59.4	S17 20.6	0.7
12 Sat	09 40	215 10.3	59.4	S17 04.6	0.7
13 SUN	09 41	214 56.7	59.4	S16 48.1	0.7
14 Mon	09 41	214 43.3	59.5	S16 31.2	0.7
15 Tu	09 42	214 30.0	59.5	S16 13.8	0.7
16 Wed	09 43	214 17.0	59.5	S15 55.9	0.8
17 Th	09 44	214 04.1	59.5	S15 37.6	0.8
18 Fri	09 45	213 51.4	59.5	S15 18.8	0.8
19 Sat	09 46	213 38.9	59.5	S14 59.6	0.8
20 SUN	09 47	213 26.6	59.5	S14 40.0	0.8
21 Mon	09 48	213 14.5	59.5	S14 19.9	0.9
22 Tu	09 49	213 02.5	59.5	S13 59.5	0.9
23 Wed	09 50	212 50.8	59.5	S13 38.6	0.9
24 Th	09 51	212 39.2	59.5	S13 17.4	0.9
25 Fri	09 52	212 27.8	59.5	S12 55.8	0.9
26 Sat	09 53	212 16.6	59.5	S12 33.8	1.0
27 SUN	09 54	212 05.6	59.6	S12 11.4	1.0
28 Mon	09 55	211 54.8	59.6	S11 48.8	1.0
29 Tu	09 55	211 44.1	59.6	S11 25.7	1.0
30 Wed	09 55	211 33.6	59.6	S11 02.4	1.0
31 Th	09 55	211 23.3	59.6	S10 38.7	1.0

VENUS, Av. Mag. −4.0
SHA March 5 54; 10 48; 15 42; 20 36; 25 30; 30 25

JUPITER

Day	GHA ° '	Mean Var/hr 15°+	Dec ° '	Mean Var/hr	Mer Pass h m
1 Tu	150 53.9	1.9	N 2 03.4	0.2	13 55
2 Wed	151 40.4	1.9	N 2 08.9	0.2	13 52
3 Th	152 26.9	1.9	N 2 14.4	0.2	13 48
4 Fri	153 13.4	1.9	N 2 19.9	0.2	13 45
5 Sat	153 59.8	1.9	N 2 25.4	0.2	13 42
6 SUN	154 46.1	1.9	N 2 31.0	0.2	13 39
7 Mon	155 32.5	1.9	N 2 36.5	0.2	13 36
8 Tu	156 18.8	1.9	N 2 42.1	0.2	13 33
9 Wed	157 05.0	1.9	N 2 47.7	0.2	13 30
10 Th	157 51.2	1.9	N 2 53.3	0.2	13 27
11 Fri	158 37.4	1.9	N 2 58.8	0.2	13 24
12 Sat	159 23.6	1.9	N 3 04.5	0.2	13 21
13 SUN	160 09.7	1.9	N 3 10.1	0.2	13 18
14 Mon	160 55.8	1.9	N 3 15.7	0.2	13 15
15 Tu	161 41.8	1.9	N 3 21.3	0.2	13 12
16 Wed	162 27.9	1.9	N 3 26.9	0.2	13 08
17 Th	163 13.9	1.9	N 3 32.5	0.2	13 05
18 Fri	163 59.9	1.9	N 3 38.2	0.2	13 02
19 Sat	164 45.8	1.9	N 3 43.8	0.2	12 59
20 SUN	165 31.7	1.9	N 3 49.4	0.2	12 56
21 Mon	166 17.7	1.9	N 3 55.1	0.2	12 53
22 Tu	167 03.5	1.9	N 4 00.7	0.2	12 50
23 Wed	167 49.4	1.9	N 4 06.3	0.2	12 47
24 Th	168 35.3	1.9	N 4 12.0	0.2	12 44
25 Fri	169 21.1	1.9	N 4 17.6	0.2	12 41
26 Sat	170 06.9	1.9	N 4 23.2	0.2	12 38
27 SUN	170 52.7	1.9	N 4 28.9	0.2	12 35
28 Mon	171 38.4	1.9	N 4 34.5	0.2	12 32
29 Tu	172 24.2	1.9	N 4 40.1	0.2	12 29
30 Wed	173 09.9	1.9	N 4 45.8	0.2	12 26
31 Th	173 55.7	1.9	N 4 51.4	0.2	12 23

JUPITER, Av. Mag. −2.1
SHA March 5 352; 10 351; 15 349; 20 348; 25 347; 30 346

MARS

Day	Mer Pass h m	GHA ° '	Mean Var/hr 15°+	Dec ° '	Mean Var/hr
1 Tu	11 54	181 30.4	0.6	S10 46.4	0.7
2 Wed	11 53	181 44.8	0.6	S10 28.8	0.7
3 Th	11 52	181 59.2	0.6	S10 11.2	0.7
4 Fri	11 51	182 13.7	0.6	S 9 53.5	0.7
5 Sat	11 50	182 28.4	0.6	S 9 35.7	0.7
6 SUN	11 49	182 43.1	0.6	S 9 17.9	0.8
7 Mon	11 48	182 57.9	0.6	S 8 59.9	0.8
8 Tu	11 47	183 12.7	0.6	S 8 41.8	0.8
9 Wed	11 46	183 27.7	0.6	S 8 23.7	0.8
10 Th	11 45	183 42.7	0.6	S 8 05.5	0.8
11 Fri	11 44	183 57.8	0.6	S 7 47.3	0.8
12 Sat	11 43	184 13.0	0.6	S 7 29.0	0.8
13 SUN	11 42	184 28.3	0.6	S 7 10.6	0.8
14 Mon	11 41	184 43.6	0.6	S 6 52.1	0.8
15 Tu	11 40	184 59.1	0.6	S 6 33.6	0.8
16 Wed	11 39	185 14.5	0.6	S 6 15.1	0.8
17 Th	11 38	185 30.1	0.6	S 5 56.5	0.8
18 Fri	11 37	185 45.7	0.7	S 5 37.9	0.8
19 Sat	11 36	186 01.4	0.7	S 5 19.2	0.8
20 SUN	11 35	186 17.1	0.7	S 5 00.5	0.8
21 Mon	11 34	186 32.9	0.7	S 4 41.7	0.8
22 Tu	11 33	186 48.8	0.7	S 4 22.9	0.8
23 Wed	11 32	187 04.7	0.7	S 4 04.1	0.8
24 Th	11 31	187 20.6	0.7	S 3 45.2	0.8
25 Fri	11 30	187 36.7	0.7	S 3 26.4	0.8
26 Sat	11 29	187 52.7	0.7	S 3 07.5	0.8
27 SUN	11 28	188 08.8	0.7	S 2 48.6	0.8
28 Mon	11 27	188 25.0	0.7	S 2 29.7	0.8
29 Tu	11 26	188 41.2	0.7	S 2 10.7	0.8
30 Wed	11 25	188 57.4	0.7	S 1 51.8	0.8
31 Th	11 24	189 13.7	0.7	S 1 32.9	0.8

MARS, Av. Mag. +1.1
SHA March 5 20; 10 16; 15 13; 20 9; 25 5; 30 2

SATURN

Day	GHA ° '	Mean Var/hr 15°+	Dec ° '	Mean Var/hr	Mer Pass h m
1 Tu	322 27.1	2.6	S 3 55.7	0.1	02 30
2 Wed	323 29.2	2.6	S 3 54.3	0.1	02 26
3 Th	324 31.4	2.6	S 3 52.8	0.1	02 21
4 Fri	325 33.7	2.6	S 3 51.3	0.1	02 17
5 Sat	326 36.0	2.6	S 3 49.8	0.1	02 13
6 SUN	327 38.4	2.6	S 3 48.2	0.1	02 09
7 Mon	328 40.9	2.6	S 3 46.6	0.1	02 05
8 Tu	329 43.4	2.6	S 3 45.0	0.1	02 01
9 Wed	330 46.0	2.6	S 3 43.4	0.1	01 57
10 Th	331 48.7	2.6	S 3 41.8	0.1	01 52
11 Fri	332 51.4	2.6	S 3 40.1	0.1	01 48
12 Sat	333 54.1	2.6	S 3 38.5	0.1	01 44
13 SUN	334 56.9	2.6	S 3 36.8	0.1	01 40
14 Mon	335 59.8	2.6	S 3 35.1	0.1	01 36
15 Tu	337 02.7	2.6	S 3 33.4	0.1	01 32
16 Wed	338 05.7	2.6	S 3 31.6	0.1	01 27
17 Th	339 08.7	2.6	S 3 29.9	0.1	01 23
18 Fri	340 11.7	2.6	S 3 28.1	0.1	01 19
19 Sat	341 14.8	2.6	S 3 26.4	0.1	01 15
20 SUN	342 17.9	2.6	S 3 24.6	0.1	01 11
21 Mon	343 21.1	2.6	S 3 22.8	0.1	01 06
22 Tu	344 24.3	2.6	S 3 21.0	0.1	01 02
23 Wed	345 27.5	2.6	S 3 19.2	0.1	00 58
24 Th	346 30.8	2.6	S 3 17.4	0.1	00 54
25 Fri	347 34.1	2.6	S 3 15.6	0.1	00 50
26 Sat	348 37.4	2.6	S 3 13.7	0.1	00 45
27 SUN	349 40.7	2.6	S 3 11.9	0.1	00 41
28 Mon	350 44.1	2.6	S 3 10.1	0.1	00 37
29 Tu	351 47.4	2.6	S 3 08.2	0.1	00 33
30 Wed	352 50.8	2.6	S 3 06.4	0.1	00 29
31 Th	353 54.2	2.6	S 3 04.6	0.1	00 24

SATURN, Av. Mag. +0.4
SHA March 5 164; 10 164; 15 165; 20 165; 25 165; 30 166

MARCH 2011

STARS

No.	Name	Mag	Transit h m	Dec ° '	SHA ° '
	0h GMT March 1				
ψ	ARIES	—	—	—	—
1	Alpheratz	2.1	13 24	N29 09.2	357 45.6
2	Ankaa	2.4	13 42	S42 14.7	353 17.6
3	Schedar	2.2	13 56	N56 36.0	349 43.0
4	Diphda	2.0	13 59	S17 55.6	348 57.8
5	Achernar	0.5	14 53	S57 10.9	335 28.3
6	POLARIS	2.0	15 59	N89 19.0	318 52.2
7	Hamal	2.0	15 22	N23 31.0	328 02.9
8	Acamar	3.2	16 13	S40 15.8	315 19.8
9	Menkar	2.5	16 17	N 4 08.0	314 16.9
10	Mirfak	1.8	16 40	N49 54.2	308 43.0
11	Aldebaran	0.9	17 51	N16 31.9	290 51.3
12	Rigel	0.1	18 29	S 8 11.5	281 13.6
13	Capella	0.1	18 32	N46 00.7	280 36.9
14	Bellatrix	1.6	18 40	N 6 21.5	278 33.7
15	Elnath	1.7	18 41	N28 37.0	278 14.7
16	Alnilam	1.7	18 51	S 1 11.9	275 48.0
17	Betelgeuse	0.1 –1.2	19 10	N 7 24.4	271 03.0
18	Canopus	−0.7	19 38	S52 42.5	263 56.8
19	Sirius	−1.5	19 59	S16 44.2	258 35.0
20	Adhara	1.5	20 13	S28 59.6	255 13.7
21	Castor	1.6	20 49	N31 51.7	246 09.8
22	Procyon	0.4	20 54	N 5 11.6	245 01.2
23	Pollux	1.1	21 00	N27 59.9	243 29.5
24	Avior	1.9	21 36	S59 33.1	234 18.3
25	Suhail	2.2	22 22	S43 29.0	222 53.3
26	Miaplacidus	1.7	22 27	S69 46.1	221 39.4
27	Alphard	2.0	22 41	S 8 42.7	217 57.4
28	Regulus	1.4	23 22	N11 54.5	207 44.9
29	Dubhe	1.8	00 21	N61 41.3	193 52.9
30	Denebola	2.1	01 07	N14 30.3	182 34.9
31	Gienah	2.6	01 33	S17 36.5	175 53.6
32	Acrux	1.3	01 44	S63 09.8	173 10.6
33	Gacrux	1.6	01 49	S57 10.7	172 02.2
34	Mimosa	1.3	02 05	S59 45.1	167 53.4
35	Alioth	1.8	02 11	N55 53.7	166 21.4
36	Spica	1.0	02 42	S11 13.4	158 32.7
37	Alkaid	1.9	03 05	N49 15.1	152 59.7
38	Hadar	0.6	03 21	S60 25.6	148 49.8
39	Menkent	2.1	03 24	S36 25.6	148 09.2
40	Arcturus	0.0	03 33	N19 07.2	145 56.9
41	Rigil Kent	−0.3	03 57	S60 52.8	139 53.6
42	Zuben'ubi	2.8	04 08	S16 05.4	137 07.0
43	Kochab	2.1	04 07	N74 06.2	137 19.0
44	Alphecca	2.2	04 51	N26 40.3	126 12.2
45	Antares	1.0	05 46	S26 27.4	112 28.1
46	Atria	1.9	06 06	S69 02.6	107 31.4
47	Sabik	2.4	06 27	S15 44.4	102 14.3
48	Shaula	1.6	06 50	S37 06.6	96 24.1
49	Rasalhague	2.1	06 51	N12 32.9	96 07.9
50	Eltanin	2.2	07 13	N51 28.9	90 46.9
51	Kaus Aust	1.9	07 41	S34 22.6	83 46.0
52	Vega	0.0	07 53	N38 47.4	80 40.2
53	Nunki	2.0	08 12	S26 16.9	76 00.4
54	Altair	0.8	09 07	N 8 53.8	62 10.0
55	Peacock	1.9	09 42	S56 41.7	53 22.1
56	Deneb	1.3	09 57	N45 19.1	49 32.9
57	Enif	2.4	11 00	N 9 55.5	33 49.0
58	Al Na'ir	1.7	11 24	S46 54.3	27 46.1
59	Fomalhaut	1.2	12 13	S29 33.7	15 26.1
60	Markab	2.5	12 20	N15 15.9	13 40.3

SUN AND MOON

SUN

Yr	Mth	Week	Transit h m	Semi-Diam	Twilight (Lat 52°N)	Sunrise	Sunset	Twilight
60	1	Tu	12 12	16.2	06 13	06 47	17 39	18 13
61	2	Wed	12 12	16.2	06 11	06 45	17 41	18 15
62	3	Th	12 12	16.2	06 09	06 42	17 43	18 16
63	4	Fri	12 12	16.2	06 06	06 40	17 44	18 18
64	5	Sat	12 11	16.2	06 04	06 38	17 46	18 20
65	6	Sun	12 11	16.1	06 02	06 36	17 48	18 22
66	7	Mon	12 11	16.1	06 00	06 33	17 50	18 23
67	8	Tu	12 11	16.1	05 58	06 31	17 51	18 25
68	9	Wed	12 10	16.1	05 55	06 29	17 53	18 27
69	10	Th	12 10	16.1	05 53	06 27	17 55	18 29
70	11	Fri	12 10	16.1	05 51	06 24	17 57	18 30
71	12	Sat	12 10	16.1	05 49	06 22	17 59	18 32
72	13	Sun	12 09	16.1	05 46	06 20	18 00	18 34
73	14	Mon	12 09	16.1	05 44	06 18	18 02	18 36
74	15	Tu	12 09	16.1	05 42	06 15	18 04	18 37
75	16	Wed	12 09	16.1	05 39	06 13	18 06	18 39
76	17	Th	12 08	16.1	05 37	06 11	18 07	18 41
77	18	Fri	12 08	16.1	05 35	06 08	18 09	18 43
78	19	Sat	12 08	16.1	05 32	06 06	18 11	18 44
79	20	Sun	12 08	16.1	05 30	06 04	18 13	18 46
80	21	Mon	12 07	16.1	05 28	06 01	18 14	18 48
81	22	Tu	12 07	16.1	05 25	05 59	18 16	18 50
82	23	Wed	12 06	16.1	05 23	05 57	18 18	18 51
83	24	Th	12 06	16.1	05 21	05 54	18 19	18 53
84	25	Fri	12 06	16.1	05 18	05 52	18 21	18 55
85	26	Sat	12 05	16.1	05 16	05 50	18 23	18 57
86	27	Sun	12 05	16.1	05 14	05 47	18 25	18 59
87	28	Mon	12 05	16.1	05 11	05 45	18 26	19 00
88	29	Tu	12 05	16.0	05 09	05 43	18 28	19 02
89	30	Wed	12 05	16.0	05 06	05 41	18 30	19 04
90	31	Th	12 04	16.0	05 04	05 38	18 31	19 06

Lat Corr to Sunrise, Sunset etc.

Lat	Twilight h m	Sunrise h m	Sunset h m	Twilight h m
N70	−0 18	+0 08	−1 06	+0 19
68	−0 15	+0 06	−1 05	+0 15
66	−0 12	+0 05	−0 04	+0 13
64	−0 09	+0 03	−0 03	+0 10
62	−0 06	+0 03	−0 02	+0 07
N60	−0 04	+0 03	−0 02	+0 05
58	−0 03	+0 02	−0 02	+0 04
56	−0 01	+0 02	−0 01	+0 04
54	−0 01	+0 01	−0 01	+0 02
50	−0 01	0 00	0 00	−0 01
N45	+0 03	−0 01	+0 02	−0 02
40	+0 05	−0 03	+0 03	−0 04
35	+0 07	−0 04	+0 04	−0 05
30	+0 07	−0 05	+0 05	−0 06
20	+0 06	−0 07	+0 06	−0 06
N10	+0 07	−0 07	+0 07	−0 05
0	+0 06	−0 08	+0 07	−0 05
S10	+0 04	−0 09	+0 09	−1 03
20	+1 01	−0 11	+0 10	0 00
30	−0 03	−0 14	+0 12	+0 04
S35	−0 06	−0 15	+0 14	+0 07
40	−0 10	−0 17	+0 15	+0 11
45	−0 13	−0 18	+0 17	+0 14
S50	−0 18	−0 20	+0 19	+0 20

NOTES
The corrections to sunrise etc. are for middle of March.

Phases of the Moon

	d	h	m
● New Moon	4	20	46
☽ First Quarter	12	23	45
○ Full Moon	19	18	10
☾ Last Quarter	26	12	07

	d	h
Apogee	6	08
Perigee	19	19

MOON

Yr	Mth	Week	Age days	Transit (Upper) h m	Diff m	Semi-diam	Hor Par	Moonrise (Lat 52°N)	Moonset
60	1	Tu	26	09 40	45	15.0	54.9	05 09	14 20
61	2	Wed	27	10 25	43	14.9	54.6	05 31	15 30
62	3	Th	28	11 08	42	14.8	54.3	05 49	16 38
63	4	Fri	29	11 50	40	14.7	54.1	06 05	17 46
64	5	Sat	01	12 30	41	14.7	54.0	06 20	18 53
65	6	Sun	02	13 11	41	14.7	53.9	06 35	20 00
66	7	Mon	03	13 52	43	14.8	54.0	06 52	21 07
67	8	Tu	04	14 35	45	14.8	54.2	07 10	22 15
68	9	Wed	05	15 20	48	14.8	54.2	07 31	23 23
69	10	Th	06	16 08	50	14.9	54.5	07 58	—
70	11	Fri	07	16 58	53	15.1	54.9	08 32	00 29
71	12	Sat	08	17 51	55	15.3	55.5	09 17	01 31
72	13	Sun	09	18 46	55	15.5	56.2	10 13	02 26
73	14	Mon	10	19 41	55	15.8	57.0	11 21	03 12
74	15	Tu	11	20 36	54	16.0	57.9	12 37	03 49
75	16	Wed	12	21 31	54	16.3	58.9	14 00	04 19
76	17	Th	13	22 25	54	16.5	59.8	15 26	04 44
77	18	Fri	14	23 19	54	16.7	60.6	16 54	05 06
78	19	Sat	15	24 13	—	16.8	61.2	18 23	05 27
79	20	Sun	16	00 13	55	16.7	61.5	19 52	05 48
80	21	Mon	17	01 08	57	16.6	61.4	21 21	06 11
81	22	Tu	18	02 05	58	16.4	61.0	22 46	06 39
82	23	Wed	19	03 04	59	16.2	60.3	—	07 14
83	24	Th	20	04 02	58	15.9	59.5	00 03	07 57
84	25	Fri	21	05 00	56	15.7	58.5	01 09	08 51
85	26	Sat	22	05 56	53	15.4	57.6	02 02	09 54
86	27	Sun	23	06 49	49	15.2	56.7	02 42	11 02
87	28	Mon	24	07 38	46	15.1	55.9	03 12	12 12
88	29	Tu	25	08 24	43	14.9	55.2	03 36	13 21
89	30	Wed	26	09 07	42	14.8	54.7	03 55	14 30
90	31	Th	27	09 49	40	14.7	54.3	04 12	15 37

SUN AND ARIES

APRIL 2011

SUN AND ARIES

GMT	SUN GHA ° ′	SUN Dec ° ′	ARIES GHA ° ′
	Friday, 1st April		
00	178 58.4	N 4 19.3	189 00.7
02	208 58.7	4 21.3	219 05.6
04	238 59.1	4 23.2	249 10.5
06	268 59.5	4 25.1	279 15.5
08	298 59.9	4 27.1	309 20.4
10	329 00.2	4 29.0	339 25.3
12	359 00.6	4 30.9	9 30.3
14	29 01.0	4 32.9	39 35.2
16	59 01.3	4 34.8	69 40.1
18	89 01.7	4 36.7	99 45.0
20	119 02.1	4 38.7	129 50.0
22	149 02.4	N 4 40.6	159 54.9
	Saturday, 2nd April		
00	179 02.8	N 4 42.5	189 59.8
02	209 03.2	4 44.4	220 04.8
04	239 03.5	4 46.4	250 09.7
06	269 03.9	4 48.3	280 14.6
08	299 04.3	4 50.2	310 19.5
10	329 04.7	4 52.1	340 24.5
12	359 05.0	4 54.1	10 29.4
14	29 05.4	4 56.0	40 34.3
16	59 05.8	4 57.9	70 39.3
18	89 06.1	4 59.8	100 44.2
20	119 06.5	5 01.8	130 49.1
22	149 06.9	N 5 03.7	160 54.0
	Sunday, 3rd April		
00	179 07.2	N 5 05.6	190 59.0
02	209 07.6	5 07.5	221 03.9
04	239 08.0	5 09.4	251 08.8
06	269 08.3	5 11.4	281 13.7
08	299 08.7	5 13.3	311 18.7
10	329 09.0	5 15.2	341 23.6
12	359 09.4	5 17.1	11 28.5
14	29 09.8	5 19.0	41 33.5
16	59 10.1	5 20.9	71 38.4
18	89 10.5	5 22.9	101 43.3
20	119 10.9	5 24.8	131 48.2
22	149 11.2	N 5 26.7	161 53.2
	Monday, 4th April		
00	179 11.6	N 5 28.6	191 58.1
02	209 12.0	5 30.5	222 03.0
04	239 12.3	5 32.4	252 08.0
06	269 12.7	5 34.3	282 12.9
08	299 13.0	5 36.2	312 17.8
10	329 13.4	5 38.1	342 22.7
12	359 13.8	5 40.1	12 27.7
14	29 14.1	5 42.0	42 32.6
16	59 14.5	5 43.9	72 37.5
18	89 14.9	5 45.8	102 42.5
20	119 15.2	5 47.7	132 47.4
22	149 15.6	N 5 49.6	162 52.3
	Tuesday, 5th April		
00	179 15.9	N 5 51.5	192 57.2
02	209 16.3	5 53.4	223 02.2
04	239 16.7	5 55.3	253 07.1
06	269 17.0	5 57.2	283 12.0
08	299 17.4	5 59.1	313 17.0
10	329 17.7	6 01.0	343 21.9
12	359 18.1	6 02.9	13 26.8
14	29 18.4	6 04.8	43 31.7
16	59 18.8	6 06.7	73 36.7
18	89 19.2	6 08.6	103 41.6
20	119 19.5	6 10.5	133 46.5
22	149 19.9	N 6 12.4	163 51.4

GMT	SUN GHA ° ′	SUN Dec ° ′	ARIES GHA ° ′
	Wednesday, 6th April		
00	179 20.2	N 6 14.3	193 56.4
02	209 20.6	6 16.2	224 01.3
04	239 20.9	6 18.1	254 06.2
06	269 21.3	6 20.0	284 11.2
08	299 21.6	6 21.9	314 16.1
10	329 22.0	6 23.8	344 21.0
12	359 22.4	6 25.6	14 25.9
14	29 22.7	6 27.5	44 30.9
16	59 23.1	6 29.4	74 35.8
18	89 23.4	6 31.3	104 40.7
20	119 23.8	6 33.2	134 45.7
22	149 24.1	N 6 35.1	164 50.6
	Thursday, 7th April		
00	179 24.5	N 6 37.0	194 55.5
02	209 24.8	6 38.9	225 00.4
04	239 25.2	6 40.7	255 05.4
06	269 25.5	6 42.6	285 10.3
08	299 25.9	6 44.5	315 15.2
10	329 26.2	6 46.4	345 20.2
12	359 26.6	6 48.3	15 25.1
14	29 26.9	6 50.2	45 30.0
16	59 27.3	6 52.0	75 34.9
18	89 27.6	6 53.9	105 39.9
20	119 28.0	6 55.8	135 44.8
22	149 28.3	N 6 57.7	165 49.7
	Friday, 8th April		
00	179 28.7	N 6 59.5	195 54.7
02	209 29.0	7 01.4	225 59.6
04	239 29.4	7 03.3	256 04.5
06	269 29.7	7 05.2	286 09.4
08	299 30.1	7 07.0	316 14.4
10	329 30.4	7 08.9	346 19.3
12	359 30.8	7 10.8	16 24.2
14	29 31.1	7 12.7	46 29.2
16	59 31.4	7 14.5	76 34.1
18	89 31.8	7 16.4	106 39.0
20	119 32.1	7 18.3	136 43.9
22	149 32.5	N 7 20.1	166 48.9
	Saturday, 9th April		
00	179 32.8	N 7 22.0	196 53.8
02	209 33.2	7 23.9	226 58.7
04	239 33.5	7 25.7	257 03.6
06	269 33.8	7 27.6	287 08.6
08	299 34.2	7 29.5	317 13.5
10	329 34.5	7 31.3	347 18.4
12	359 34.9	7 33.2	17 23.4
14	29 35.2	7 35.0	47 28.3
16	59 35.5	7 36.9	77 33.2
18	89 35.9	7 38.8	107 38.1
20	119 36.2	7 40.6	137 43.1
22	149 36.6	N 7 42.5	167 48.0
	Sunday, 10th April		
00	179 36.9	N 7 44.3	197 52.9
02	209 37.2	7 46.2	227 57.9
04	239 37.6	7 48.0	258 02.8
06	269 37.9	7 49.9	288 07.7
08	299 38.3	7 51.7	318 12.6
10	329 38.6	7 53.6	348 17.6
12	359 38.9	7 55.4	18 22.5
14	29 39.3	7 57.3	48 27.4
16	59 39.6	7 59.1	78 32.4
18	89 39.9	8 01.0	108 37.3
20	119 40.3	8 02.8	138 42.2
22	149 40.6	N 8 04.7	168 47.1

GMT	SUN GHA ° ′	SUN Dec ° ′	ARIES GHA ° ′
	Monday, 11th April		
00	179 40.9	N 8 06.5	198 52.1
02	209 41.3	8 08.4	228 57.0
04	239 41.6	8 10.2	259 01.9
06	269 41.9	8 12.1	289 06.9
08	299 42.3	8 13.9	319 11.8
10	329 42.6	8 15.7	349 16.7
12	359 42.9	8 17.6	19 21.6
14	29 43.2	8 19.4	49 26.6
16	59 43.6	8 21.2	79 31.5
18	89 43.9	8 23.1	109 36.4
20	119 44.2	8 24.9	139 41.4
22	149 44.6	N 8 26.7	169 46.3
	Tuesday, 12th April		
00	179 44.9	N 8 28.6	199 51.2
02	209 45.2	8 30.4	229 56.1
04	239 45.5	8 32.2	260 01.1
06	269 45.9	8 34.1	290 06.0
08	299 46.2	8 35.9	320 10.9
10	329 46.5	8 37.7	350 15.9
12	359 46.8	8 39.6	20 20.8
14	29 47.2	8 41.4	50 25.7
16	59 47.5	8 43.2	80 30.6
18	89 47.8	8 45.0	110 35.6
20	119 48.1	8 46.9	140 40.5
22	149 48.5	N 8 48.7	170 45.4
	Wednesday, 13th April		
00	179 48.8	N 8 50.5	200 50.4
02	209 49.1	8 52.3	230 55.3
04	239 49.4	8 54.1	261 00.2
06	269 49.7	8 56.0	291 05.1
08	299 50.1	8 57.8	321 10.1
10	329 50.4	8 59.6	351 15.0
12	359 50.7	9 01.4	21 19.9
14	29 51.0	9 03.2	51 24.8
16	59 51.3	9 05.0	81 29.8
18	89 51.7	9 06.8	111 34.7
20	119 52.0	9 08.6	141 39.6
22	149 52.3	N 9 10.5	171 44.6
	Thursday, 14th April		
00	179 52.6	N 9 12.3	201 49.5
02	209 52.9	9 14.1	231 54.4
04	239 53.2	9 15.9	261 59.3
06	269 53.5	9 17.7	292 04.3
08	299 53.9	9 19.5	322 09.2
10	329 54.2	9 21.3	352 14.1
12	359 54.5	9 23.1	22 19.1
14	29 54.8	9 24.9	52 24.0
16	59 55.1	9 26.7	82 28.9
18	89 55.4	9 28.5	112 33.8
20	119 55.7	9 30.3	142 38.8
22	149 56.0	N 9 32.1	172 43.7
	Friday, 15th April		
00	179 56.3	N 9 33.9	202 48.6
02	209 56.6	9 35.7	232 53.6
04	239 57.0	9 37.5	262 58.5
06	269 57.3	9 39.2	293 03.4
08	299 57.6	9 41.0	323 08.3
10	329 57.9	9 42.8	353 13.3
12	359 58.2	9 44.6	23 18.2
14	29 58.5	9 46.4	53 23.1
16	59 58.8	9 48.2	83 28.1
18	89 59.1	9 50.0	113 33.0
20	119 59.4	9 51.8	143 37.9
22	149 59.7	N 9 53.5	173 42.8

GMT	SUN GHA ° ′	SUN Dec ° ′	ARIES GHA ° ′
	Saturday, 16th April		
00	180 00.0	N 9 55.3	203 47.8
02	210 00.3	9 57.1	233 52.7
04	240 00.6	9 58.9	263 57.6
06	270 00.9	10 00.7	294 02.5
08	300 01.2	10 02.4	324 07.5
10	330 01.5	10 04.2	354 12.4
12	0 01.8	10 06.0	24 17.3
14	30 02.1	10 07.8	54 22.3
16	60 02.4	10 09.5	84 27.2
18	90 02.7	10 11.3	114 32.1
20	120 03.0	10 13.1	144 37.0
22	150 03.3	N10 14.8	174 42.0
	Sunday, 17th April		
00	180 03.6	N10 16.6	204 46.9
02	210 03.9	10 18.4	234 51.8
04	240 04.2	10 20.1	264 56.8
06	270 04.4	10 21.9	295 01.7
08	300 04.7	10 23.7	325 06.6
10	330 05.0	10 25.4	355 11.5
12	0 05.3	10 27.2	25 16.5
14	30 05.6	10 28.9	55 21.4
16	60 05.9	10 30.7	85 26.3
18	90 06.2	10 32.5	115 31.3
20	120 06.5	10 34.2	145 36.2
22	150 06.8	N10 36.0	175 41.1
	Monday, 18th April		
00	180 07.0	N10 37.7	205 46.0
02	210 07.3	10 39.5	235 51.0
04	240 07.6	10 41.2	265 55.9
06	270 07.9	10 43.0	296 00.8
08	300 08.2	10 44.7	326 05.7
10	330 08.5	10 46.5	356 10.7
12	0 08.8	10 48.2	26 15.6
14	30 09.0	10 50.0	56 20.5
16	60 09.3	10 51.7	86 25.5
18	90 09.6	10 53.4	116 30.4
20	120 09.9	10 55.2	146 35.3
22	150 10.2	N10 56.9	176 40.2
	Tuesday, 19th April		
00	180 10.4	N10 58.7	206 45.2
02	210 10.7	11 00.4	236 50.1
04	240 11.0	11 02.1	266 55.0
06	270 11.3	11 03.9	297 00.0
08	300 11.5	11 05.6	327 04.9
10	330 11.8	11 07.3	357 09.8
12	0 12.1	11 09.1	27 14.7
14	30 12.4	11 10.8	57 19.7
16	60 12.6	11 12.5	87 24.6
18	90 12.9	11 14.2	117 29.5
20	120 13.2	11 16.0	147 34.5
22	150 13.4	N11 17.7	177 39.4
	Wednesday, 20th April		
00	180 13.7	N11 19.4	207 44.3
02	210 14.0	11 21.1	237 49.2
04	240 14.2	11 22.9	267 54.2
06	270 14.5	11 24.6	297 59.1
08	300 14.8	11 26.3	328 04.0
10	330 15.0	11 28.0	358 09.0
12	0 15.3	11 29.7	28 13.9
14	30 15.6	11 31.4	58 18.8
16	60 15.8	11 33.2	88 23.7
18	90 16.1	11 34.9	118 28.7
20	120 16.4	11 36.6	148 33.6
22	150 16.6	N11 38.3	178 38.5

GMT	SUN GHA ° ′	SUN Dec ° ′	ARIES GHA ° ′
	Thursday, 21st April		
00	180 16.9	N11 40.0	208 43.5
02	210 17.1	11 41.7	238 48.4
04	240 17.4	11 43.4	268 53.3
06	270 17.7	11 45.1	298 58.2
08	300 17.9	11 46.8	329 03.2
10	330 18.2	11 48.5	359 08.1
12	0 18.4	11 50.2	29 13.0
14	30 18.7	11 51.9	59 18.0
16	60 18.9	11 53.6	89 22.9
18	90 19.2	11 55.3	119 27.8
20	120 19.4	11 57.0	149 32.7
22	150 19.7	N11 58.7	179 37.7
	Friday, 22nd April		
00	180 19.9	N12 00.4	209 42.6
02	210 20.2	12 02.1	239 47.5
04	240 20.4	12 03.8	269 52.5
06	270 20.7	12 05.5	299 57.4
08	300 20.9	12 07.1	330 02.3
10	330 21.2	12 08.8	0 07.2
12	0 21.4	12 10.5	30 12.2
14	30 21.7	12 12.2	60 17.1
16	60 21.9	12 13.9	90 22.0
18	90 22.2	12 15.5	120 26.9
20	120 22.4	12 17.2	150 31.9
22	150 22.7	N12 18.9	180 36.8
	Saturday, 23rd April		
00	180 22.9	N12 20.6	210 41.7
02	210 23.1	12 22.2	240 46.7
04	240 23.4	12 23.9	270 51.6
06	270 23.6	12 25.6	300 56.5
08	300 23.9	12 27.3	331 01.4
10	330 24.1	12 28.9	1 06.4
12	0 24.3	12 30.6	31 11.3
14	30 24.6	12 32.3	61 16.2
16	60 24.8	12 33.9	91 21.2
18	90 25.0	12 35.6	121 26.1
20	120 25.3	12 37.2	151 31.0
22	150 25.5	N12 38.9	181 35.9
	Sunday, 24th April		
00	180 25.7	N12 40.6	211 40.9
02	210 26.0	12 42.2	241 45.8
04	240 26.2	12 43.9	271 50.7
06	270 26.4	12 45.5	301 55.7
08	300 26.6	12 47.2	332 00.6
10	330 26.9	12 48.8	2 05.5
12	0 27.1	12 50.5	32 10.4
14	30 27.3	12 52.1	62 15.4
16	60 27.6	12 53.8	92 20.3
18	90 27.8	12 55.4	122 25.2
20	120 28.0	12 57.1	152 30.2
22	150 28.2	N12 58.7	182 35.1
	Monday, 25th April		
00	180 28.4	N13 00.3	212 40.0
02	210 28.7	13 02.0	242 44.9
04	240 28.9	13 03.6	272 49.9
06	270 29.1	13 05.3	302 54.8
08	300 29.3	13 06.9	332 59.7
10	330 29.5	13 08.5	3 04.7
12	0 29.8	13 10.2	33 09.6
14	30 30.0	13 11.8	63 14.5
16	60 30.2	13 13.4	93 19.4
18	90 30.4	13 15.0	123 24.4
20	120 30.6	13 16.7	153 29.3
22	150 30.8	N13 18.3	183 34.2

GMT	SUN GHA ° ′	SUN Dec ° ′	ARIES GHA ° ′
	Tuesday, 26th April		
00	180 31.0	N13 19.9	213 39.2
02	210 31.2	13 21.5	243 44.1
04	240 31.4	13 23.2	273 49.0
06	270 31.7	13 24.8	303 53.9
08	300 31.9	13 26.4	333 58.9
10	330 32.1	13 28.0	4 03.8
12	0 32.3	13 29.6	34 08.7
14	30 32.5	13 31.2	64 13.6
16	60 32.7	13 32.8	94 18.6
18	90 32.9	13 34.4	124 23.5
20	120 33.1	13 36.1	154 28.4
22	150 33.3	N13 37.7	184 33.4
	Wednesday, 27th April		
00	180 33.5	N13 39.3	214 38.3
02	210 33.7	13 40.9	244 43.2
04	240 33.9	13 42.5	274 48.1
06	270 34.1	13 44.1	304 53.1
08	300 34.3	13 45.7	334 58.0
10	330 34.5	13 47.3	5 02.9
12	0 34.7	13 48.9	35 07.9
14	30 34.9	13 50.4	65 12.8
16	60 35.1	13 52.0	95 17.7
18	90 35.3	13 53.6	125 22.6
20	120 35.4	13 55.2	155 27.6
22	150 35.6	N13 56.8	185 32.5
	Thursday, 28th April		
00	180 35.8	N13 58.4	215 37.4
02	210 36.0	14 00.0	245 42.4
04	240 36.2	14 01.6	275 47.3
06	270 36.4	14 03.1	305 52.2
08	300 36.6	14 04.7	335 57.1
10	330 36.8	14 06.3	6 02.1
12	0 36.9	14 07.9	36 07.0
14	30 37.1	14 09.4	66 11.9
16	60 37.3	14 11.0	96 16.9
18	90 37.5	14 12.6	126 21.8
20	120 37.7	14 14.1	156 26.7
22	150 37.8	N14 15.7	186 31.6
	Friday, 29th April		
00	180 38.0	N14 17.3	216 36.6
02	210 38.2	14 18.8	246 41.5
04	240 38.4	14 20.4	276 46.4
06	270 38.6	14 22.0	306 51.3
08	300 38.7	14 23.5	336 56.3
10	330 38.9	14 25.1	7 01.2
12	0 39.1	14 26.6	37 06.1
14	30 39.2	14 28.2	67 11.1
16	60 39.4	14 29.7	97 16.0
18	90 39.6	14 31.3	127 20.9
20	120 39.8	14 32.8	157 25.8
22	150 39.9	N14 34.4	187 30.8
	Saturday, 30th April		
00	180 40.1	N14 35.9	217 35.7
02	210 40.3	14 37.5	247 40.6
04	240 40.4	14 39.0	277 45.6
06	270 40.6	14 40.6	307 50.5
08	300 40.8	14 42.1	337 55.4
10	330 40.9	14 43.6	8 00.3
12	0 41.1	14 45.2	38 05.3
14	30 41.2	14 46.7	68 10.2
16	60 41.4	14 48.2	98 15.1
18	90 41.6	14 49.8	128 20.1
20	120 41.7	14 51.3	158 25.0
22	150 41.9	N14 52.8	188 29.9

APRIL

MOON

Day	GMT hr	GHA ° '	Mean Var/hr 14°+	Dec ° '	Mean Var/hr '
1 Fri	0	206 56.0	35.4	S 2 21.3	11.9
	6	294 28.1	35.4	S 1 10.2	11.9
	12	22 00.5	35.4	N 0 00.9	11.8
	18	109 32.9	35.3	N 1 12.0	11.8
2 Sat	0	197 05.2	35.4	N 2 22.9	11.7
	6	284 37.2	35.2	N 3 33.5	11.7
	12	12 08.8	35.2	N 4 43.6	11.6
	18	99 39.7	35.0	N 5 53.0	11.4
3 Sun	0	187 09.9	34.9	N 7 01.7	11.3
	6	274 39.2	34.7	N 8 09.5	11.1
	12	2 07.4	34.5	N 9 16.1	10.9
	18	89 34.4	34.3	N10 21.5	10.7
4 Mon	0	177 00.1	34.0	N11 25.5	10.4
	6	264 24.2	33.8	N12 28.0	10.1
	12	351 46.8	33.4	N13 28.7	9.7
	18	79 07.7	33.2	N14 27.5	9.4
5 Tu	0	166 26.7	32.8	N15 24.3	9.1
	6	253 43.8	32.5	N16 18.8	8.7
	12	340 59.0	32.2	N17 10.9	8.2
	18	68 12.1	31.8	N18 00.4	7.8
6 Wed	0	155 23.1	31.5	N18 47.2	7.2
	6	242 32.0	31.1	N19 31.0	6.7
	12	329 38.8	30.7	N20 11.7	6.2
	18	56 43.4	30.4	N20 49.2	5.7
7 Th	0	143 46.0	30.0	N21 23.2	5.0
	6	230 46.5	29.7	N21 53.6	4.4
	12	317 45.0	29.4	N22 20.3	3.7
	18	44 41.7	29.1	N22 43.1	3.0
8 Fri	0	131 36.5	28.9	N23 01.8	2.4
	6	218 29.7	28.6	N23 16.4	1.6
	12	305 21.4	28.4	N23 26.7	0.9
	18	32 11.8	28.2	N23 32.6	0.2
9 Sat	0	119 01.0	28.0	N23 34.1	0.6
	6	205 49.2	27.8	N23 31.0	1.3
	12	292 36.5	27.8	N23 23.3	2.1
	18	19 23.3	27.7	N23 11.0	2.9
10 Sun	0	106 09.6	27.6	N22 54.0	3.7
	6	192 55.7	27.7	N22 32.4	4.4
	12	279 41.6	27.7	N22 06.1	5.2
	18	6 27.6	27.7	N21 35.2	6.0
11 Mon	0	93 13.8	27.7	N20 59.8	6.7
	6	180 00.4	27.9	N20 19.9	7.5
	12	266 47.4	27.9	N19 35.6	8.2
	18	353 34.9	28.1	N18 47.0	8.9
12 Tu	0	80 22.9	28.1	N17 54.2	9.5
	6	167 11.6	28.2	N16 57.4	10.1
	12	254 00.9	28.4	N15 56.7	10.8
	18	340 50.8	28.4	N14 52.3	11.3
13 Wed	0	67 41.2	28.5	N13 44.5	11.9
	6	154 32.1	28.6	N12 33.3	12.4
	12	241 23.5	28.6	N11 19.0	12.9
	18	328 15.1	28.7	N10 01.9	13.3
14 Th	0	55 07.0	28.6	N 8 42.1	13.7
	6	141 58.9	28.6	N 7 20.1	14.0
	12	228 50.8	28.6	N 5 55.9	14.3
	18	315 42.4	28.5	N 4 30.1	14.6
15 Fri	0	42 33.6	28.4	N 3 02.7	14.7
	6	129 24.3	28.3	N 1 34.3	14.9
	12	216 14.1	28.2	N 0 05.1	14.9
	18	303 03.1	27.9	S 1 24.5	14.9
16 Sat	0	29 51.0	27.8	S 2 54.2	14.8
	6	116 37.5	27.5	S 4 23.6	14.7
	12	203 22.7	27.2	S 5 52.2	14.5
	18	290 06.3	27.0	S 7 19.7	14.3
17 Sun	0	16 48.1	26.6	S 8 45.6	14.0
	6	103 28.2	26.3	S10 09.6	13.6
	12	190 06.3	26.0	S11 31.3	13.1
	18	276 42.5	25.7	S12 50.2	12.6
18 Mon	0	3 16.7	25.4	S14 06.0	12.0
	6	89 48.9	25.0	S15 18.2	11.3
	12	176 19.2	24.7	S16 26.6	10.6
	18	262 47.7	24.5	S17 30.7	9.9
19 Tu	0	349 14.5	24.2	S18 30.2	9.1
	6	75 39.9	24.0	S19 24.9	8.2
	12	162 03.9	23.8	S20 14.4	7.3
	18	248 27.0	23.7	S20 58.7	6.4
20 Wed	0	334 49.4	23.7	S21 37.4	5.4
	6	61 11.4	23.7	S22 10.5	4.5
	12	147 33.5	23.7	S22 37.8	3.6
	18	233 55.9	23.9	S22 59.4	2.6
21 Th	0	320 19.0	24.1	S23 15.2	1.6
	6	46 43.3	24.4	S23 25.2	0.6
	12	133 09.1	24.7	S23 29.6	0.3
	18	219 36.8	25.0	S23 28.4	1.2
22 Fri	0	306 06.7	25.4	S23 21.8	2.0
	6	32 39.0	25.8	S23 10.0	2.9
	12	119 14.0	26.4	S22 53.1	3.7
	18	205 52.0	26.8	S22 31.5	4.4
23 Sat	0	292 33.1	27.5	S22 05.3	5.1
	6	19 17.4	28.0	S21 34.7	5.8
	12	106 05.0	28.6	S21 00.1	6.4
	18	192 56.0	29.1	S20 21.7	7.0
24 Sun	0	279 50.4	29.7	S19 39.7	7.6
	6	6 48.1	30.2	S18 54.4	8.1
	12	93 49.0	30.7	S18 06.0	8.6
	18	180 53.2	31.3	S17 14.9	9.0
25 Mon	0	268 00.4	31.8	S16 21.1	9.4
	6	355 10.6	32.2	S15 25.0	9.8
	12	82 23.5	32.6	S14 26.7	10.1
	18	169 39.1	33.0	S13 26.5	10.4
26 Tu	0	256 57.2	33.4	S12 24.5	10.6
	6	344 17.5	33.7	S11 20.9	10.9
	12	71 39.9	34.0	S10 15.9	11.1
	18	159 04.2	34.4	S 9 09.8	11.3
27 Wed	0	246 30.2	34.6	S 8 02.5	11.3
	6	333 57.7	34.8	S 6 54.4	11.5
	12	61 26.5	35.0	S 5 45.4	11.6
	18	148 56.4	35.2	S 4 35.9	11.7
28 Th	0	236 27.1	35.2	S 3 25.9	11.7
	6	323 58.6	35.3	S 2 15.6	11.7
	12	51 30.5	35.4	S 1 05.1	11.8
	18	139 02.7	35.4	N 0 05.5	11.8
29 Fri	0	226 35.1	35.4	N 1 16.1	11.7
	6	314 07.3	35.4	N 2 26.4	11.7
	12	41 39.3	35.3	N 3 36.4	11.6
	18	129 10.8	35.2	N 4 45.9	11.5
30 Sat	0	216 41.7	35.0	N 5 54.8	11.4
	6	304 11.8	34.9	N 7 02.9	11.2
	12	31 40.8	34.6	N 8 10.1	11.0
	18	119 08.8	34.4	N 9 16.3	10.9

PLANETS

VENUS Mer Pass h m	VENUS GHA ° '	VENUS Mean Var/hr 14°+	VENUS Dec ° '	VENUS Mean Var/hr '	Day	JUPITER GHA ° '	JUPITER Mean Var/hr 15°+	JUPITER Dec ° '	JUPITER Mean Var/hr '	JUPITER Mer Pass h m
09 55	211 13.2	59.6	S10 14.7	1.0	1 Fri	174 41.4	1.9	N 4 57.0	0.2	12 20
09 56	211 03.3	59.6	S 9 50.5	1.0	2 Sat	175 27.1	1.9	N 5 02.6	0.2	12 17
09 57	210 53.5	59.6	S 9 26.0	1.0	3 SUN	176 12.8	1.9	N 5 08.2	0.2	12 14
09 57	210 43.8	59.6	S 9 01.1	1.0	4 Mon	176 58.5	1.9	N 5 13.8	0.2	12 11
09 58	210 34.3	59.6	S 8 36.1	1.1	5 Tu	177 44.2	1.9	N 5 19.4	0.2	12 08
09 59	210 25.0	59.6	S 8 10.8	1.1	6 Wed	178 29.9	1.9	N 5 25.0	0.2	12 04
09 59	210 15.8	59.6	S 7 45.2	1.1	7 Th	179 15.5	1.9	N 5 30.6	0.2	12 01
10 00	210 06.8	59.6	S 7 19.4	1.1	8 Fri	180 01.2	1.9	N 5 36.1	0.2	11 58
10 00	209 57.9	59.6	S 6 53.4	1.1	9 Sat	180 46.9	1.9	N 5 41.7	0.2	11 55
10 01	209 49.1	59.6	S 6 27.2	1.1	10 SUN	181 32.5	1.9	N 5 47.2	0.2	11 52
10 02	209 40.4	59.6	S 6 00.9	1.1	11 Mon	182 18.2	1.9	N 5 52.8	0.2	11 49
10 02	209 31.8	59.7	S 5 34.3	1.1	12 Tu	183 03.9	1.9	N 5 58.3	0.2	11 46
10 03	209 23.4	59.7	S 5 07.5	1.1	13 Wed	183 49.6	1.9	N 6 03.8	0.2	11 43
10 03	209 15.0	59.7	S 4 40.6	1.1	14 Th	184 35.2	1.9	N 6 09.3	0.2	11 40
10 04	209 06.7	59.7	S 4 13.6	1.1	15 Fri	185 20.9	1.9	N 6 14.8	0.2	11 37
10 04	208 58.6	59.7	S 3 46.4	1.1	16 Sat	186 06.6	1.9	N 6 20.3	0.2	11 34
10 05	208 50.4	59.7	S 3 19.1	1.1	17 SUN	186 52.3	1.9	N 6 25.7	0.2	11 31
10 05	208 42.4	59.7	S 2 51.6	1.1	18 Mon	187 38.0	1.9	N 6 31.2	0.2	11 28
10 06	208 34.4	59.7	S 2 24.1	1.2	19 Tu	188 23.7	1.9	N 6 36.6	0.2	11 25
10 06	208 26.4	59.7	S 1 56.4	1.2	20 Wed	189 09.4	1.9	N 6 42.1	0.2	11 22
10 07	208 18.5	59.7	S 1 28.7	1.2	21 Th	189 55.1	1.9	N 6 47.5	0.2	11 19
10 08	208 10.6	59.7	S 1 00.9	1.2	22 Fri	190 40.9	1.9	N 6 52.8	0.2	11 16
10 08	208 02.7	59.7	S 0 33.0	1.2	23 Sat	191 26.6	1.9	N 6 58.2	0.2	11 13
10 09	207 54.9	59.7	S 0 05.1	1.2	24 SUN	192 12.4	1.9	N 7 03.6	0.2	11 10
10 09	207 47.0	59.7	N 0 22.8	1.2	25 Mon	192 58.1	1.9	N 7 08.9	0.2	11 07
10 10	207 39.2	59.7	N 0 50.8	1.2	26 Tu	193 43.9	1.9	N 7 14.2	0.2	11 04
10 10	207 31.3	59.7	N 1 18.8	1.2	27 Wed	194 29.7	1.9	N 7 19.5	0.2	11 01
10 11	207 23.4	59.7	N 1 46.8	1.2	28 Th	195 15.6	1.9	N 7 24.8	0.2	10 58
10 11	207 15.4	59.7	N 2 14.8	1.2	29 Fri	196 01.4	1.9	N 7 30.1	0.2	10 55
10 12	207 07.5	59.7	N 2 42.8	1.2	30 Sat	196 47.3	1.9	N 7 35.3	0.2	10 51

VENUS, Av. Mag. –3.9
SHA April 5 18; 10 12; 15 6; 20 1; 25 355; 30 350

JUPITER, Av. Mag. –2.1
SHA April 5 345; 10 344; 15 343; 20 341; 25 340; 30 339

MARS Mer Pass h m	MARS GHA ° '	MARS Mean Var/hr 15°+	MARS Dec ° '	MARS Mean Var/hr '	Day	SATURN GHA ° '	SATURN Mean Var/hr 15°+	SATURN Dec ° '	SATURN Mean Var/hr '	SATURN Mer Pass h m
11 21	189 30.0	0.7	S 1 13.9	0.8	1 Fri	354 57.7	2.6	S 3 02.7	0.1	00 20
11 20	189 46.3	0.7	S 0 55.0	0.8	2 Sat	356 01.1	2.6	S 3 00.9	0.1	00 16
11 19	190 02.7	0.7	S 0 36.1	0.8	3 SUN	357 04.5	2.6	S 2 59.1	0.1	00 12
11 18	190 19.1	0.7	S 0 17.2	0.8	4 Mon	358 07.9	2.6	S 2 57.3	0.1	00 07
11 17	190 35.5	0.7	N 0 01.7	0.8	5 Tu	359 11.4	2.6	S 2 55.4	0.1	00 03
11 16	190 51.9	0.7	N 0 20.6	0.8	6 Wed	0 14.8	2.6	S 2 53.6	0.1	23 55
11 15	191 08.4	0.7	N 0 39.5	0.8	7 Th	1 18.3	2.6	S 2 51.8	0.1	23 51
11 14	191 24.9	0.7	N 0 58.3	0.8	8 Fri	2 21.7	2.6	S 2 50.0	0.1	23 46
11 13	191 41.4	0.7	N 1 17.1	0.8	9 Sat	3 25.1	2.6	S 2 48.2	0.1	23 42
11 12	191 58.0	0.7	N 1 35.9	0.8	10 SUN	4 28.5	2.6	S 2 46.4	0.1	23 38
11 11	192 14.6	0.7	N 1 54.7	0.8	11 Mon	5 31.9	2.6	S 2 44.7	0.1	23 34
11 09	192 31.2	0.7	N 2 13.4	0.8	12 Tu	6 35.3	2.6	S 2 42.9	0.1	23 30
11 08	192 47.8	0.7	N 2 32.1	0.8	13 Wed	7 38.7	2.6	S 2 41.2	0.1	23 25
11 07	193 04.4	0.7	N 2 50.7	0.8	14 Th	8 42.0	2.6	S 2 39.4	0.1	23 21
11 06	193 21.0	0.7	N 3 09.3	0.8	15 Fri	9 45.4	2.6	S 2 37.7	0.1	23 17
11 05	193 37.6	0.7	N 3 27.8	0.8	16 Sat	10 48.7	2.6	S 2 36.0	0.1	23 13
11 04	193 54.3	0.7	N 3 46.3	0.8	17 SUN	11 51.9	2.6	S 2 34.3	0.1	23 08
11 03	194 10.9	0.7	N 4 04.8	0.8	18 Mon	12 55.2	2.6	S 2 32.7	0.1	23 04
11 02	194 27.6	0.7	N 4 23.2	0.8	19 Tu	13 58.4	2.6	S 2 31.0	0.1	23 00
11 01	194 44.2	0.7	N 4 41.5	0.8	20 Wed	15 01.6	2.6	S 2 29.4	0.1	22 56
10 59	195 00.8	0.7	N 4 59.8	0.8	21 Th	16 04.8	2.6	S 2 27.8	0.1	22 52
10 58	195 17.5	0.7	N 5 18.0	0.8	22 Fri	17 07.9	2.6	S 2 26.2	0.1	22 47
10 57	195 34.1	0.7	N 5 36.1	0.8	23 Sat	18 11.0	2.6	S 2 24.6	0.1	22 43
10 56	195 50.7	0.7	N 5 54.2	0.8	24 SUN	19 14.0	2.6	S 2 23.0	0.1	22 39
10 55	196 07.3	0.7	N 6 12.2	0.7	25 Mon	20 17.0	2.6	S 2 21.5	0.1	22 35
10 54	196 23.9	0.7	N 6 30.1	0.7	26 Tu	21 20.0	2.6	S 2 20.0	0.1	22 31
10 53	196 40.5	0.7	N 6 47.9	0.7	27 Wed	22 22.9	2.6	S 2 18.5	0.1	22 27
10 52	196 57.1	0.7	N 7 05.7	0.7	28 Th	23 25.8	2.6	S 2 17.1	0.1	22 22
10 51	197 13.6	0.7	N 7 23.4	0.7	29 Fri	24 28.6	2.6	S 2 15.6	0.1	22 18
10 49	197 30.2	0.7	N 7 41.0	0.7	30 Sat	25 31.4	2.6	S 2 14.2	0.1	22 14

MARS, Av. Mag. +1.2
SHA April 5 358; 10 354; 15 351; 20 347; 25 343; 30 340

SATURN, Av. Mag. +0.4
SHA April 5 166; 10 167; 15 167; 20 167; 25 168; 30 168.

APRIL 2011

STARS — SUN AND MOON

STARS

0h GMT April 1

No.	Name	Mag	Transit (h m)	Dec (° ′)	SHA (° ′)
	γ ARIES	–	11 22	–	
1	Alpheratz	2.1	11 31	N29 09.1	357 45.5
2	Ankaa	2.4	11 49	S42 14.6	353 17.6
3	Schedar	2.2	12 03	N56 35.9	349 43.0
4	Diphda	2.0	12 06	S17 55.5	348 57.7
5	Achernar	0.5	13 00	S57 10.7	335 28.4
6	POLARIS	2.0	14 06	N89 18.9	318 58.3
7	Hamal	2.0	13 30	N23 30.9	328 02.9
8	Acamar	3.2	14 20	S40 15.7	315 19.9
9	Menkar	2.5	14 24	N 4 08.0	314 17.0
10	Mirfak	1.8	14 47	N49 54.1	308 43.1
11	Aldebaran	0.9	15 58	N16 31.9	290 51.4
12	Rigel	0.1	16 36	S 8 11.5	281 13.7
13	Capella	0.1	16 39	N46 00.6	280 37.0
14	Bellatrix	1.6	16 47	N 6 21.5	278 33.9
15	Elnath	1.7	16 48	N28 37.0	278 14.8
16	Alnilam	1.7	16 58	S 1 11.8	275 48.1
17	Betelgeuse	0.1–1.2	17 17	N 7 24.4	271 03.1
18	Canopus	-0.7	17 45	S52 42.5	263 57.0
19	Sirius	-1.5	18 07	S16 44.2	258 35.2
20	Adhara	1.5	18 20	S28 59.6	255 13.8
21	Procyon	0.4	19 01	N 5 11.6	245 01.4
22	Pollux	1.1	19 07	N27 59.9	243 29.6
23	Avior	1.9	19 43	S59 33.2	234 18.6
24	Suhail	2.2	20 29	S43 29.1	222 53.5
25	Miaplacidus	1.7	20 34	S69 46.2	221 39.8
26	Alphard	2.0	20 49	S 8 42.7	217 57.5
27	Regulus	1.4	21 29	N11 54.5	207 45.0
28	Dubhe	1.8	22 25	N61 41.4	193 53.3
29	Denebola	2.1	23 10	N14 30.3	182 34.9
30	Gienah	2.6	23 36	S17 36.5	175 53.6
31	Acrux	1.3	23 47	S63 10.0	173 10.5
32	Gacrux	1.6	23 52	S57 10.8	172 02.2
33	Mimosa	1.3	0 12	S59 45.2	167 53.3
34	Alioth	1.8	0 18	N55 53.8	166 21.4
35	Spica	1.0	0 50	S11 13.4	158 32.6
36	Alkaid	1.9	1 12	N49 15.2	152 59.6
37	Hadar	0.6	1 28	S60 25.7	148 49.6
38	Menkent	2.1	1 31	S36 25.7	148 09.1
39	Arcturus	0.0	1 40	N19 07.2	145 56.8
40	Rigil Kent	-0.3	2 04	S60 52.9	139 53.3
41	Zuben'ubi	2.8	2 15	S16 05.5	137 06.8
42	Kochab	2.1	2 14	N74 06.4	137 18.7
43	Alphecca	2.2	2 59	N26 40.4	126 12.0
44	Antares	1.0	3 53	S26 27.4	112 27.9
45	Atria	1.9	4 13	S69 02.7	107 30.8
46	Sabik	2.4	4 34	S15 44.4	102 14.1
47	Shaula	1.6	4 58	S37 06.6	96 23.8
48	Rasalhague	2.1	4 59	N12 33.0	96 05.7
49	Eltanin	2.2	5 20	N51 29.0	90 46.6
50	Kaus Aust	1.9	5 48	S34 22.6	83 45.7
51	Vega	0.0	6 00	N38 47.4	80 39.9
52	Nunki	2.0	6 19	S26 16.9	76 00.2
53	Altair	0.8	7 14	N 8 53.8	62 09.8
54	Peacock	1.9	7 49	S56 41.6	53 21.7
55	Deneb	1.3	8 04	N45 19.0	49 32.7
56	Enif	2.4	9 07	N 9 55.6	33 48.8
57	Al Na'ir	1.7	9 31	S46 54.2	27 45.9
58	Fomalhaut	1.2	10 21	S29 33.6	15 25.9
59	Markab	2.5	10 28	N15 15.9	13 40.2

SUN — Lat 52°N

Yr	Day of Mth	Week	Transit (h m)	Semi-Diam	Twilight (h m)	Sunrise (h m)	Sunset (h m)	Twilight (h m)
91	1	Fri	12 04	16.0	05 02	05 36	18 33	19 07
92	2	Sat	12 04	16.0	04 59	05 34	18 35	19 09
93	3	Sun	12 03	16.0	04 57	05 31	18 37	19 11
94	4	Mon	12 03	16.0	04 55	05 29	18 38	19 13
95	5	Tu	12 03	16.0	04 52	05 27	18 40	19 15
96	6	Wed	12 03	16.0	04 50	05 24	18 42	19 17
97	7	Th	12 02	16.0	04 47	05 22	18 43	19 18
98	8	Fri	12 02	16.0	04 45	05 20	18 45	19 20
99	9	Sat	12 02	16.0	04 43	05 18	18 47	19 22
100	10	Sun	12 01	16.0	04 40	05 15	18 49	19 24
101	11	Mon	12 01	16.0	04 38	05 13	18 50	19 26
102	12	Tu	12 01	16.0	04 36	05 11	18 52	19 27
103	13	Wed	12 01	16.0	04 33	05 09	18 54	19 29
104	14	Th	12 00	16.0	04 31	05 07	18 55	19 31
105	15	Fri	12 00	16.0	04 29	05 04	18 57	19 33
106	16	Sat	12 00	16.0	04 26	05 02	18 59	19 35
107	17	Sun	12 00	16.0	04 24	05 00	19 00	19 37
108	18	Mon	11 59	16.0	04 22	04 58	19 02	19 39
109	19	Tu	11 59	16.0	04 19	04 56	19 04	19 40
110	20	Wed	11 59	15.9	04 17	04 54	19 06	19 42
111	21	Th	11 59	15.9	04 15	04 51	19 07	19 44
112	22	Fri	11 59	15.9	04 12	04 49	19 09	19 46
113	23	Sat	11 58	15.9	04 10	04 47	19 11	19 48
114	24	Sun	11 58	15.9	04 08	04 45	19 12	19 50
115	25	Mon	11 58	15.9	04 06	04 43	19 14	19 52
116	26	Tu	11 58	15.9	04 03	04 41	19 16	19 54
117	27	Wed	11 58	15.9	04 01	04 39	19 17	19 56
118	28	Th	11 57	15.9	03 59	04 37	19 19	19 57
119	29	Fri	11 57	15.9	03 57	04 35	19 21	19 59
120	30	Sat	11 57	15.9	03 55	04 33	19 23	20 01

Lat Corr to Sunrise, Sunset etc.

Lat	Twilight	Sunrise	Sunset	Twilight
N70	-1 50	-1 05	+1 09	+1 50
68	-1 26	-0 53	+0 55	+1 27
66	-1 07	-0 42	+0 44	+1 08
64	-0 52	-0 33	+0 35	+0 53
62	-0 40	-0 26	+0 27	+0 41
N60	-0 30	-0 19	+0 20	+0 30
58	-0 21	-0 13	+0 14	+0 21
56	-0 13	-0 08	+0 09	+0 13
54	-0 06	-0 04	+0 04	+0 06
50	+0 05	+0 04	-0 04	-0 05
N45	+0 17	+0 12	-0 11	-0 17
40	+0 26	+0 19	-0 18	-0 25
35	+0 33	+0 25	-0 24	-0 33
30	+0 40	+0 30	-0 29	-0 40
20	+0 51	+0 38	-0 38	-0 50
N10	+0 59	+0 45	-0 42	-0 59
0	+1 07	+0 52	-0 53	-1 07
S10	+1 12	+0 59	-0 59	-1 13
20	+1 18	+1 05	-1 06	-1 18
30	+1 24	+1 13	-1 15	-1 24
S35	+1 27	+1 17	-1 19	-1 27
40	+1 30	+1 22	-1 24	-1 30
45	+1 34	+1 28	-1 30	-1 34
S50	+1 38	+1 35	-1 36	-1 38

NOTES
The corrections to sunrise etc. are for middle of April.

MOON — Lat 52°N

Yr	Day of Mth	Week	Age (days)	Transit (Upper) (h m)	Diff (m)	Semi-diam	Hor Par	Moonrise (h m)	Moonset (h m)
91	1	Fri	28	10 29	41	14.7	54.0	04 28	16 44
92	2	Sat	29	11 10	41	14.7	53.9	04 43	17 51
93	3	Sun	30	11 51	43	14.7	54.0	04 59	18 58
94	4	Mon	01	12 34	44	14.8	54.1	05 17	20 06
95	5	Tu	02	13 18	47	14.8	54.3	05 37	21 14
96	6	Wed	03	14 05	50	14.9	54.7	06 03	22 20
97	7	Th	04	14 55	52	15.0	55.1	06 35	23 23
98	8	Fri	05	15 47	53	15.1	55.6	07 16	–
99	9	Sat	06	16 40	54	15.3	56.2	08 07	00 19
100	10	Sun	07	17 33	54	15.5	56.9	09 09	01 07
101	11	Mon	08	18 27	52	15.7	57.7	10 21	01 46
102	12	Tu	09	19 19	53	16.0	58.5	11 38	02 18
103	13	Wed	10	20 12	52	16.2	59.4	12 59	02 44
104	14	Th	11	21 04	52	16.4	60.2	14 23	03 06
105	15	Fri	12	21 56	54	16.6	60.8	15 49	03 27
106	16	Sat	13	22 50	56	16.7	61.1	17 16	03 48
107	17	Sun	14	23 46	59	16.7	61.2	18 45	04 10
108	18	Mon	15	24 45	–	16.6	61.0	20 13	04 36
109	19	Tu	16	00 45	60	16.5	60.4	21 36	05 08
110	20	Wed	17	01 45	60	16.3	59.7	22 50	05 48
111	21	Th	18	02 45	59	16.0	58.8	23 51	06 39
112	22	Fri	19	03 44	56	15.8	57.8	–	07 40
113	23	Sat	20	04 40	52	15.5	56.9	00 37	08 48
114	24	Sun	21	05 32	48	15.3	56.0	01 12	09 59
115	25	Mon	22	06 20	45	15.1	55.3	01 39	11 10
116	26	Tu	23	07 05	42	14.9	54.8	02 00	12 20
117	27	Wed	24	07 47	41	14.8	54.4	02 18	13 28
118	28	Th	25	08 28	41	14.7	54.2	02 34	14 35
119	29	Fri	26	09 09	41	14.7	54.1	02 50	15 41
120	30	Sat	27	09 50	42	14.7	54.0	03 05	16 48

Phases of the Moon

		d	h	m
●	New Moon	3	14	32
☽	First Quarter	11	12	05
○	Full Moon	18	02	44
☽	Last Quarter	25	02	47

	d	h
Apogee	2	09
Perigee	17	06
Apogee	29	18

MAY 2011

SUN AND ARIES

SUN AND ARIES

GMT	SUN GHA	SUN Dec.	ARIES GHA	GMT	SUN GHA	SUN Dec.	ARIES GHA

Sunday, 1st May

Monday, 2nd May

Tuesday, 3rd May

Wednesday, 4th May

Thursday, 5th May

Friday, 6th May

Saturday, 7th May

Sunday, 8th May

Monday, 9th May

Tuesday, 10th May

Wednesday, 11th May

Thursday, 12th May

Friday, 13th May

Saturday, 14th May

Sunday, 15th May

SUN AND ARIES

Monday, 16th May

Tuesday, 17th May

Wednesday, 18th May

Thursday, 19th May

Friday, 20th May

Saturday, 21st May

Sunday, 22nd May

Monday, 23rd May

Tuesday, 24th May

Wednesday, 25th May

Thursday, 26th May

Friday, 27th May

Saturday, 28th May

Sunday, 29th May

Monday, 30th May

Tuesday, 31st May

20

MOON MAY 2011 PLANETS

Day	GMT hr	GHA ° ′	Mean Var/hr 14°+	Dec ° ′	Mean Var/hr ′
1 Sun	0	206 35.4	34.2	N10 21.3	10.6
	6	294 00.6	33.9	N11 24.9	10.3
	12	21 24.3	33.7	N12 26.9	10.1
	18	108 46.2	33.3	N13 27.3	9.7
2 Mon	0	196 06.2	33.0	N14 25.8	9.4
	6	283 24.4	32.6	N15 22.3	9.0
	12	10 40.5	32.3	N16 16.5	8.6
	18	97 54.5	32.0	N17 08.4	8.1
3 Tu	0	185 06.3	31.6	N17 57.7	7.7
	6	272 15.9	31.2	N18 44.2	7.3
	12	359 23.3	30.9	N19 27.7	6.7
	18	86 28.5	30.5	N20 08.2	6.1
4 Wed	0	173 31.6	30.1	N20 45.3	5.6
	6	260 32.4	29.8	N21 19.0	4.9
	12	347 31.2	29.5	N21 49.0	4.3
	18	74 28.1	29.1	N22 15.2	3.7
5 Th	0	161 23.1	28.9	N22 37.5	2.9
	6	248 16.4	28.6	N22 55.6	2.3
	12	335 08.1	28.4	N23 09.6	1.5
	18	61 58.5	28.2	N23 19.2	0.8
6 Fri	0	148 47.8	28.0	N23 24.3	0.1
	6	235 36.0	27.9	N23 25.0	0.8
	12	322 23.6	27.8	N23 21.1	1.5
	18	49 10.6	27.7	N23 12.7	2.2
7 Sat	0	135 57.3	27.8	N22 59.6	3.0
	6	222 43.9	27.8	N22 41.9	3.8
	12	309 30.6	27.8	N22 19.7	4.5
	18	36 17.6	27.9	N21 52.9	5.3
8 Sun	0	123 05.1	28.0	N21 21.6	6.0
	6	209 53.2	28.2	N20 46.0	6.7
	12	296 42.0	28.3	N20 06.0	7.4
	18	23 31.7	28.4	N19 21.9	8.1
9 Mon	0	110 22.3	28.6	N18 33.8	8.7
	6	197 13.8	28.7	N17 41.7	9.3
	12	284 06.4	28.9	N16 45.9	9.9
	18	10 59.9	29.1	N15 46.5	10.5
10 Tu	0	97 54.3	29.2	N14 43.7	11.0
	6	184 49.7	29.3	N13 37.8	11.5
	12	271 45.8	29.5	N12 28.8	12.0
	18	358 42.7	29.6	N11 17.0	12.4
11 Wed	0	85 40.2	29.7	N10 02.7	12.8
	6	172 38.1	29.7	N 8 45.9	13.2
	12	259 36.3	29.7	N 7 27.1	13.5
	18	346 34.7	29.7	N 6 06.4	13.7
12 Th	0	73 33.0	29.7	N 4 44.1	14.0
	6	160 31.0	29.6	N 3 20.4	14.1
	12	247 28.6	29.5	N 1 55.6	14.3
	18	334 25.6	29.3	N 0 30.1	14.3
13 Fri	0	61 21.7	29.1	S 0 56.0	14.4
	6	148 16.8	29.0	S 2 22.2	14.4
	12	235 10.6	28.7	S 3 48.3	14.2
	18	322 02.9	28.4	S 5 14.0	14.1
14 Sat	0	48 53.6	28.1	S 6 38.9	14.0
	6	135 42.5	27.8	S 8 02.6	13.7
	12	222 29.4	27.5	S 9 24.9	13.4
	18	309 14.1	27.0	S10 45.2	13.0
15 Sun	0	35 56.7	26.6	S12 03.3	12.5
	6	122 36.9	26.3	S13 18.9	12.1
	12	209 14.7	25.9	S14 31.4	11.5
	18	295 50.1	25.5	S15 40.6	10.9
16 Mon	0	22 23.2	25.1	S16 46.1	10.2
	6	108 54.0	24.8	S17 47.5	9.5
	12	195 22.6	24.4	S18 44.6	8.6
	18	281 49.2	24.1	S19 36.9	7.8
17 Tu	0	8 14.0	23.8	S20 24.4	7.0
	6	94 37.3	23.7	S21 06.6	6.1
	12	180 59.3	23.6	S21 43.4	5.1
	18	267 20.6	23.4	S22 14.6	4.2
18 Wed	0	353 41.3	23.4	S22 40.0	3.2
	6	80 02.0	23.5	S22 59.7	2.2
	12	166 23.1	23.6	S23 13.6	1.2
	18	252 44.9	23.9	S23 21.7	0.3
19 Th	0	339 07.9	24.1	S23 24.0	0.6
	6	65 32.6	24.5	S23 20.6	1.6
	12	151 59.2	24.9	S23 11.7	2.5
	18	238 28.1	25.3	S22 57.5	3.3
20 Fri	0	324 59.7	25.8	S22 38.2	4.1
	6	51 34.2	26.3	S22 13.9	4.9
	12	138 11.8	26.9	S21 44.9	5.6
	18	224 52.7	27.5	S21 11.6	6.3
21 Sat	0	311 37.0	28.0	S20 34.1	7.0
	6	38 24.8	28.6	S19 52.7	7.6
	12	125 16.1	29.2	S19 07.8	8.1
	18	212 10.9	29.7	S18 19.6	8.6
22 Sun	0	299 09.1	30.3	S17 28.3	9.0
	6	26 10.7	30.9	S16 34.3	9.4
	12	113 15.5	31.3	S15 37.8	9.8
	18	200 23.5	31.8	S14 39.0	10.2
23 Mon	0	287 34.5	32.4	S13 38.1	10.4
	6	14 48.3	32.8	S12 35.5	10.7
	12	102 04.7	33.1	S11 31.3	10.9
	18	189 23.5	33.6	S10 25.7	11.2
24 Tu	0	276 44.5	33.9	S 9 18.8	11.3
	6	4 07.6	34.2	S 8 11.0	11.5
	12	91 32.5	34.5	S 7 02.2	11.6
	18	178 59.0	34.7	S 5 52.8	11.7
25 Wed	0	266 26.8	34.9	S 4 42.8	11.8
	6	353 55.9	35.0	S 3 32.4	11.7
	12	81 25.8	35.1	S 2 21.8	11.8
	18	168 56.5	35.2	S 1 11.0	11.8
26 Th	0	256 27.8	35.3	S 0 00.2	11.8
	6	343 59.3	35.2	N 1 10.4	11.7
	12	71 31.0	35.2	N 2 20.8	11.7
	18	159 02.5	35.2	N 3 30.7	11.6
27 Fri	0	246 33.8	35.1	N 4 40.2	11.4
	6	334 04.5	35.0	N 5 49.0	11.3
	12	61 34.6	34.8	N 6 57.0	11.2
	18	149 03.7	34.7	N 8 04.1	11.0
28 Sat	0	236 31.8	34.5	N 9 10.2	10.8
	6	323 58.7	34.2	N10 15.0	10.5
	12	51 24.1	34.0	N11 18.5	10.4
	18	138 48.0	33.7	N12 20.5	10.1
29 Sun	0	226 10.1	33.4	N13 20.9	9.7
	6	313 30.3	33.0	N14 19.4	9.4
	12	40 48.6	32.7	N15 16.0	9.1
	18	128 04.7	32.2	N16 10.4	8.6
30 Mon	0	215 18.6	31.9	N17 02.4	8.2
	6	302 30.3	31.5	N17 51.9	7.7
	12	29 39.6	31.1	N18 38.7	7.3
	18	116 46.5	30.7	N19 22.6	6.8
31 Tu	0	203 51.0	30.3	N20 03.4	6.2
	6	290 53.1	29.9	N20 41.0	5.6
	12	17 52.9	29.5	N21 15.1	5.1
	18	104 50.3	29.1	N21 45.5	4.4

VENUS					JUPITER					
Mer Pass h m	GHA ° ′	Mean Var/hr 14°+	Dec ° ′	Mean Var/hr ′	Day	GHA ° ′	Mean Var/hr 15°+	Dec ° ′	Mean Var/hr ′	Mer Pass h m
10 12	206 59.4	59.7	N 3 10.8	1.2	1 SUN	197 33.1	1.9	N 7 40.5	0.2	10 48
10 13	206 51.3	59.7	N 3 38.7	1.2	2 Mon	198 19.0	1.9	N 7 45.8	0.2	10 45
10 13	206 43.2	59.7	N 4 06.6	1.2	3 Tu	199 05.0	1.9	N 7 50.9	0.2	10 42
10 14	206 35.0	59.7	N 4 34.4	1.2	4 Wed	199 50.9	1.9	N 7 56.1	0.2	10 39
10 14	206 26.7	59.6	N 5 02.1	1.2	5 Th	200 36.9	1.9	N 8 01.2	0.2	10 36
10 15	206 18.3	59.6	N 5 29.8	1.1	6 Fri	201 22.9	1.9	N 8 06.3	0.2	10 33
10 16	206 09.8	59.6	N 5 57.3	1.1	7 Sat	202 08.9	1.9	N 8 11.4	0.2	10 30
10 16	206 01.2	59.6	N 6 24.8	1.1	8 SUN	202 55.0	1.9	N 8 16.5	0.2	10 27
10 17	205 52.4	59.6	N 6 52.1	1.1	9 Mon	203 41.0	1.9	N 8 21.5	0.2	10 24
10 17	205 43.6	59.6	N 7 19.3	1.1	10 Tu	204 27.2	1.9	N 8 26.6	0.2	10 21
10 18	205 34.6	59.6	N 7 46.4	1.1	11 Wed	205 13.3	1.9	N 8 31.6	0.2	10 18
10 19	205 25.5	59.6	N 8 13.3	1.1	12 Th	205 59.5	1.9	N 8 36.5	0.2	10 15
10 19	205 16.2	59.6	N 8 40.0	1.1	13 Fri	206 45.7	1.9	N 8 41.5	0.2	10 12
10 20	205 06.8	59.6	N 9 06.6	1.1	14 Sat	207 32.0	1.9	N 8 46.4	0.2	10 09
10 20	204 57.2	59.6	N 9 32.9	1.1	15 SUN	208 18.3	1.9	N 8 51.3	0.2	10 05
10 21	204 47.4	59.6	N 9 59.1	1.1	16 Mon	209 04.6	1.9	N 8 56.1	0.2	10 02
10 22	204 37.4	59.6	N10 25.1	1.1	17 Tu	209 51.0	1.9	N 9 00.9	0.2	09 59
10 22	204 27.3	59.6	N10 50.8	1.1	18 Wed	210 37.4	1.9	N 9 05.7	0.2	09 56
10 23	204 16.9	59.6	N11 16.3	1.1	19 Th	211 23.8	1.9	N 9 10.5	0.2	09 53
10 24	204 06.3	59.6	N11 41.6	1.0	20 Fri	212 10.3	1.9	N 9 15.3	0.2	09 50
10 25	203 55.6	59.5	N12 06.6	1.0	21 Sat	212 56.9	1.9	N 9 20.0	0.2	09 47
10 25	203 44.6	59.5	N12 31.4	1.0	22 SUN	213 43.5	1.9	N 9 24.7	0.2	09 44
10 26	203 33.3	59.5	N12 55.8	1.0	23 Mon	214 30.1	1.9	N 9 29.3	0.2	09 41
10 27	203 21.8	59.5	N13 20.0	1.0	24 Tu	215 16.8	1.9	N 9 34.0	0.2	09 38
10 28	203 10.1	59.5	N13 43.9	1.0	25 Wed	216 03.5	1.9	N 9 38.6	0.2	09 35
10 28	202 58.2	59.5	N14 07.4	1.0	26 Th	216 50.3	2.0	N 9 43.1	0.2	09 31
10 29	202 46.0	59.5	N14 30.6	1.0	27 Fri	217 37.1	2.0	N 9 47.7	0.2	09 28
10 30	202 33.5	59.5	N14 53.5	0.9	28 Sat	218 24.0	2.0	N 9 52.2	0.2	09 25
10 31	202 20.8	59.5	N15 16.1	0.9	29 SUN	219 10.9	2.0	N 9 56.6	0.2	09 22
10 32	202 07.8	59.5	N15 38.2	0.9	30 Mon	219 57.9	2.0	N10 01.1	0.2	09 19
10 33	201 54.6	59.4	N16 00.0	0.9	31 Tu	220 44.9	2.0	N10 05.5	0.2	09 16

VENUS, Av. Mag. –3.8
SHA May 5 344; 10 338; 15 333; 20 327; 25 321; 30 315

JUPITER, Av. Mag. –2.1
SHA May 5 338; 10 337; 15 336; 20 335; 25 334; 30 333

MARS					SATURN					
Mer Pass h m	GHA ° ′	Mean Var/hr 15°+	Dec ° ′	Mean Var/hr	Day	GHA ° ′	Mean Var/hr 15°+	Dec ° ′	Mean Var/hr ′	Mer Pass h m
10 48	197 46.7	0.7	N 7 58.5	0.7	1 SUN	26 34.1	2.6	S 2 12.9	0.1	22 10
10 47	198 03.2	0.7	N 8 15.9	0.7	2 Mon	27 36.7	2.6	S 2 11.5	0.1	22 06
10 46	198 19.7	0.7	N 8 33.2	0.7	3 Tu	28 39.3	2.6	S 2 10.2	0.1	22 02
10 45	198 36.1	0.7	N 8 50.4	0.7	4 Wed	29 41.9	2.6	S 2 08.9	0.1	21 57
10 44	198 52.6	0.7	N 9 07.5	0.7	5 Th	30 44.4	2.6	S 2 07.7	0.1	21 53
10 43	199 09.0	0.7	N 9 24.6	0.7	6 Fri	31 46.8	2.6	S 2 06.4	0.1	21 49
10 42	199 25.4	0.7	N 9 41.5	0.7	7 Sat	32 49.1	2.6	S 2 05.2	0.0	21 45
10 41	199 41.7	0.7	N 9 58.3	0.7	8 SUN	33 51.4	2.6	S 2 04.1	0.0	21 41
10 40	199 58.1	0.7	N10 15.0	0.7	9 Mon	34 53.7	2.6	S 2 02.9	0.0	21 37
10 39	200 14.4	0.7	N10 31.5	0.7	10 Tu	35 55.8	2.6	S 2 01.8	0.0	21 33
10 37	200 30.6	0.7	N10 48.0	0.7	11 Wed	36 57.9	2.6	S 2 00.8	0.0	21 28
10 36	200 46.9	0.7	N11 04.3	0.7	12 Th	37 59.9	2.6	S 1 59.7	0.0	21 24
10 35	201 03.1	0.7	N11 20.6	0.7	13 Fri	39 01.9	2.6	S 1 58.7	0.0	21 20
10 34	201 19.3	0.7	N11 36.7	0.7	14 Sat	40 03.8	2.6	S 1 57.8	0.0	21 16
10 33	201 35.4	0.7	N11 52.6	0.7	15 SUN	41 05.6	2.6	S 1 56.8	0.0	21 12
10 32	201 51.6	0.7	N12 08.5	0.7	16 Mon	42 07.3	2.6	S 1 55.9	0.0	21 08
10 31	202 07.6	0.7	N12 24.2	0.6	17 Tu	43 08.9	2.6	S 1 55.1	0.0	21 04
10 30	202 23.7	0.7	N12 39.8	0.6	18 Wed	44 10.5	2.6	S 1 54.3	0.0	21 00
10 29	202 39.7	0.7	N12 55.2	0.6	19 Th	45 12.0	2.6	S 1 53.5	0.0	20 56
10 28	202 55.6	0.7	N13 10.6	0.6	20 Fri	46 13.4	2.6	S 1 52.7	0.0	20 52
10 27	203 11.5	0.7	N13 25.7	0.6	21 Sat	47 14.7	2.6	S 1 52.0	0.0	20 47
10 26	203 27.4	0.7	N13 40.8	0.6	22 SUN	48 16.0	2.5	S 1 51.4	0.0	20 43
10 25	203 43.3	0.7	N13 55.7	0.6	23 Mon	49 17.2	2.5	S 1 50.7	0.0	20 39
10 24	203 59.1	0.7	N14 10.4	0.6	24 Tu	50 18.2	2.5	S 1 50.1	0.0	20 35
10 23	204 14.8	0.7	N14 25.0	0.6	25 Wed	51 19.2	2.5	S 1 49.6	0.0	20 31
10 22	204 30.5	0.7	N14 39.4	0.6	26 Th	52 20.2	2.5	S 1 49.0	0.0	20 27
10 20	204 46.2	0.6	N14 53.7	0.6	27 Fri	53 21.0	2.5	S 1 48.6	0.0	20 23
10 19	205 01.8	0.6	N15 07.9	0.6	28 Sat	54 21.7	2.5	S 1 48.1	0.0	20 19
10 18	205 17.3	0.7	N15 21.9	0.6	29 SUN	55 22.4	2.5	S 1 47.7	0.0	20 15
10 17	205 32.9	0.6	N15 35.7	0.6	30 Mon	56 23.0	2.5	S 1 47.3	0.0	20 11
10 16	205 48.4	0.6	N15 49.4	0.6	31 Tu	57 23.5	2.5	S 1 47.0	0.0	20 07

MARS, Av. Mag. +1.3
SHA May 5 336; 10 333; 15 329; 20 326; 25 322; 30 318

SATURN, Av. Mag. +0.6
SHA May 5 168; 10 168; 15 169; 20 169; 25 169; 30 169

MAY 2011

STARS   SUN AND MOON

STARS

0h GMT May 1

No.	Name	Mag	Transit h m	Dec ° '	SHA ° '
ψ	ARIES	—	9 24	—	357 45.3
1	Alpheratz	2.1	9 33	N29 09.1	357 45.3
2	Ankaa	2.4	9 51	S42 14.4	353 17.4
3	Schedar	2.2	10 05	N56 35.8	349 42.7
4	Diphda	2.0	10 08	S17 55.4	348 57.6
5	Achernar	0.5	11 02	S57 10.6	335 28.3
6	POLARIS	2.0	12 08	N89 18.8	318 56.6
7	Hamal	2.0	11 32	N23 30.9	328 02.8
8	Acamar	3.2	12 22	S40 15.5	315 19.9
9	Menkar	2.5	12 26	N 4 08.0	314 17.0
10	Mirfak	1.8	12 49	N49 54.0	308 43.1
11	Aldebaran	0.9	14 00	N16 31.8	290 51.4
12	Rigel	0.1	14 38	S 8 11.4	281 13.8
13	Capella	0.1	14 41	N46 00.6	280 37.1
14	Bellatrix	1.6	14 49	N 6 21.5	278 33.9
15	Elnath	1.7	14 50	N28 37.0	278 14.9
16	Alnilam	1.7	15 00	S 1 11.8	275 48.2
17	Betelgeuse	0.1–1.2	15 19	N 7 24.4	271 03.2
18	Canopus	–0.7	15 47	S52 42.4	263 57.2
19	Sirius	–1.5	16 09	S16 44.1	258 35.3
20	Adhara	1.6	16 22	S28 59.5	255 14.0
21	Castor	1.6	16 58	N31 51.8	246 01.1
22	Procyon	0.4	17 03	N 5 11.6	245 01.5
23	Pollux	1.1	17 09	N27 59.9	243 29.7
24	Avior	1.9	17 45	S59 33.2	234 18.9
25	Suhail	2.2	18 31	S43 29.1	222 53.7
26	Miaplacidus	1.7	18 36	S69 46.3	221 40.2
27	Alphard	2.0	18 51	S 8 42.7	217 57.6
28	Regulus	1.4	19 31	N11 54.6	207 45.1
29	Dubhe	1.8	20 27	N61 41.5	193 53.2
30	Denebola	2.1	21 12	N14 30.4	182 35.0
31	Gienah	2.6	21 39	S17 36.6	175 53.7
32	Acrux	1.3	21 49	S63 10.1	173 10.6
33	Gacrux	1.6	21 54	S57 11.0	172 02.3
34	Mimosa	1.3	22 10	S59 45.4	167 53.3
35	Alioth	1.8	22 17	N55 53.9	166 21.5
36	Spica	1.0	22 48	S11 13.4	158 32.6
37	Alkaid	1.9	23 10	N49 15.4	152 59.6
38	Hadar	0.6	23 27	S60 25.9	148 49.5
39	Menkent	2.1	23 29	S36 25.7	148 09.0
40	Arcturus	0.0	23 38	N19 07.3	145 56.8
41	Rigil Kent	–0.3	0 06	S60 53.1	139 53.2
42	Zuben'ubi	2.8	0 17	S16 05.5	137 06.8
43	Kochab	2.1	0 16	N74 06.5	137 18.6
44	Alphecca	2.2	1 01	N26 40.5	126 11.9
45	Antares	1.0	1 56	S26 27.5	112 27.8
46	Atria	1.9	2 15	S69 02.8	107 30.4
47	Sabik	2.4	2 36	S15 44.3	102 13.9
48	Shaula	1.6	3 00	S37 06.6	96 23.6
49	Rasalhague	2.1	3 01	N12 33.0	96 07.5
50	Eltanin	2.2	3 22	N51 29.1	90 46.4
51	Kaus Aust	1.9	3 50	S34 22.6	83 45.5
52	Vega	0.0	4 02	N38 47.5	80 39.7
53	Nunki	2.0	4 21	S26 16.8	75 59.9
54	Altair	0.8	5 16	N 8 53.9	62 09.5
55	Peacock	1.9	5 51	S56 41.6	53 21.3
56	Deneb	1.3	6 07	N45 19.1	49 32.4
57	Enif	2.4	7 09	N 9 55.6	33 48.6
58	Al Na'ir	1.7	7 33	S46 54.0	27 45.6
59	Fomalhaut	1.2	8 23	S29 33.5	15 25.7
60	Markab	2.5	8 30	N15 15.9	13 40.0

SUN — Lat 52°N

Yr	Day of Mth/Week	Transit h m	Semi-Diam	Twilight h m	Sunrise h m	Sunset h m	Twilight h m
121	1 Sun	11 57	15.9	03 52	04 31	19 24	20 03
122	2 Mon	11 57	15.9	03 50	04 29	19 26	20 05
123	3 Tu	11 57	15.9	03 48	04 27	19 28	20 07
124	4 Wed	11 57	15.9	03 46	04 26	19 29	20 09
125	5 Th	11 57	15.9	03 44	04 24	19 31	20 11
126	6 Fri	11 57	15.9	03 42	04 22	19 33	20 13
127	7 Sat	11 57	15.9	03 40	04 20	19 34	20 14
128	8 Sun	11 57	15.9	03 38	04 18	19 36	20 16
129	9 Mon	11 56	15.9	03 36	04 17	19 37	20 18
130	10 Tu	11 56	15.9	03 34	04 15	19 39	20 20
131	11 Wed	11 56	15.9	03 32	04 13	19 41	20 22
132	12 Th	11 56	15.9	03 30	04 12	19 42	20 24
133	13 Fri	11 56	15.9	03 28	04 10	19 44	20 26
134	14 Sat	11 56	15.9	03 27	04 08	19 45	20 27
135	15 Sun	11 56	15.8	03 25	04 07	19 47	20 29
136	16 Mon	11 56	15.8	03 23	04 05	19 48	20 31
137	17 Tu	11 56	15.8	03 21	04 04	19 50	20 33
138	18 Wed	11 56	15.8	03 20	04 02	19 51	20 35
139	19 Th	11 56	15.8	03 18	04 01	19 53	20 36
140	20 Fri	11 57	15.8	03 16	04 00	19 54	20 38
141	21 Sat	11 57	15.8	03 15	03 58	19 56	20 40
142	22 Sun	11 57	15.8	03 13	03 57	19 57	20 41
143	23 Mon	11 57	15.8	03 12	03 56	19 59	20 43
144	24 Tu	11 57	15.8	03 10	03 54	20 00	20 45
145	25 Wed	11 57	15.8	03 09	03 53	20 01	20 46
146	26 Th	11 57	15.8	03 07	03 52	20 03	20 48
147	27 Fri	11 57	15.8	03 06	03 51	20 04	20 49
148	28 Sat	11 57	15.8	03 05	03 50	20 05	20 51
149	29 Sun	11 57	15.8	03 03	03 49	20 06	20 52
150	30 Mon	11 57	15.8	03 03	03 48	20 08	20 54
151	31 Tu	11 58	15.8	03 01	03 47	20 09	20 55

Lat Corr to Sunrise, Sunset etc.

Lat °	Twilight h m	Sunrise h m	Sunset h m	Twilight h m
N70	TAN	–3 17	+3 32	TAN
68	TAN	–2 16	+2 20	TAN
66	TAN	–1 42	+1 45	TAN
64	–2 12	–1 17	+1 20	+2 13
62	–1 31	–0 58	+1 01	+1 31
N60	–1 04	–0 43	+0 44	+1 03
58	–0 42	–0 30	+0 30	+0 45
56	–0 27	–0 18	+0 19	+0 27
54	–0 12	–0 08	+0 09	+0 12
52	+0 01	+0 08	–0 08	–0 11
50	+0 33	+0 25	–0 24	–0 32
N45	+0 50	+0 39	–0 38	–0 49
40	+1 04	+0 50	–0 49	–1 03
35	+1 17	+1 00	–1 00	–1 16
30	+1 35	+1 17	–1 17	–1 35
20	+1 51	+1 32	–1 32	–1 51
N10	+2 06	+1 46	–1 46	–2 06
0	+2 19	+1 59	–2 00	–2 19
S10	+2 32	+2 13	–2 15	–2 32
20	+2 46	+2 30	–2 32	–2 47
30	+3 03	+2 50	–2 41	–3 03
S35	+3 13	+3 03	–2 52	–3 13
40	+3 24	+3 18	–3 04	–3 24
S50				

NOTES
The corrections to sunrise etc. are for middle of May. TAN means Twilight all night.

MOON — Lat 52°N

Yr	Day of Mth/Week	Age days	Transit (Upper) h m	Diff m	Semi-diam	Hor Par	Moonrise h m	Moonset h m
121	1 Sun	28	10 32	44	14.8	54.2	03 23	17 55
122	2 Mon	29	11 16	47	14.8	54.4	03 43	19 04
123	3 Tu	00	12 03	49	14.9	54.7	04 07	20 11
124	4 Wed	01	12 52	51	15.0	55.1	04 37	21 16
125	5 Th	02	13 43	53	15.1	55.5	05 15	22 15
126	6 Fri	03	14 36	53	15.3	56.0	06 04	23 05
127	7 Sat	04	15 29	53	15.4	56.5	07 03	23 46
128	8 Sun	05	16 22	52	15.6	57.1	08 12	—
129	9 Mon	06	17 14	51	15.7	57.7	09 26	00 20
130	10 Tu	07	18 05	51	15.9	58.4	10 44	00 47
131	11 Wed	08	18 56	50	16.1	59.0	12 04	01 10
132	12 Th	09	19 46	51	16.2	59.6	13 26	01 30
133	13 Fri	10	20 37	54	16.4	60.1	14 49	01 50
134	14 Sat	11	21 31	56	16.5	60.4	16 15	02 11
135	15 Sun	12	22 27	59	16.5	60.6	17 41	02 34
136	16 Mon	13	23 26	60	16.4	60.4	19 06	03 02
137	17 Tu	14	24 26	—	16.2	59.4	20 25	03 38
138	18 Wed	15	00 26	61	16.0	58.7	21 33	04 24
139	19 Th	16	01 27	59	15.7	57.8	22 27	05 21
140	20 Fri	17	02 26	55	15.5	56.9	23 08	06 28
141	21 Sat	18	03 21	51	15.3	56.1	23 39	07 40
142	22 Sun	19	04 12	47	15.1	55.4	—	08 53
143	23 Mon	20	04 59	44	14.9	54.8	00 03	10 05
144	24 Tu	21	05 43	42	14.8	54.4	00 22	11 15
145	25 Wed	22	06 25	41	14.8	54.4	00 40	12 23
146	26 Th	23	07 06	41	14.8	54.1	00 55	13 29
147	27 Fri	24	07 47	41	14.8	54.1	01 11	14 36
148	28 Sat	25	08 28	44	14.8	54.2	01 28	15 43
149	29 Sun	26	09 12	45	14.9	54.5	01 47	16 51
150	30 Mon	27	09 57	49	14.8	54.8	02 09	17 59
151	31 Tu	28	10 46	51	15.0	55.2	02 37	19 05

Phases of the Moon

Phase	d	h	m
● New Moon	3	06	51
☽ First Quarter	10	20	33
○ Full Moon	17	11	09
☾ Last Quarter	24	18	52

	d	h
Perigee	15	11
Apogee	27	10

SUN AND ARIES JUNE 2011 SUN AND ARIES

GMT	SUN GHA ° '	SUN Dec ° '	ARIES GHA ° '
Wednesday, 1st June			
00	180 34.2	**N21 58.5**	249 08.1
02	210 34.0	**21 59.2**	279 13.1
04	240 33.8	**21 59.9**	309 18.0
06	270 33.6	**22 00.6**	339 22.9
08	300 33.4	**22 01.3**	9 27.9
10	330 33.3	**22 02.0**	39 32.8
12	0 33.1	**22 02.7**	69 37.7
14	30 32.9	**22 03.3**	99 42.6
16	60 32.7	**22 04.0**	129 47.6
18	90 32.5	**22 04.7**	159 52.5
20	120 32.3	**22 05.4**	189 57.4
22	150 32.1	**N22 06.0**	220 02.4
Thursday, 2nd June			
00	180 31.9	**N22 06.7**	250 07.3
02	210 31.7	**22 07.4**	280 12.2
04	240 31.5	**22 08.0**	310 17.1
06	270 31.3	**22 08.7**	340 22.1
08	300 31.1	**22 09.3**	10 27.0
10	330 30.9	**22 10.0**	40 31.9
12	0 30.7	**22 10.6**	70 36.9
14	30 30.5	**22 11.3**	100 41.8
16	60 30.3	**22 11.9**	130 46.7
18	90 30.1	**22 12.6**	160 51.6
20	120 29.9	**22 13.2**	190 56.6
22	150 29.7	**N22 13.8**	221 01.5
Friday, 3rd June			
00	180 29.5	**N22 14.5**	251 06.4
02	210 29.3	**22 15.1**	281 11.4
04	240 29.1	**22 15.7**	311 16.3
06	270 28.9	**22 16.4**	341 21.2
08	300 28.7	**22 17.0**	11 26.1
10	330 28.5	**22 17.6**	41 31.1
12	0 28.3	**22 18.2**	71 36.0
14	30 28.1	**22 18.8**	101 40.9
16	60 27.9	**22 19.5**	131 45.9
18	90 27.6	**22 20.1**	161 50.8
20	120 27.4	**22 20.7**	191 55.7
22	150 27.2	**N22 21.3**	222 00.6
Saturday, 4th June			
00	180 27.0	**N22 21.9**	252 05.6
02	210 26.8	**22 22.5**	282 10.5
04	240 26.6	**22 23.1**	312 15.4
06	270 26.4	**22 23.7**	342 20.4
08	300 26.2	**22 24.3**	12 25.3
10	330 25.9	**22 24.8**	42 30.2
12	0 25.7	**22 25.4**	72 35.1
14	30 25.5	**22 26.0**	102 40.1
16	60 25.3	**22 26.6**	132 45.0
18	90 25.1	**22 27.2**	162 49.9
20	120 24.9	**22 27.7**	192 54.9
22	150 24.7	**N22 28.3**	222 59.8
Sunday, 5th June			
00	180 24.4	**N22 28.9**	253 04.7
02	210 24.2	**22 29.4**	283 09.6
04	240 24.0	**22 30.0**	313 14.6
06	270 23.8	**22 30.6**	343 19.5
08	300 23.6	**22 31.1**	13 24.4
10	330 23.3	**22 31.7**	43 29.4
12	0 23.1	**22 32.2**	73 34.3
14	30 22.9	**22 32.8**	103 39.2
16	60 22.7	**22 33.3**	133 44.1
18	90 22.4	**22 33.9**	163 49.1
20	120 22.2	**22 34.4**	193 54.0
22	150 22.0	**N22 35.0**	223 58.9

GMT	SUN GHA ° '	SUN Dec ° '	ARIES GHA ° '
Monday, 6th June			
00	180 21.8	**N22 35.5**	254 03.9
02	210 21.5	**22 36.0**	284 08.8
04	240 21.3	**22 36.6**	314 13.7
06	270 21.1	**22 37.1**	344 18.6
08	300 20.9	**22 37.6**	14 23.6
10	330 20.6	**22 38.1**	44 28.5
12	0 20.4	**22 38.6**	74 33.4
14	30 20.2	**22 39.2**	104 38.3
16	60 20.0	**22 39.7**	134 43.3
18	90 19.7	**22 40.2**	164 48.2
20	120 19.5	**22 40.7**	194 53.1
22	150 19.3	**N22 41.2**	224 58.1
Tuesday, 7th June			
00	180 19.0	**N22 41.7**	255 03.0
02	210 18.8	**22 42.2**	285 07.9
04	240 18.6	**22 42.7**	315 12.8
06	270 18.3	**22 43.2**	345 17.8
08	300 18.1	**22 43.7**	15 22.7
10	330 17.9	**22 44.2**	45 27.6
12	0 17.6	**22 44.7**	75 32.6
14	30 17.4	**22 45.1**	105 37.5
16	60 17.2	**22 45.6**	135 42.4
18	90 16.9	**22 46.1**	165 47.3
20	120 16.7	**22 46.6**	195 52.3
22	150 16.5	**N22 47.0**	225 57.2
Wednesday, 8th June			
00	180 16.2	**N22 47.5**	256 02.1
02	210 16.0	**22 48.0**	286 07.1
04	240 15.8	**22 48.4**	316 12.0
06	270 15.5	**22 48.9**	346 16.9
08	300 15.3	**22 49.4**	16 21.8
10	330 15.1	**22 49.8**	46 26.8
12	0 14.8	**22 50.3**	76 31.7
14	30 14.6	**22 50.7**	106 36.6
16	60 14.3	**22 51.2**	136 41.6
18	90 14.1	**22 51.6**	166 46.5
20	120 13.9	**22 52.1**	196 51.4
22	150 13.6	**N22 52.5**	226 56.3
Thursday, 9th June			
00	180 13.4	**N22 52.9**	257 01.3
02	210 13.1	**22 53.4**	287 06.2
04	240 12.9	**22 53.8**	317 11.1
06	270 12.6	**22 54.2**	347 16.1
08	300 12.4	**22 54.6**	17 21.0
10	330 12.2	**22 55.1**	47 25.9
12	0 11.9	**22 55.5**	77 30.8
14	30 11.7	**22 55.9**	107 35.8
16	60 11.4	**22 56.3**	137 40.7
18	90 11.2	**22 56.7**	167 45.6
20	120 10.9	**22 57.1**	197 50.5
22	150 10.7	**N22 57.5**	227 55.5
Friday, 10th June			
00	180 10.4	**N22 57.9**	258 00.4
02	210 10.2	**22 58.3**	288 05.3
04	240 10.0	**22 58.7**	318 10.3
06	270 09.7	**22 59.1**	348 15.2
08	300 09.5	**22 59.5**	18 20.1
10	330 09.2	**22 59.9**	48 25.0
12	0 09.0	**23 00.3**	78 30.0
14	30 08.7	**23 00.7**	108 34.9
16	60 08.5	**23 01.1**	138 39.8
18	90 08.2	**23 01.4**	168 44.8
20	120 08.0	**23 01.8**	198 49.7
22	150 07.7	**N23 02.2**	228 54.6

GMT	SUN GHA ° '	SUN Dec ° '	ARIES GHA ° '
Saturday, 11th June			
00	180 07.5	**N23 02.6**	258 59.5
02	210 07.2	**23 02.9**	289 04.5
04	240 07.0	**23 03.3**	319 09.4
06	270 06.7	**23 03.6**	349 14.3
08	300 06.5	**23 04.0**	19 19.3
10	330 06.2	**23 04.4**	49 24.2
12	0 06.0	**23 04.7**	79 29.1
14	30 05.7	**23 05.1**	109 34.0
16	60 05.5	**23 05.4**	139 39.0
18	90 05.2	**23 05.7**	169 43.9
20	120 05.0	**23 06.1**	199 48.8
22	150 04.7	**N23 06.4**	229 53.8
Sunday, 12th June			
00	180 04.4	**N23 06.8**	259 58.7
02	210 04.2	**23 07.1**	290 03.6
04	240 03.9	**23 07.4**	320 08.5
06	270 03.7	**23 07.7**	350 13.5
08	300 03.4	**23 08.1**	20 18.4
10	330 03.2	**23 08.4**	50 23.3
12	0 02.9	**23 08.7**	80 28.2
14	30 02.7	**23 09.0**	110 33.2
16	60 02.4	**23 09.3**	140 38.1
18	90 02.1	**23 09.6**	170 43.0
20	120 01.9	**23 09.9**	200 48.0
22	150 01.6	**N23 10.2**	230 52.9
Monday, 13th June			
00	180 01.4	**N23 10.5**	260 57.8
02	210 01.1	**23 10.8**	291 02.7
04	240 00.9	**23 11.1**	321 07.7
06	270 00.6	**23 11.4**	351 12.6
08	300 00.3	**23 11.7**	21 17.5
10	330 00.1	**23 12.0**	51 22.5
12	359 59.8	**23 12.3**	81 27.4
14	29 59.6	**23 12.6**	111 32.3
16	59 59.3	**23 12.9**	141 37.2
18	89 59.0	**23 13.1**	171 42.2
20	119 58.8	**23 13.4**	201 47.1
22	149 58.5	**N23 13.7**	231 52.0
Tuesday, 14th June			
00	179 58.3	**N23 13.9**	261 57.0
02	209 58.0	**23 14.2**	292 01.9
04	239 57.7	**23 14.5**	322 06.8
06	269 57.5	**23 14.7**	352 11.7
08	299 57.2	**23 15.0**	22 16.7
10	329 57.0	**23 15.2**	52 21.6
12	359 56.7	**23 15.5**	82 26.5
14	29 56.4	**23 15.7**	112 31.5
16	59 56.2	**23 16.0**	142 36.4
18	89 55.9	**23 16.2**	172 41.3
20	119 55.6	**23 16.4**	202 46.2
22	149 55.4	**N23 16.7**	232 51.2
Wednesday, 15th June			
00	179 55.1	**N23 16.9**	262 56.1
02	209 54.9	**23 17.1**	293 01.0
04	239 54.6	**23 17.4**	323 06.0
06	269 54.3	**23 17.6**	353 10.9
08	299 54.1	**23 17.8**	23 15.8
10	329 53.8	**23 18.0**	53 20.7
12	359 53.5	**23 18.2**	83 25.7
14	29 53.3	**23 18.5**	113 30.6
16	59 53.0	**23 18.7**	143 35.5
18	89 52.7	**23 18.9**	173 40.5
20	119 52.5	**23 19.1**	203 45.4
22	149 52.2	**N23 19.3**	233 50.3

GMT	SUN GHA ° '	SUN Dec ° '	ARIES GHA ° '
Thursday, 16th June			
00	179 51.9	**N23 19.5**	263 55.2
02	209 51.7	**23 19.7**	294 00.2
04	239 51.4	**23 19.9**	324 05.1
06	269 51.1	**23 20.1**	354 10.0
08	299 50.9	**23 20.2**	24 15.0
10	329 50.6	**23 20.4**	54 19.9
12	359 50.3	**23 20.6**	84 24.8
14	29 50.1	**23 20.8**	114 29.7
16	59 49.8	**23 21.0**	144 34.7
18	89 49.5	**23 21.1**	174 39.6
20	119 49.3	**23 21.3**	204 44.5
22	149 49.0	**N23 21.5**	234 49.5
Friday, 17th June			
00	179 48.7	**N23 21.6**	264 54.4
02	209 48.5	**23 21.8**	294 59.3
04	239 48.2	**23 22.0**	325 04.2
06	269 47.9	**23 22.1**	355 09.2
08	299 47.7	**23 22.3**	25 14.1
10	329 47.4	**23 22.4**	55 19.0
12	359 47.1	**23 22.6**	85 24.0
14	29 46.9	**23 22.7**	115 28.9
16	59 46.6	**23 22.8**	145 33.8
18	89 46.3	**23 23.0**	175 38.7
20	119 46.1	**23 23.1**	205 43.7
22	149 45.8	**N23 23.3**	235 48.6
Saturday, 18th June			
00	179 45.5	**N23 23.4**	265 53.5
02	209 45.3	**23 23.5**	295 58.5
04	239 45.0	**23 23.6**	326 03.4
06	269 44.7	**23 23.8**	356 08.3
08	299 44.4	**23 23.9**	26 13.2
10	329 44.2	**23 24.0**	56 18.2
12	359 43.9	**23 24.1**	86 23.1
14	29 43.6	**23 24.2**	116 28.0
16	59 43.4	**23 24.3**	146 33.0
18	89 43.1	**23 24.4**	176 37.9
20	119 42.8	**23 24.5**	206 42.8
22	149 42.6	**N23 24.6**	236 47.7
Sunday, 19th June			
00	179 42.3	**N23 24.7**	266 52.7
02	209 42.0	**23 24.8**	296 57.6
04	239 41.7	**23 24.9**	327 02.5
06	269 41.5	**23 25.0**	357 07.4
08	299 41.2	**23 25.1**	27 12.4
10	329 40.9	**23 25.1**	57 17.3
12	359 40.7	**23 25.2**	87 22.2
14	29 40.4	**23 25.3**	117 27.2
16	59 40.1	**23 25.4**	147 32.1
18	89 39.9	**23 25.4**	177 37.0
20	119 39.6	**23 25.5**	207 41.9
22	149 39.3	**N23 25.6**	237 46.9
Monday, 20th June			
00	179 39.0	**N23 25.6**	267 51.8
02	209 38.8	**23 25.7**	297 56.7
04	239 38.5	**23 25.7**	328 01.7
06	269 38.2	**23 25.8**	358 06.6
08	299 38.0	**23 25.8**	28 11.5
10	329 37.7	**23 25.9**	58 16.4
12	359 37.4	**23 25.9**	88 21.4
14	29 37.1	**23 26.0**	118 26.3
16	59 36.9	**23 26.0**	148 31.2
18	89 36.6	**23 26.1**	178 36.2
20	119 36.3	**23 26.1**	208 41.1
22	149 36.1	**N23 26.1**	238 46.0

GMT	SUN GHA ° '	SUN Dec ° '	ARIES GHA ° '
Tuesday, 21st June			
00	179 35.8	**N23 26.1**	268 50.9
02	209 35.5	**23 26.2**	298 55.9
04	239 35.2	**23 26.2**	329 00.8
06	269 35.0	**23 26.2**	359 05.7
08	299 34.7	**23 26.2**	29 10.7
10	329 34.4	**23 26.2**	59 15.6
12	359 34.2	**23 26.2**	89 20.5
14	29 33.9	**23 26.2**	119 25.4
16	59 33.6	**23 26.2**	149 30.4
18	89 33.3	**23 26.2**	179 35.3
20	119 33.1	**23 26.2**	209 40.2
22	149 32.8	**N23 26.2**	239 45.2
Wednesday, 22nd June			
00	179 32.5	**N23 26.2**	269 50.1
02	209 32.3	**23 26.2**	299 55.0
04	239 32.0	**23 26.2**	329 59.9
06	269 31.7	**23 26.2**	0 04.9
08	299 31.4	**23 26.2**	30 09.8
10	329 31.2	**23 26.1**	60 14.7
12	359 30.9	**23 26.1**	90 19.6
14	29 30.6	**23 26.1**	120 24.6
16	59 30.4	**23 26.1**	150 29.5
18	89 30.1	**23 26.0**	180 34.4
20	119 29.8	**23 26.0**	210 39.4
22	149 29.6	**N23 26.0**	240 44.3
Thursday, 23rd June			
00	179 29.3	**N23 25.9**	270 49.2
02	209 29.0	**23 25.9**	300 54.1
04	239 28.7	**23 25.8**	330 59.1
06	269 28.5	**23 25.8**	1 04.0
08	299 28.2	**23 25.7**	31 08.9
10	329 27.9	**23 25.7**	61 13.9
12	359 27.7	**23 25.6**	91 18.8
14	29 27.4	**23 25.5**	121 23.7
16	59 27.1	**23 25.5**	151 28.6
18	89 26.9	**23 25.4**	181 33.6
20	119 26.6	**23 25.3**	211 38.5
22	149 26.3	**N23 25.2**	241 43.4
Friday, 24th June			
00	179 26.0	**N23 25.2**	271 48.4
02	209 25.8	**23 25.1**	301 53.3
04	239 25.5	**23 25.0**	331 58.2
06	269 25.2	**23 24.9**	2 03.1
08	299 25.0	**23 24.8**	32 08.1
10	329 24.7	**23 24.7**	62 13.0
12	359 24.4	**23 24.6**	92 17.9
14	29 24.2	**23 24.6**	122 22.9
16	59 23.9	**23 24.5**	152 27.8
18	89 23.6	**23 24.3**	182 32.7
20	119 23.4	**23 24.2**	212 37.6
22	149 23.1	**N23 24.1**	242 42.6
Saturday, 25th June			
00	179 22.8	**N23 24.0**	272 47.5
02	209 22.6	**23 23.9**	302 52.4
04	239 22.3	**23 23.8**	332 57.4
06	269 22.0	**23 23.7**	3 02.3
08	299 21.8	**23 23.5**	33 07.2
10	329 21.5	**23 23.4**	63 12.1
12	359 21.2	**23 23.3**	93 17.1
14	29 21.0	**23 23.2**	123 22.0
16	59 20.7	**23 23.0**	153 26.9
18	89 20.4	**23 22.9**	183 31.8
20	119 20.2	**23 22.8**	213 36.8
22	149 19.9	**N23 22.6**	243 41.7

GMT	SUN GHA ° '	SUN Dec ° '	ARIES GHA ° '
Sunday, 26th June			
00	179 19.6	**N23 22.5**	273 46.6
02	209 19.4	**23 22.3**	303 51.6
04	239 19.1	**23 22.2**	333 56.5
06	269 18.8	**23 22.0**	4 01.4
08	299 18.6	**23 21.8**	34 06.3
10	329 18.3	**23 21.7**	64 11.3
12	359 18.0	**23 21.5**	94 16.2
14	29 17.8	**23 21.4**	124 21.1
16	59 17.5	**23 21.2**	154 26.1
18	89 17.2	**23 21.0**	184 31.0
20	119 17.0	**23 20.8**	214 35.9
22	149 16.7	**N23 20.7**	244 40.8
Monday, 27th June			
00	179 16.5	**N23 20.5**	274 45.8
02	209 16.2	**23 20.3**	304 50.7
04	239 15.9	**23 20.1**	334 55.6
06	269 15.7	**23 19.9**	5 00.6
08	299 15.4	**23 19.7**	35 05.5
10	329 15.1	**23 19.5**	65 10.4
12	359 14.9	**23 19.3**	95 15.3
14	29 14.6	**23 19.1**	125 20.3
16	59 14.4	**23 18.9**	155 25.2
18	89 14.1	**23 18.7**	185 30.1
20	119 13.8	**23 18.5**	215 35.1
22	149 13.6	**N23 18.3**	245 40.0
Tuesday, 28th June			
00	179 13.3	**N23 18.1**	275 44.9
02	209 13.1	**23 17.9**	305 49.8
04	239 12.8	**23 17.7**	335 54.8
06	269 12.5	**23 17.4**	5 59.7
08	299 12.3	**23 17.2**	36 04.6
10	329 12.0	**23 17.0**	66 09.6
12	359 11.8	**23 16.8**	96 14.5
14	29 11.5	**23 16.5**	126 19.4
16	59 11.3	**23 16.3**	156 24.3
18	89 11.0	**23 16.0**	186 29.3
20	119 10.7	**23 15.8**	216 34.2
22	149 10.5	**N23 15.6**	246 39.1
Wednesday, 29th June			
00	179 10.2	**N23 15.3**	276 44.1
02	209 10.0	**23 15.1**	306 49.0
04	239 09.7	**23 14.8**	336 53.9
06	269 09.5	**23 14.5**	6 58.8
08	299 09.2	**23 14.3**	37 03.8
10	329 09.0	**23 14.0**	67 08.7
12	359 08.7	**23 13.8**	97 13.6
14	29 08.4	**23 13.5**	127 18.6
16	59 08.2	**23 13.2**	157 23.5
18	89 07.9	**23 12.9**	187 28.4
20	119 07.7	**23 12.7**	217 33.3
22	149 07.4	**N23 12.4**	247 38.3
Thursday, 30th June			
00	179 07.2	**N23 12.1**	277 43.2
02	209 06.9	**23 11.8**	307 48.1
04	239 06.7	**23 11.5**	337 53.0
06	269 06.4	**23 11.2**	7 58.0
08	299 06.2	**23 10.9**	38 02.9
10	329 05.9	**23 10.7**	68 07.8
12	359 05.7	**23 10.4**	98 12.8
14	29 05.4	**23 10.0**	128 17.7
16	59 05.2	**23 09.7**	158 22.6
18	89 04.9	**23 09.4**	188 27.5
20	119 04.7	**23 09.1**	218 32.5
22	149 04.4	**N23 08.8**	248 37.4

23

JUNE

23

JUNE 2011

MOON

Day	GMT hr	GHA ° '	Mean Var/hr 14°+	Dec ° '	Mean Var/hr '
1 Wed	0	191 45.6	28.8	N22 12.1	3.7
	6	278 38.9	28.6	N22 34.7	3.0
	12	5 30.2	28.2	N22 53.2	2.3
	18	Eclipse of the Sun occurs today			
2 Th	0	179 07.8	27.7	N23 17.2	0.9
	6	265 54.5	27.6	N23 22.5	0.1
	12	352 40.1	27.4	N23 23.2	0.7
	18	79 24.9	27.4	N23 19.2	1.6
3 Fri	0	166 09.0	27.3	N23 10.5	2.3
	6	252 52.9	27.3	N22 57.1	3.1
	12	339 36.6	27.3	N22 39.0	3.8
	18	66 20.6	27.3	N22 16.1	4.7
4 Sat	0	153 04.9	27.5	N21 48.7	5.4
	6	239 49.8	27.6	N21 16.7	6.2
	12	326 35.5	27.7	N20 40.2	6.8
	18	53 22.1	28.0	N19 59.4	7.6
5 Sun	0	140 09.8	28.2	N19 14.4	8.3
	6	226 58.7	28.4	N18 25.3	8.9
	12	313 48.9	28.6	N17 32.4	9.5
	18	40 40.4	28.8	N16 35.8	10.0
6 Mon	0	127 33.1	29.1	N15 35.8	10.6
	6	214 27.2	29.2	N14 32.4	11.1
	12	301 22.6	29.5	N13 26.0	11.6
	18	28 19.2	29.7	N12 16.8	12.0
7 Tu	0	115 16.8	29.8	N11 05.0	12.4
	6	202 15.4	29.9	N 9 50.8	12.8
	12	289 14.9	30.1	N 8 34.5	13.1
	18	16 15.1	30.1	N 7 16.3	13.4
8 Wed	0	103 15.8	30.2	N 5 56.5	13.5
	6	190 16.9	30.3	N 4 35.3	13.7
	12	277 18.1	30.1	N 3 13.0	13.9
	18	4 19.2	30.1	N 1 49.8	14.0
9 Th	0	91 20.1	30.1	N 0 26.0	14.0
	6	178 20.5	29.9	S 0 58.0	14.0
	12	265 20.3	29.8	S 2 22.2	14.0
	18	352 19.1	29.6	S 3 46.0	13.9
10 Fri	0	79 16.8	29.4	S 5 09.4	13.8
	6	166 13.3	29.1	S 6 31.9	13.6
	12	253 08.1	28.8	S 7 53.3	13.3
	18	340 01.3	28.6	S 9 13.2	13.0
11 Sat	0	66 52.7	28.2	S10 31.4	12.6
	6	153 42.0	27.8	S11 47.6	12.2
	12	240 29.1	27.5	S13 01.3	11.8
	18	327 14.0	27.0	S14 12.4	11.3
12 Sun	0	53 56.6	26.7	S15 20.4	10.7
	6	140 36.7	26.2	S16 25.1	10.1
	12	227 14.6	25.9	S17 26.1	9.5
	18	313 50.1	25.5	S18 23.2	8.7
13 Mon	0	40 23.4	25.2	S19 16.0	7.9
	6	126 54.6	24.9	S20 04.2	7.2
	12	213 23.8	24.6	S20 47.7	6.3
	18	299 51.4	24.3	S21 26.2	5.4
14 Tu	0	26 17.6	24.2	S21 59.4	4.5
	6	112 42.7	24.1	S22 27.3	3.7
	12	199 07.0	24.0	S22 49.7	2.7
	18	285 31.0	24.0	S23 06.5	1.7
15 Wed	0	11 55.0	24.1	S23 17.6	0.8
	6	98 19.4	24.2	S23 23.1	0.1
	12	184 44.6	24.4	S23 23.0	1.0
	18	271 11.1	24.7	S23 17.3	1.9
16 Th	0	357 39.2	25.0	S23 06.2	2.8
	6	84 09.3	25.4	S22 49.9	3.7
	12	170 41.6	25.9	S22 28.4	4.4
	18	257 16.5	26.3	S22 02.1	5.2

Day	GMT hr	GHA ° '	Mean Var/hr 14°+	Dec ° '	Mean Var/hr '
17 Fri	0	343 54.2	26.8	S21 31.1	6.0
	6	70 35.0	27.3	S20 55.7	6.6
	12	157 18.8	27.9	S20 16.1	7.3
	18	244 05.9	28.5	S19 32.7	7.9
18 Sat	0	330 56.4	29.0	S18 45.7	8.4
	6	57 50.1	29.6	S17 55.4	8.9
	12	144 47.2	30.1	S17 02.1	9.4
	18	231 47.5	30.6	S16 06.0	9.8
19 Sun	0	318 50.9	31.2	S15 07.5	10.2
	6	45 57.5	31.6	S14 06.6	10.5
	12	133 06.9	32.1	S13 03.8	10.8
	18	220 19.2	32.5	S11 59.3	11.0
20 Mon	0	307 34.1	32.9	S10 53.2	11.3
	6	34 51.4	33.2	S 9 45.8	11.4
	12	122 10.9	33.7	S 8 37.3	11.6
	18	209 32.5	33.9	S 7 27.9	11.7
21 Tu	0	296 55.9	34.2	S 6 17.8	11.8
	6	24 21.0	34.4	S 5 07.0	11.8
	12	111 47.5	34.6	S 3 55.9	11.9
	18	199 15.2	34.8	S 2 44.6	11.9
22 Wed	0	286 43.9	34.9	S 1 33.1	11.9
	6	14 13.4	35.0	S 0 21.7	11.8
	12	101 43.5	35.0	N 0 49.5	11.9
	18	189 13.9	35.1	N 2 00.5	11.7
23 Th	0	276 44.5	35.1	N 3 10.9	11.6
	6	4 15.1	35.0	N 4 20.8	11.6
	12	91 45.4	35.0	N 5 30.0	11.4
	18	179 15.2	34.8	N 6 38.4	11.3
24 Fri	0	266 44.4	34.7	N 7 45.8	11.1
	6	354 12.7	34.5	N 8 52.2	10.8
	12	81 40.0	34.3	N 9 57.3	10.6
	18	169 06.1	34.1	N11 01.0	10.3
25 Sat	0	256 30.7	33.8	N12 03.3	10.1
	6	343 53.8	33.5	N13 03.9	9.8
	12	71 15.2	33.2	N14 02.8	9.4
	18	158 34.7	32.9	N14 59.7	9.1
26 Sun	0	245 52.2	32.5	N15 54.5	8.8
	6	333 07.5	32.1	N16 47.0	8.3
	12	60 20.5	31.7	N17 37.1	7.8
	18	147 31.3	31.4	N18 24.6	7.5
27 Mon	0	234 39.5	30.9	N19 09.2	7.0
	6	321 45.4	30.5	N19 50.9	6.3
	12	48 48.7	30.1	N20 29.4	5.8
	18	135 49.6	29.7	N21 04.6	5.2
28 Tu	0	222 48.0	29.3	N21 36.2	4.6
	6	309 43.9	29.0	N22 04.1	4.0
	12	36 37.6	28.5	N22 28.0	3.3
	18	123 29.0	28.2	N22 47.9	2.6
29 Wed	0	210 18.4	27.9	N23 03.6	1.8
	6	297 05.9	27.6	N23 14.9	1.0
	12	23 51.8	27.4	N23 21.7	0.3
	18	110 36.1	27.2	N23 23.8	0.5
30 Th	0	197 19.2	27.0	N23 21.3	1.3
	6	284 01.4	26.9	N23 13.9	2.1
	12	10 42.9	26.8	N23 01.7	2.9
	18	97 23.9	26.8	N22 44.7	3.7

PLANETS

VENUS					JUPITER					
Mer Pass h m	GHA ° '	Mean Var/hr 14°+	Dec ° '	Mean Var/hr '	Day	GHA ° '	Mean Var/hr 15°+	Dec ° '	Mean Var/hr '	Mer Pass h m
10 34	201 41.0	59.4	N16 21.4	0.9	1 Wed	221 32.0	2.0	N10 09.8	0.2	09 13
10 35	201 27.3	59.4	N16 42.4	0.9	2 Th	222 19.2	2.0	N10 14.2	0.2	09 10
10 36	201 13.2	59.4	N17 03.0	0.8	3 Fri	223 06.4	2.0	N10 18.4	0.2	09 06
10 37	200 58.9	59.4	N17 23.2	0.8	4 Sat	223 53.7	2.0	N10 22.7	0.2	09 03
10 37	200 44.3	59.4	N17 42.9	0.8	5 SUN	224 41.0	2.0	N10 26.9	0.2	09 00
10 38	200 29.4	59.4	N18 02.2	0.8	6 Mon	225 28.5	2.0	N10 31.1	0.2	08 57
10 40	200 14.2	59.4	N18 21.0	0.8	7 Tu	226 15.9	2.0	N10 35.3	0.2	08 54
10 41	199 58.8	59.3	N18 39.4	0.7	8 Wed	227 03.5	2.0	N10 39.4	0.2	08 51
10 42	199 43.1	59.3	N18 57.3	0.7	9 Th	227 51.1	2.0	N10 43.5	0.2	08 47
10 43	199 27.1	59.3	N19 14.7	0.7	10 Fri	228 38.8	2.0	N10 47.5	0.2	08 44
10 44	199 10.9	59.3	N19 31.6	0.7	11 Sat	229 26.6	2.0	N10 51.5	0.2	08 41
10 45	198 54.4	59.3	N19 48.0	0.7	12 SUN	230 14.4	2.0	N10 55.5	0.2	08 38
10 46	198 37.6	59.3	N20 03.8	0.6	13 Mon	231 02.3	2.0	N10 59.4	0.2	08 35
10 47	198 20.5	59.3	N20 19.1	0.6	14 Tu	231 50.3	2.0	N11 03.3	0.2	08 32
10 48	198 03.2	59.3	N20 33.9	0.6	15 Wed	232 38.4	2.0	N11 07.1	0.2	08 28
10 49	197 45.7	59.3	N20 48.2	0.6	16 Th	233 26.5	2.0	N11 10.9	0.2	08 25
10 51	197 27.8	59.2	N21 01.8	0.5	17 Fri	234 14.8	2.0	N11 14.7	0.2	08 22
10 52	197 09.8	59.2	N21 14.9	0.5	18 Sat	235 03.1	2.0	N11 18.4	0.2	08 19
10 53	196 51.5	59.2	N21 27.5	0.5	19 SUN	235 51.5	2.0	N11 22.1	0.2	08 15
10 54	196 33.0	59.2	N21 39.4	0.5	20 Mon	236 40.0	2.0	N11 25.8	0.2	08 12
10 56	196 14.2	59.2	N21 50.7	0.4	21 Tu	237 28.5	2.0	N11 29.4	0.1	08 09
10 57	195 55.2	59.2	N22 01.4	0.4	22 Wed	238 17.2	2.0	N11 32.9	0.1	08 06
10 58	195 36.0	59.2	N22 11.5	0.4	23 Th	239 05.9	2.0	N11 36.4	0.1	08 03
10 59	195 16.6	59.2	N22 21.0	0.4	24 Fri	239 54.8	2.0	N11 39.9	0.1	07 59
11 01	194 57.0	59.2	N22 29.9	0.3	25 Sat	240 43.7	2.0	N11 43.3	0.1	07 56
11 02	194 37.3	59.2	N22 38.1	0.3	26 SUN	241 32.7	2.0	N11 46.7	0.1	07 53
11 03	194 17.3	59.2	N22 45.7	0.3	27 Mon	242 21.8	2.1	N11 50.1	0.1	07 49
11 05	193 57.2	59.2	N22 52.6	0.3	28 Tu	243 11.0	2.1	N11 53.4	0.1	07 46
11 06	193 37.0	59.2	N22 58.9	0.2	29 Wed	244 00.4	2.1	N11 56.6	0.1	07 43
11 08	193 16.6	59.1	N23 04.5	0.2	30 Th	244 49.8	2.1	N11 59.8	0.1	07 40

VENUS, Av. Mag. –3.8
SHA June 5 308; 10 301; 15 295; 20 289; 25 282; 30 276

JUPITER, Av. Mag. –2.2
SHA June 5 332; 10 331; 15 330; 20 329; 25 328; 30 327

MARS					SATURN					
Mer Pass h m	GHA ° '	Mean Var/hr 15°+	Dec ° '	Mean Var/hr '	Day	GHA ° '	Mean Var/hr 15°+	Dec ° '	Mean Var/hr '	Mer Pass h m
10 15	206 03.8	0.6	N16 02.9	0.6	1 Wed	58 23.9	2.5	S 1 46.7	0.0	20 03
10 14	206 19.2	0.6	N16 16.2	0.5	2 Th	59 24.2	2.5	S 1 46.5	0.0	19 59
10 13	206 34.6	0.6	N16 29.4	0.5	3 Fri	60 24.4	2.5	S 1 46.3	0.0	19 55
10 12	206 49.9	0.6	N16 42.4	0.5	4 Sat	61 24.5	2.5	S 1 46.1	0.0	19 51
10 11	207 05.1	0.6	N16 55.2	0.5	5 SUN	62 24.5	2.5	S 1 46.0	0.0	19 47
10 10	207 20.4	0.6	N17 07.8	0.5	6 Mon	63 24.5	2.5	S 1 45.9	0.0	19 43
10 09	207 35.6	0.6	N17 20.3	0.5	7 Tu	64 24.3	2.5	S 1 45.9	0.0	19 39
10 08	207 50.7	0.6	N17 32.6	0.5	8 Wed	65 24.1	2.5	S 1 45.9	0.0	19 35
10 07	208 05.8	0.6	N17 44.8	0.5	9 Th	66 23.8	2.5	S 1 45.9	0.0	19 31
10 06	208 20.9	0.6	N17 56.7	0.5	10 Fri	67 23.3	2.5	S 1 46.0	0.0	19 27
10 05	208 35.9	0.6	N18 08.5	0.5	11 Sat	68 22.8	2.5	S 1 46.1	0.0	19 23
10 04	208 50.9	0.6	N18 20.1	0.5	12 SUN	69 22.2	2.5	S 1 46.3	0.0	19 19
10 03	209 05.9	0.6	N18 31.5	0.5	13 Mon	70 21.5	2.5	S 1 46.5	0.0	19 15
10 02	209 20.8	0.6	N18 42.7	0.5	14 Tu	71 20.7	2.5	S 1 46.7	0.0	19 11
10 01	209 35.7	0.6	N18 53.7	0.4	15 Wed	72 19.8	2.5	S 1 47.0	0.0	19 08
10 00	209 50.6	0.6	N19 04.5	0.4	16 Th	73 18.8	2.5	S 1 47.3	0.0	19 04
09 59	210 05.4	0.6	N19 15.2	0.4	17 Fri	74 17.8	2.5	S 1 47.7	0.0	19 00
09 58	210 20.2	0.6	N19 25.7	0.4	18 Sat	75 16.6	2.4	S 1 48.1	0.0	18 56
09 57	210 34.9	0.6	N19 35.9	0.4	19 SUN	76 15.3	2.4	S 1 48.5	0.0	18 52
09 56	210 49.6	0.6	N19 46.0	0.4	20 Mon	77 14.0	2.4	S 1 49.0	0.0	18 48
09 55	211 04.3	0.6	N19 55.9	0.4	21 Tu	78 12.6	2.4	S 1 49.5	0.0	18 44
09 54	211 18.9	0.6	N20 05.5	0.4	22 Wed	79 11.0	2.4	S 1 50.0	0.0	18 40
09 53	211 33.6	0.6	N20 15.0	0.4	23 Th	80 09.4	2.4	S 1 50.6	0.0	18 36
09 52	211 48.1	0.6	N20 24.3	0.4	24 Fri	81 07.7	2.4	S 1 51.3	0.0	18 32
09 51	212 02.7	0.6	N20 33.4	0.4	25 Sat	82 05.9	2.4	S 1 51.9	0.0	18 29
09 50	212 17.2	0.6	N20 42.3	0.4	26 SUN	83 04.0	2.4	S 1 52.6	0.0	18 25
09 49	212 31.7	0.6	N20 51.0	0.4	27 Mon	84 02.0	2.4	S 1 53.4	0.0	18 21
09 49	212 46.2	0.6	N20 59.4	0.3	28 Tu	84 59.9	2.4	S 1 54.2	0.0	18 17
09 48	213 00.7	0.6	N21 07.7	0.3	29 Wed	85 57.7	2.4	S 1 55.0	0.0	18 13
09 47	213 15.1	0.6	N21 15.8	0.3	30 Th	86 55.5	2.4	S 1 55.9	0.0	18 09

MARS, Av. Mag. +1.4
SHA June 5 314; 10 310; 15 307; 20 303; 25 299; 30 296

SATURN, Av. Mag. +0.8
SHA June 5 169; 10 169; 15 169; 20 169; 25 169; 30 169

JUNE 2011 — SUN AND MOON

STARS

No.	Name	Mag	Transit (h m)	Dec (° ')	SHA (° ')
	0h GMT June 1				
ψ	ARIES	—	7 22		
1	Alpheratz	2.1	7 31	N29 09.2	357 45.1
2	Ankaa	2.4	7 49	S42 14.3	353 17.2
3	Schedar	2.2	8 03	N56 35.8	349 42.4
4	Diphda	2.0	8 06	S17 55.2	348 57.4
5	Achernar	0.5	9 00	S57 10.4	335 28.1
6	POLARIS	2.0	10 07	N89 18.6	318 48.2
7	Hamal	2.0	9 30	N23 30.9	328 02.6
8	Acamar	3.2	10 20	S40 15.4	315 19.7
9	Menkar	2.5	10 25	N 4 08.1	314 16.8
10	Mirfak	1.8	10 47	N49 54.0	308 42.9
11	Aldebaran	0.9	11 58	N16 31.9	290 51.4
12	Rigel	0.1	12 36	S 8 11.4	281 13.8
13	Capella	0.1	12 39	N46 00.5	280 37.1
14	Bellatrix	1.6	12 47	N 6 21.5	278 33.9
15	Elnath	1.7	12 48	N28 36.9	278 14.8
16	Alnilam	1.7	12 58	S 1 11.8	275 48.2
17	Betelgeuse	0.1–1.2	13 17	N 7 24.5	271 03.2
18	Canopus	-0.7	13 45	S52 42.2	263 57.3
19	Sirius	-1.5	14 07	S16 44.0	258 35.3
20	Adhara	1.5	14 20	S28 59.4	255 14.0
21	Castor	1.6	14 56	N31 51.7	246 10.1
22	Procyon	0.4	15 01	N 5 11.6	245 01.5
23	Pollux	1.1	15 07	N27 59.9	243 29.8
24	Avior	1.9	15 44	S59 33.1	234 19.1
25	Suhail	2.2	16 29	S43 29.0	222 53.8
26	Miaplacidus	1.7	16 34	S69 46.2	221 40.6
27	Alphard	2.0	16 49	S 8 42.7	217 57.7
28	Regulus	1.4	17 30	N11 54.6	207 45.2
29	Dubhe	1.8	18 25	N61 41.5	193 53.5
30	Denebola	2.1	19 10	N14 30.4	182 35.1
31	Gienah	2.6	19 37	S17 36.6	175 53.7
32	Acrux	1.3	19 47	S63 10.2	173 10.8
33	Gacrux	1.6	19 52	S57 11.0	172 02.4
34	Mimosa	1.3	20 09	S59 45.5	167 53.5
35	Alioth	1.8	20 15	N55 54.0	166 21.7
36	Spica	1.0	20 46	S11 13.4	158 32.6
37	Alkaid	1.9	21 08	N49 15.5	152 59.7
38	Hadar	0.6	21 25	S60 26.0	148 49.6
39	Menkent	2.1	21 27	S36 25.8	148 09.0
40	Arcturus	0.0	21 36	N19 07.4	145 56.8
41	Rigil Kent	-0.3	22 00	S60 53.2	139 53.2
42	Zuben'ubi	2.8	22 11	S16 05.5	137 06.7
43	Kochab	2.1	22 11	N74 06.7	137 18.9
44	Alphecca	2.2	22 55	N26 40.6	126 11.9
45	Antares	1.0	23 50	S26 27.5	112 27.7
46	Atria	1.9	0 13	S69 02.9	107 30.2
47	Sabik	2.4	0 34	S15 44.3	102 13.8
48	Shaula	1.6	0 58	S37 06.7	96 23.4
49	Rasalhague	2.1	0 59	N12 33.1	96 07.4
50	Eltanin	2.2	1 20	N51 29.3	90 46.3
51	Kaus Aust	1.9	1 48	S34 22.6	83 45.3
52	Vega	0.0	2 00	N38 47.7	80 39.5
53	Nunki	2.0	2 19	S26 16.8	75 59.7
54	Altair	0.8	3 14	N 8 54.0	62 09.3
55	Peacock	1.9	3 49	S56 41.6	53 21.0
56	Deneb	1.3	4 05	N45 19.2	49 32.1
57	Enif	2.4	5 07	N 9 55.7	33 48.4
58	Al Na'ir	1.7	5 32	S46 54.0	27 45.3
59	Fomalhaut	1.2	6 21	S29 33.4	15 25.5
60	Markab	2.5	6 28	N15 16.0	13 39.7

SUN

Yr	Day of Mth / Week	Transit (h m)	Semi-Diam	Lat 52°N Twilight	Sunrise	Sunset	Twilight
152	1 Wed	11 58	15.8	03 00	03 46	20 10	20 56
153	2 Th	11 58	15.8	02 59	03 46	20 11	20 58
154	3 Fri	11 58	15.8	02 58	03 45	20 12	20 59
155	4 Sat	11 58	15.8	02 57	03 44	20 13	21 00
156	5 Sun	11 58	15.8	02 56	03 44	20 14	21 01
157	6 Mon	11 59	15.8	02 55	03 43	20 15	21 03
158	7 Tu	11 59	15.8	02 55	03 42	20 16	21 04
159	8 Wed	11 59	15.8	02 54	03 42	20 17	21 05
160	9 Th	11 59	15.8	02 53	03 41	20 18	21 06
161	10 Fri	11 59	15.8	02 53	03 41	20 18	21 07
162	11 Sat	12 00	15.8	02 52	03 40	20 19	21 07
163	12 Sun	12 00	15.8	02 52	03 40	20 20	21 08
164	13 Mon	12 00	15.8	02 51	03 40	20 20	21 09
165	14 Tu	12 00	15.8	02 51	03 40	20 21	21 09
166	15 Wed	12 00	15.8	02 51	03 39	20 22	21 10
167	16 Th	12 01	15.8	02 51	03 39	20 22	21 11
168	17 Fri	12 01	15.8	02 51	03 39	20 23	21 11
169	18 Sat	12 01	15.8	02 51	03 39	20 23	21 12
170	19 Sun	12 01	15.8	02 51	03 40	20 23	21 12
171	20 Mon	12 02	15.8	02 51	03 40	20 24	21 12
172	21 Tu	12 02	15.8	02 51	03 40	20 24	21 13
173	22 Wed	12 02	15.8	02 52	03 41	20 24	21 13
174	23 Th	12 02	15.8	02 52	03 41	20 24	21 13
175	24 Fri	12 02	15.8	02 52	03 41	20 24	21 13
176	25 Sat	12 03	15.8	02 52	03 41	20 24	21 13
177	26 Sun	12 03	15.8	02 52	03 42	20 24	21 13
178	27 Mon	12 03	15.8	02 53	03 42	20 24	21 13
179	28 Tu	12 03	15.8	02 54	03 42	20 24	21 12
180	29 Wed	12 03	15.8	02 54	03 43	20 24	21 12
181	30 Th	12 04	15.8	02 55	03 43	20 24	21 12

MOON

Yr	Day of Mth / Week	Age (days)	Transit (Upper)	Diff (m)	Semi-diam	Hor Par	Lat 52°N Moonrise	Moonset
152	1 Wed	29	11 37	53	15.2	55.7	03 13	20 07
153	2 Th	01	12 30	55	15.3	56.2	03 59	21 01
154	3 Fri	02	13 25	54	15.4	56.7	04 56	21 46
155	4 Sat	03	14 19	52	15.6	57.2	06 02	22 22
156	5 Sun	04	15 11	52	15.7	57.7	07 16	22 51
157	6 Mon	05	16 03	50	15.8	58.1	08 34	23 15
158	7 Tu	06	16 53	49	16.0	58.6	09 53	23 36
159	8 Wed	07	17 42	50	16.1	59.0	11 13	23 56
160	9 Th	08	18 32	51	16.2	59.3	12 34	– –
161	10 Fri	09	19 23	53	16.2	59.5	13 56	00 16
162	11 Sat	10	20 16	56	16.3	59.7	15 19	00 37
163	12 Sun	11	21 12	58	16.3	59.7	16 42	01 02
164	13 Mon	12	22 10	60	16.2	59.6	18 03	01 33
165	14 Tu	13	23 10	60	16.1	59.2	19 15	02 13
166	15 Wed	14	24 10	–	16.0	58.7	20 15	03 04
167	16 Th	15	00 10	57	15.8	58.1	21 02	04 07
168	17 Fri	16	01 07	53	15.6	57.4	21 37	05 17
169	18 Sat	17	02 00	50	15.4	56.7	22 05	06 31
170	19 Sun	18	02 50	46	15.3	56.0	22 26	07 45
171	20 Mon	19	03 36	44	15.1	55.4	22 45	08 57
172	21 Tu	20	04 20	41	14.9	54.8	23 01	10 07
173	22 Wed	21	05 01	42	14.8	54.5	23 17	11 15
174	23 Th	22	05 43	41	14.8	54.3	23 33	12 22
175	24 Fri	23	06 24	42	14.8	54.2	23 51	13 29
176	25 Sat	24	07 06	45	14.8	54.4	– –	14 36
177	26 Sun	25	07 51	47	14.9	54.7	00 12	15 44
178	27 Mon	26	08 38	50	15.0	55.1	00 37	16 51
179	28 Tu	27	09 28	53	15.2	55.7	01 10	17 55
180	29 Wed	28	10 21	55	15.3	56.3	01 51	18 53
181	30 Th	29	11 16	55	15.5	56.9	02 44	19 42

Lat Corr to Sunrise, Sunset etc.

Lat	Twilight	Sunrise	Sunset	Twilight
N70	SAH	SAH	SAH	SAH
68	SAH	SAH	SAH	SAH
66	SAH	SAH	SAH	SAH
64	TAN	-2 06	+2 07	TAN
62	TAN	-1 29	+1 30	TAN
N60	-1 57	-1 03	+1 04	+1 58
58	-1 09	-0 43	+0 43	+1 11
56	-0 39	-0 26	+0 26	+0 40
54	-0 17	-0 12	+0 12	+0 18
50	+0 16	+0 11	-0 11	-0 15
N45	+0 45	+0 33	-0 34	-0 46
40	+1 07	+0 51	-0 51	-1 07
35	+1 25	+1 06	-1 06	-1 25
30	+1 40	+1 19	-1 20	-1 40
20	+2 05	+1 41	-1 41	-2 05
N10	+2 26	+2 00	-2 00	-2 26
0	+2 44	+2 18	-2 18	-2 44
S10	+3 01	+2 35	-2 36	-3 01
20	+3 18	+2 54	-2 54	-3 15
30	+3 48	+3 15	-3 15	-3 37
S35	+4 00	+3 27	-3 28	-3 48
40	+4 13	+3 41	-3 42	-3 59
45	+4 29	+3 58	-3 58	-4 13
S50		+4 19	-4 19	-4 29

NOTES

The corrections to sunrise etc. are for middle of June. SAH means Sun above Horizon. TAN means Twilight all night.

Phases of the Moon

	d	h	m
● New Moon	1	21	03
☾ First Quarter	9	02	11
○ Full Moon	15	20	14
☽ Last Quarter	23	11	48

	d	h
Perigee	12	02
Apogee	24	04

JULY

SUN AND ARIES

JULY 2011

SUN AND ARIES

GMT	SUN GHA ° '	SUN Dec ° '	ARIES GHA ° '
Friday, 1st July			
00	179 04.2	**N23 08.5**	278 42.3
02	209 03.9	**23 08.2**	308 47.3
04	239 03.7	**23 07.9**	338 52.2
06	269 03.5	**23 07.5**	8 57.1
08	299 03.2	**23 07.2**	39 02.0
10	329 03.0	**23 06.9**	69 07.0
12	359 02.7	**23 06.5**	99 11.9
14	29 02.5	**23 06.2**	129 16.8
16	59 02.2	**23 05.9**	159 21.8
18	89 02.0	**23 05.5**	189 26.7
20	119 01.8	**23 05.2**	219 31.6
22	149 01.5	**N23 04.8**	249 36.5
Saturday, 2nd July			
00	179 01.3	**N23 04.5**	279 41.5
02	209 01.0	**23 04.1**	309 46.4
04	239 00.8	**23 03.8**	339 51.3
06	269 00.5	**23 03.4**	9 56.3
08	299 00.3	**23 03.1**	40 01.2
10	329 00.1	**23 02.7**	70 06.1
12	358 59.8	**23 02.3**	100 11.0
14	28 59.6	**23 02.0**	130 16.0
16	58 59.4	**23 01.6**	160 20.9
18	88 59.1	**23 01.2**	190 25.8
20	118 58.9	**23 00.8**	220 30.8
22	148 58.6	**N23 00.4**	250 35.7
Sunday, 3rd July			
00	178 58.4	**N23 00.1**	280 40.6
02	208 58.2	**22 59.7**	310 45.5
04	238 57.9	**22 59.3**	340 50.5
06	268 57.7	**22 58.9**	10 55.4
08	298 57.5	**22 58.5**	41 00.3
10	328 57.2	**22 58.1**	71 05.3
12	358 57.0	**22 57.7**	101 10.2
14	28 56.8	**22 57.3**	131 15.1
16	58 56.5	**22 56.9**	161 20.0
18	88 56.3	**22 56.5**	191 25.0
20	118 56.1	**22 56.1**	221 29.9
22	148 55.9	**N22 55.7**	251 34.8
Monday, 4th July			
00	178 55.6	**N22 55.2**	281 39.8
02	208 55.4	**22 54.8**	311 44.7
04	238 55.2	**22 54.4**	341 49.6
06	268 54.9	**22 54.0**	11 54.5
08	298 54.7	**22 53.6**	41 59.5
10	328 54.5	**22 53.1**	72 04.4
12	358 54.3	**22 52.7**	102 09.3
14	28 54.0	**22 52.3**	132 14.3
16	58 53.8	**22 51.8**	162 19.2
18	88 53.6	**22 51.4**	192 24.1
20	118 53.4	**22 50.9**	222 29.0
22	148 53.1	**N22 50.5**	252 34.0
Tuesday, 5th July			
00	178 52.9	**N22 50.0**	282 38.9
02	208 52.7	**22 49.6**	312 43.8
04	238 52.5	**22 49.1**	342 48.7
06	268 52.3	**22 48.7**	12 53.7
08	298 52.0	**22 48.2**	42 58.6
10	328 51.8	**22 47.7**	73 03.5
12	358 51.6	**22 47.3**	103 08.5
14	28 51.4	**22 46.8**	133 13.4
16	58 51.2	**22 46.3**	163 18.3
18	88 50.9	**22 45.9**	193 23.2
20	118 50.7	**22 45.4**	223 28.2
22	148 50.5	**N22 44.9**	253 33.1

GMT	SUN GHA ° '	SUN Dec ° '	ARIES GHA ° '
Wednesday, 6th July			
00	178 50.3	**N22 44.4**	283 38.0
02	208 50.1	**22 43.9**	313 43.0
04	238 49.9	**22 43.4**	343 47.9
06	268 49.7	**22 43.0**	13 52.8
08	298 49.4	**22 42.5**	43 57.7
10	328 49.2	**22 42.0**	74 02.7
12	358 49.0	**22 41.5**	104 07.6
14	28 48.8	**22 41.0**	134 12.5
16	58 48.6	**22 40.5**	164 17.5
18	88 48.4	**22 39.9**	194 22.4
20	118 48.2	**22 39.4**	224 27.3
22	148 48.0	**N22 38.9**	254 32.2
Thursday, 7th July			
00	178 47.8	**N22 38.4**	284 37.2
02	208 47.6	**22 37.9**	314 42.1
04	238 47.4	**22 37.4**	344 47.0
06	268 47.2	**22 36.8**	14 52.0
08	298 47.0	**22 36.3**	44 56.9
10	328 46.7	**22 35.8**	75 01.8
12	358 46.5	**22 35.3**	105 06.7
14	28 46.3	**22 34.7**	135 11.7
16	58 46.1	**22 34.2**	165 16.6
18	88 45.9	**22 33.6**	195 21.5
20	118 45.7	**22 33.1**	225 26.5
22	148 45.5	**N22 32.6**	255 31.4
Friday, 8th July			
00	178 45.3	**N22 32.0**	285 36.3
02	208 45.1	**22 31.5**	315 41.2
04	238 44.9	**22 30.9**	345 46.2
06	268 44.8	**22 30.4**	15 51.1
08	298 44.6	**22 29.8**	45 56.0
10	328 44.4	**22 29.2**	76 00.9
12	358 44.2	**22 28.7**	106 05.9
14	28 44.0	**22 28.1**	136 10.8
16	58 43.8	**22 27.5**	166 15.7
18	88 43.6	**22 27.0**	196 20.7
20	118 43.4	**22 26.4**	226 25.6
22	148 43.2	**N22 25.8**	256 30.5
Saturday, 9th July			
00	178 43.0	**N22 25.2**	286 35.4
02	208 42.8	**22 24.6**	316 40.4
04	238 42.6	**22 24.1**	346 45.3
06	268 42.5	**22 23.5**	16 50.2
08	298 42.3	**22 22.9**	46 55.2
10	328 42.1	**22 22.3**	77 00.1
12	358 41.9	**22 21.7**	107 05.0
14	28 41.7	**22 21.1**	137 09.9
16	58 41.5	**22 20.5**	167 14.9
18	88 41.3	**22 19.9**	197 19.8
20	118 41.2	**22 19.3**	227 24.7
22	148 41.0	**N22 18.7**	257 29.7
Sunday, 10th July			
00	178 40.8	**N22 18.0**	287 34.6
02	208 40.6	**22 17.4**	317 39.5
04	238 40.4	**22 16.8**	347 44.4
06	268 40.3	**22 16.2**	17 49.4
08	298 40.1	**22 15.6**	47 54.3
10	328 39.9	**22 14.9**	77 59.2
12	358 39.7	**22 14.3**	108 04.2
14	28 39.6	**22 13.7**	138 09.1
16	58 39.4	**22 13.1**	168 14.0
18	88 39.2	**22 12.4**	198 18.9
20	118 39.0	**22 11.8**	228 23.9
22	148 38.9	**N22 11.1**	258 28.8

GMT	SUN GHA ° '	SUN Dec ° '	ARIES GHA ° '
Monday, 11th July			
00	178 38.7	**N22 10.5**	288 33.7
02	208 38.5	**22 09.8**	318 38.7
04	238 38.3	**22 09.2**	348 43.6
06	268 38.2	**22 08.5**	18 48.5
08	298 38.0	**22 07.9**	48 53.4
10	328 37.8	**22 07.2**	78 58.4
12	358 37.7	**22 06.6**	109 03.3
14	28 37.5	**22 05.9**	139 08.2
16	58 37.3	**22 05.2**	169 13.2
18	88 37.2	**22 04.6**	199 18.1
20	118 37.0	**22 03.9**	229 23.0
22	148 36.9	**N22 03.2**	259 27.9
Tuesday, 12th July			
00	178 36.7	**N22 02.6**	289 32.9
02	208 36.5	**22 01.9**	319 37.8
04	238 36.4	**22 01.2**	349 42.7
06	268 36.2	**22 00.5**	19 47.6
08	298 36.0	**21 59.8**	49 52.6
10	328 35.9	**21 59.1**	79 57.5
12	358 35.7	**21 58.4**	110 02.4
14	28 35.6	**21 57.7**	140 07.4
16	58 35.4	**21 57.1**	170 12.3
18	88 35.3	**21 56.4**	200 17.2
20	118 35.1	**21 55.6**	230 22.1
22	148 35.0	**N21 54.9**	260 27.1
Wednesday, 13th July			
00	178 34.8	**N21 54.2**	290 32.0
02	208 34.7	**21 53.5**	320 36.9
04	238 34.5	**21 52.8**	350 41.9
06	268 34.4	**21 52.1**	20 46.8
08	298 34.2	**21 51.4**	50 51.7
10	328 34.1	**21 50.7**	80 56.6
12	358 33.9	**21 49.9**	111 01.6
14	28 33.8	**21 49.2**	141 06.5
16	58 33.6	**21 48.5**	171 11.4
18	88 33.5	**21 47.8**	201 16.4
20	118 33.3	**21 47.0**	231 21.3
22	148 33.2	**N21 46.3**	261 26.2
Thursday, 14th July			
00	178 33.0	**N21 45.5**	291 31.1
02	208 32.9	**21 44.8**	321 36.1
04	238 32.8	**21 44.1**	351 41.0
06	268 32.6	**21 43.3**	21 45.9
08	298 32.5	**21 42.6**	51 50.9
10	328 32.4	**21 41.8**	81 55.8
12	358 32.2	**21 41.1**	112 00.7
14	28 32.1	**21 40.3**	142 05.6
16	58 31.9	**21 39.5**	172 10.6
18	88 31.8	**21 38.8**	202 15.5
20	118 31.7	**21 38.0**	232 20.4
22	148 31.5	**N21 37.3**	262 25.4
Friday, 15th July			
00	178 31.4	**N21 36.5**	292 30.3
02	208 31.3	**21 35.7**	322 35.2
04	238 31.2	**21 34.9**	352 40.1
06	268 31.0	**21 34.2**	22 45.1
08	298 30.9	**21 33.4**	52 50.0
10	328 30.8	**21 32.6**	82 54.9
12	358 30.6	**21 31.8**	112 59.9
14	28 30.5	**21 31.0**	143 04.8
16	58 30.4	**21 30.2**	173 09.7
18	88 30.3	**21 29.5**	203 14.6
20	118 30.1	**21 28.7**	233 19.6
22	148 30.0	**N21 27.9**	263 24.5

GMT	SUN GHA ° '	SUN Dec ° '	ARIES GHA ° '
Saturday, 16th July			
00	178 29.9	**N21 27.1**	293 29.4
02	208 29.8	**21 26.3**	323 34.4
04	238 29.7	**21 25.5**	353 39.3
06	268 29.5	**21 24.6**	23 44.2
08	298 29.4	**21 23.8**	53 49.1
10	328 29.3	**21 23.0**	83 54.1
12	358 29.2	**21 22.2**	113 59.0
14	28 29.1	**21 21.4**	144 03.9
16	58 29.0	**21 20.6**	174 08.9
18	88 28.8	**21 19.7**	204 13.8
20	118 28.7	**21 18.9**	234 18.7
22	148 28.6	**N21 18.1**	264 23.6
Sunday, 17th July			
00	178 28.5	**N21 17.3**	294 28.6
02	208 28.4	**21 16.4**	324 33.5
04	238 28.3	**21 15.6**	354 38.4
06	268 28.2	**21 14.8**	24 43.4
08	298 28.1	**21 13.9**	54 48.3
10	328 28.0	**21 13.1**	84 53.2
12	358 27.9	**21 12.2**	114 58.1
14	28 27.8	**21 11.4**	145 03.1
16	58 27.7	**21 10.5**	175 08.0
18	88 27.6	**21 09.7**	205 12.9
20	118 27.5	**21 08.8**	235 17.8
22	148 27.4	**N21 08.0**	265 22.8
Monday, 18th July			
00	178 27.3	**N21 07.1**	295 27.7
02	208 27.2	**21 06.2**	325 32.6
04	238 27.1	**21 05.4**	355 37.6
06	268 27.0	**21 04.5**	25 42.5
08	298 26.9	**21 03.6**	55 47.4
10	328 26.8	**21 02.8**	85 52.3
12	358 26.7	**21 01.9**	115 57.3
14	28 26.6	**21 01.0**	146 02.2
16	58 26.5	**21 00.1**	176 07.1
18	88 26.4	**20 59.3**	206 12.1
20	118 26.3	**20 58.4**	236 17.0
22	148 26.2	**N20 57.5**	266 21.9
Tuesday, 19th July			
00	178 26.1	**N20 56.6**	296 26.8
02	208 26.0	**20 55.7**	326 31.8
04	238 25.9	**20 54.8**	356 36.7
06	268 25.9	**20 53.9**	26 41.6
08	298 25.8	**20 53.0**	56 46.6
10	328 25.7	**20 52.1**	86 51.5
12	358 25.6	**20 51.2**	116 56.4
14	28 25.5	**20 50.3**	147 01.3
16	58 25.4	**20 49.4**	177 06.3
18	88 25.4	**20 48.5**	207 11.2
20	118 25.3	**20 47.6**	237 16.1
22	148 25.2	**N20 46.7**	267 21.1
Wednesday, 20th July			
00	178 25.1	**N20 45.7**	297 26.0
02	208 25.1	**20 44.8**	327 30.9
04	238 25.0	**20 43.9**	357 35.8
06	268 24.9	**20 43.0**	27 40.8
08	298 24.8	**20 42.0**	57 45.7
10	328 24.8	**20 41.1**	87 50.6
12	358 24.7	**20 40.2**	117 55.5
14	28 24.6	**20 39.2**	148 00.5
16	58 24.5	**20 38.3**	178 05.4
18	88 24.5	**20 37.4**	208 10.3
20	118 24.4	**20 36.4**	238 15.3
22	148 24.3	**N20 35.5**	268 20.2

GMT	SUN GHA ° '	SUN Dec ° '	ARIES GHA ° '
Thursday, 21st July			
00	178 24.3	**N20 34.5**	298 25.1
02	208 24.2	**20 33.6**	328 30.0
04	238 24.1	**20 32.6**	358 35.0
06	268 24.1	**20 31.7**	28 39.9
08	298 24.0	**20 30.7**	58 44.8
10	328 24.0	**20 29.7**	88 49.8
12	358 23.9	**20 28.8**	118 54.7
14	28 23.8	**20 27.8**	148 59.6
16	58 23.8	**20 26.8**	179 04.5
18	88 23.7	**20 25.9**	209 09.5
20	118 23.7	**20 24.9**	239 14.4
22	148 23.6	**N20 23.9**	269 19.3
Friday, 22nd July			
00	178 23.5	**N20 23.0**	299 24.3
02	208 23.5	**20 22.0**	329 29.2
04	238 23.4	**20 21.0**	359 34.1
06	268 23.4	**20 20.0**	29 39.0
08	298 23.3	**20 19.0**	59 44.0
10	328 23.3	**20 18.0**	89 48.9
12	358 23.2	**20 17.0**	119 53.8
14	28 23.2	**20 16.0**	149 58.8
16	58 23.1	**20 15.1**	180 03.7
18	88 23.1	**20 14.1**	210 08.6
20	118 23.0	**20 13.1**	240 13.5
22	148 23.0	**N20 12.0**	270 18.5
Saturday, 23rd July			
00	178 23.0	**N20 11.0**	300 23.4
02	208 22.9	**20 10.0**	330 28.3
04	238 22.9	**20 09.0**	0 33.3
06	268 22.8	**20 08.0**	30 38.2
08	298 22.8	**20 07.0**	60 43.1
10	328 22.8	**20 06.0**	90 48.0
12	358 22.7	**20 05.0**	120 53.0
14	28 22.7	**20 03.9**	150 57.9
16	58 22.6	**20 02.9**	181 02.8
18	88 22.6	**20 01.9**	211 07.7
20	118 22.6	**20 00.9**	241 12.7
22	148 22.5	**N19 59.8**	271 17.6
Sunday, 24th July			
00	178 22.5	**N19 58.8**	301 22.5
02	208 22.5	**19 57.8**	331 27.5
04	238 22.5	**19 56.7**	1 32.4
06	268 22.4	**19 55.7**	31 37.3
08	298 22.4	**19 54.6**	61 42.2
10	328 22.4	**19 53.6**	91 47.2
12	358 22.3	**19 52.5**	121 52.1
14	28 22.3	**19 51.5**	151 57.0
16	58 22.3	**19 50.4**	182 02.0
18	88 22.3	**19 49.4**	212 06.9
20	118 22.3	**19 48.3**	242 11.8
22	148 22.2	**N19 47.3**	272 16.7
Monday, 25th July			
00	178 22.2	**N19 46.2**	302 21.7
02	208 22.2	**19 45.2**	332 26.6
04	238 22.2	**19 44.1**	2 31.5
06	268 22.2	**19 43.0**	32 36.5
08	298 22.1	**19 42.0**	62 41.4
10	328 22.1	**19 40.9**	92 46.3
12	358 22.1	**19 39.8**	122 51.2
14	28 22.1	**19 38.7**	152 56.2
16	58 22.1	**19 37.6**	183 01.1
18	88 22.1	**19 36.6**	213 06.0
20	118 22.1	**19 35.5**	243 11.0
22	148 22.1	**N19 34.4**	273 15.9

GMT	SUN GHA ° '	SUN Dec ° '	ARIES GHA ° '
Tuesday, 26th July			
00	178 22.0	N19 33.3	303 20.8
02	208 22.0	19 32.2	333 25.7
04	238 22.0	19 31.1	3 30.7
06	268 22.0	19 30.0	33 35.6
08	298 22.0	19 28.9	63 40.5
10	328 22.0	19 27.8	93 45.5
12	358 22.0	19 26.7	123 50.4
14	28 22.0	19 25.6	153 55.3
16	58 22.0	19 24.5	184 00.2
18	88 22.0	19 23.4	214 05.2
20	118 22.0	19 22.3	244 10.1
22	148 22.0	N19 21.2	274 15.0
Wednesday, 27th July			
00	178 22.0	N19 20.1	304 20.0
02	208 22.0	19 19.0	334 24.9
04	238 22.0	19 17.8	4 29.8
06	268 22.1	19 16.7	34 34.7
08	298 22.1	19 15.6	64 39.7
10	328 22.1	19 14.5	94 44.6
12	358 22.1	19 13.3	124 49.5
14	28 22.1	19 12.2	154 54.5
16	58 22.1	19 11.1	184 59.4
18	88 22.1	19 09.9	215 04.3
20	118 22.1	19 08.8	245 09.2
22	148 22.1	N19 07.7	275 14.2
Thursday, 28th July			
00	178 22.2	N19 06.5	305 19.1
02	208 22.2	19 05.4	335 24.0
04	238 22.2	19 04.2	5 28.9
06	268 22.2	19 03.1	35 33.9
08	298 22.2	19 01.9	65 38.8
10	328 22.3	19 00.8	95 43.7
12	358 22.3	18 59.6	125 48.7
14	28 22.3	18 58.5	155 53.6
16	58 22.3	18 57.3	185 58.5
18	88 22.4	18 56.1	216 03.4
20	118 22.4	18 55.0	246 08.4
22	148 22.4	N18 53.8	276 13.3
Friday, 29th July			
00	178 22.4	N18 52.6	306 18.2
02	208 22.5	18 51.5	336 23.2
04	238 22.5	18 50.3	6 28.1
06	268 22.5	18 49.1	36 33.0
08	298 22.6	18 48.0	66 37.9
10	328 22.6	18 46.8	96 42.9
12	358 22.6	18 45.6	126 47.8
14	28 22.7	18 44.4	156 52.7
16	58 22.7	18 43.2	186 57.7
18	88 22.7	18 42.0	217 02.6
20	118 22.8	18 40.8	247 07.5
22	148 22.8	N18 39.7	277 12.4
Saturday, 30th July			
00	178 22.9	N18 38.5	307 17.4
02	208 22.9	18 37.3	337 22.3
04	238 23.0	18 36.1	7 27.2
06	268 23.0	18 34.9	37 32.2
08	298 23.0	18 33.7	67 37.1
10	328 23.1	18 32.5	97 42.0
12	358 23.1	18 31.3	127 46.9
14	28 23.2	18 30.0	157 51.9
16	58 23.2	18 28.8	187 56.8
18	88 23.3	18 27.6	218 01.7
20	118 23.3	18 26.4	248 06.7
22	148 23.4	N18 25.2	278 11.6

GMT	SUN GHA ° '	SUN Dec ° '	ARIES GHA ° '
Sunday, 31st July			
00	178 23.4	**N18 24.0**	308 16.5
02	208 23.5	**18 22.8**	338 21.4
04	238 23.6	**18 21.5**	8 26.4
06	268 23.6	**N18 20.3**	38 31.3
08	298 23.7	**N18 19.1**	68 36.2
10	328 23.7	**18 17.8**	98 41.2
12	358 23.8	**18 16.6**	128 46.1
14	28 23.9	**N18 15.4**	158 51.0
16	58 23.9	N18 14.1	188 55.9
18	88 24.0	18 12.9	219 00.9
20	118 24.0	18 11.7	249 05.8
22	148 24.1	N18 10.4	279 10.7

MOON

Day	GMT hr	GHA ° ′	Mean Var/hr 14°+	Dec ° ′	Mean Var/hr ′
1 Fri	0	184 04.7	26.9	N22 22.8	4.5
	6	Eclipse of the Sun occurs today			
	12	357 26.9	27.0	N21 24.6	6.1
	18	84 08.7	27.1	N20 48.5	6.8
2 Sat	0	170 51.3	27.2	N20 07.8	7.6
	6	257 34.8	27.5	N19 22.7	8.3
	12	344 19.4	27.7	N18 33.4	9.0
	18	71 05.2	27.9	N17 40.0	9.6
3 Sun	0	157 52.3	28.1	N16 42.7	10.2
	6	244 40.8	28.3	N15 41.8	10.8
	12	331 30.7	28.6	N14 37.5	11.2
	18	58 21.9	28.8	N13 30.0	11.8
4 Mon	0	145 14.6	29.0	N12 19.6	12.2
	6	232 08.5	29.2	N11 06.6	12.6
	12	319 03.6	29.3	N 9 51.1	13.0
	18	45 59.8	29.5	N 8 33.6	13.3
5 Tu	0	132 57.0	29.7	N 7 14.2	13.5
	6	219 55.0	29.8	N 5 53.3	13.7
	12	306 53.6	29.8	N 4 31.1	13.9
	18	33 52.8	29.9	N 3 07.9	14.0
6 Wed	0	120 52.3	29.9	N 1 44.0	14.0
	6	207 51.9	30.0	N 0 19.8	14.1
	12	294 51.4	29.9	S 1 04.6	14.0
	18	21 50.6	29.8	S 2 28.9	14.0
7 Th	0	108 49.3	29.6	S 3 52.6	13.8
	6	195 47.4	29.5	S 5 15.7	13.7
	12	282 44.5	29.4	S 6 37.7	13.4
	18	9 40.6	29.2	S 7 58.3	13.2
8 Fri	0	96 35.5	28.9	S 9 17.4	12.8
	6	183 28.9	28.7	S10 34.6	12.5
	12	270 20.8	28.3	S11 49.6	12.0
	18	357 10.9	28.0	S13 02.1	11.6
9 Sat	0	83 59.2	27.7	S14 11.9	11.1
	6	170 45.7	27.4	S15 18.7	10.5
	12	257 30.1	27.0	S16 22.1	9.9
	18	344 12.6	26.7	S17 22.0	9.3
10 Sun	0	70 53.1	26.4	S18 18.0	8.6
	6	157 31.6	26.1	S19 09.9	7.9
	12	244 08.3	25.8	S19 57.5	7.1
	18	330 43.3	25.5	S20 40.6	6.3
11 Mon	0	57 16.7	25.4	S21 18.9	5.5
	6	143 48.7	25.1	S21 52.3	4.7
	12	230 19.6	25.0	S22 20.6	3.8
	18	316 49.7	24.9	S22 43.7	2.9
12 Tu	0	43 19.2	24.9	S23 01.6	2.0
	6	129 48.5	24.9	S23 14.0	1.1
	12	216 17.9	25.0	S23 21.1	0.2
	18	302 47.7	25.1	S23 22.8	0.7
13 Wed	0	29 18.3	25.4	S23 19.2	1.6
	6	115 50.1	25.6	S23 10.3	2.4
	12	202 23.3	25.8	S22 56.3	3.3
	18	288 58.3	26.2	S22 37.2	4.1
14 Th	0	15 35.3	26.5	S22 13.3	4.8
	6	102 14.6	27.0	S21 44.8	5.6
	12	188 56.4	27.5	S21 11.8	6.3
	18	275 40.8	27.9	S20 34.5	6.9
15 Fri	0	2 28.0	28.4	S19 53.3	7.5
	6	89 18.1	28.8	S19 08.4	8.1
	12	176 11.2	29.4	S18 20.0	8.6
	18	263 07.2	29.8	S17 28.3	9.2
16 Sat	0	350 06.1	30.3	S16 33.7	9.6
	6	77 08.0	30.8	S15 36.4	10.0
	12	164 12.7	31.3	S14 36.7	10.3
	18	251 20.2	31.7	S13 34.8	10.7
17 Sun	0	338 30.3	32.1	S12 30.9	10.9
	6	65 43.0	32.6	S11 25.3	11.2
	12	152 58.1	32.9	S10 18.3	11.4
	18	240 15.3	33.3	S 9 10.0	11.6
18 Mon	0	327 34.6	33.6	S 8 00.6	11.7
	6	54 55.8	33.8	S 6 50.3	11.8
	12	142 18.7	34.1	S 5 39.4	11.9
	18	229 43.2	34.3	S 4 28.0	12.0
19 Tu	0	317 08.9	34.5	S 3 16.3	12.0
	6	44 35.8	34.7	S 2 04.4	11.9
	12	132 03.6	34.8	S 0 52.5	11.9
	18	219 32.2	34.8	N 0 19.2	11.9
20 Wed	0	307 01.3	34.9	N 1 30.6	11.8
	6	34 30.8	34.9	N 2 41.6	11.7
	12	122 00.4	34.9	N 3 52.0	11.6
	18	209 30.0	34.9	N 5 01.6	11.4
21 Th	0	296 59.4	34.8	N 6 10.4	11.3
	6	24 28.3	34.7	N 7 18.3	11.1
	12	111 56.7	34.5	N 8 25.0	10.9
	18	199 24.2	34.4	N 9 30.5	10.7
22 Fri	0	286 50.8	34.3	N10 34.6	10.4
	6	14 16.3	34.0	N11 37.2	10.2
	12	101 40.5	33.8	N12 38.2	9.8
	18	189 03.2	33.5	N13 37.4	9.6
23 Sat	0	276 24.2	33.2	N14 34.6	9.2
	6	3 43.6	32.9	N15 29.8	8.8
	12	91 01.0	32.5	N16 22.8	8.4
	18	178 16.3	32.2	N17 13.4	7.9
24 Sun	0	265 29.6	31.8	N18 01.5	7.5
	6	352 40.6	31.4	N18 46.8	7.0
	12	79 49.3	31.0	N19 29.3	6.5
	18	166 55.7	30.6	N20 08.8	6.0
25 Mon	0	253 59.6	30.2	N20 45.1	5.4
	6	341 01.2	29.8	N21 18.0	4.8
	12	68 00.4	29.4	N21 47.3	4.2
	18	154 57.2	29.1	N22 12.9	3.6
26 Tu	0	241 51.7	28.7	N22 34.6	2.8
	6	328 44.0	28.3	N22 52.3	2.2
	12	55 34.2	28.0	N23 05.7	1.5
	18	142 22.5	27.7	N23 14.8	0.7
27 Wed	0	229 09.0	27.5	N23 19.4	0.0
	6	315 54.0	27.3	N23 19.4	0.8
	12	42 37.6	27.0	N23 14.8	1.6
	18	129 20.0	26.9	N23 05.3	2.5
28 Th	0	216 01.5	26.8	N22 51.0	3.3
	6	302 42.3	26.7	N22 31.9	4.1
	12	29 22.8	26.7	N22 07.9	4.9
	18	116 03.0	26.7	N21 39.0	5.7
29 Fri	0	202 43.2	26.7	N21 05.4	6.5
	6	289 23.8	26.8	N20 27.0	7.2
	12	16 04.7	27.0	N19 44.0	8.0
	18	102 46.3	27.1	N18 56.5	8.7
30 Sat	0	189 28.7	27.2	N18 04.6	9.4
	6	276 12.0	27.4	N17 08.6	10.1
	12	2 56.3	27.6	N16 08.6	10.7
	18	89 41.6	27.8	N15 04.9	11.2
31 Sun	0	176 28.1	28.0	N13 57.7	11.7
	6	263 15.7	28.1	N12 47.2	12.3
	12	350 04.3	28.3	N11 33.8	12.7
	18	76 54.1	28.5	N10 17.7	13.1

PLANETS

VENUS Mer Pass h m	VENUS GHA ° ′	VENUS Mean Var/hr 14°+	VENUS Dec ° ′	VENUS Mean Var/hr ′	Day	JUPITER GHA ° ′	JUPITER Mean Var/hr 15°+	JUPITER Dec ° ′	JUPITER Mean Var/hr ′	JUPITER Mer Pass h m
11 09	192 56.1	59.1	N23 09.4	0.2	1 Fri	245 39.3	2.1	N12 03.0	0.1	07 36
11 10	192 35.5	59.1	N23 13.7	0.1	2 Sat	246 29.0	2.1	N12 06.1	0.1	07 33
11 12	192 14.8	59.1	N23 17.3	0.1	3 SUN	247 18.7	2.1	N12 09.2	0.1	07 30
11 13	191 54.0	59.1	N23 20.2	0.1	4 Mon	248 08.6	2.1	N12 12.2	0.1	07 26
11 14	191 33.2	59.1	N23 22.4	0.1	5 Tu	248 58.5	2.1	N12 15.2	0.1	07 23
11 16	191 12.3	59.1	N23 24.0	0.0	6 Wed	249 48.6	2.1	N12 18.1	0.1	07 20
11 17	190 51.3	59.1	N23 24.9	0.0	7 Th	250 38.8	2.1	N12 21.0	0.1	07 16
11 19	190 30.3	59.1	N23 25.0	0.0	8 Fri	251 29.1	2.1	N12 23.8	0.1	07 13
11 20	190 09.4	59.1	N23 24.5	0.1	9 Sat	252 19.5	2.1	N12 26.6	0.1	07 10
11 21	189 48.4	59.1	N23 23.3	0.1	10 SUN	253 10.1	2.1	N12 29.4	0.1	07 06
11 23	189 27.4	59.1	N23 21.4	0.1	11 Mon	254 00.7	2.1	N12 32.1	0.1	07 03
11 24	189 06.5	59.1	N23 18.9	0.1	12 Tu	254 51.5	2.1	N12 34.7	0.1	07 00
11 26	188 45.6	59.1	N23 15.6	0.2	13 Wed	255 42.5	2.1	N12 37.3	0.1	06 56
11 27	188 24.7	59.1	N23 11.7	0.2	14 Th	256 33.5	2.1	N12 39.8	0.1	06 53
11 28	188 03.9	59.1	N23 07.0	0.2	15 Fri	257 24.7	2.1	N12 42.3	0.1	06 49
11 30	187 43.3	59.1	N23 01.7	0.3	16 Sat	258 16.0	2.1	N12 44.8	0.1	06 46
11 31	187 22.7	59.1	N22 55.7	0.3	17 SUN	259 07.4	2.2	N12 47.1	0.1	06 43
11 33	187 02.2	59.1	N22 49.0	0.3	18 Mon	259 59.0	2.2	N12 49.5	0.1	06 39
11 34	186 41.8	59.2	N22 41.7	0.3	19 Tu	260 50.7	2.2	N12 51.8	0.1	06 36
11 35	186 21.5	59.2	N22 33.7	0.4	20 Wed	261 42.5	2.2	N12 54.0	0.1	06 32
11 37	186 01.4	59.2	N22 25.0	0.4	21 Th	262 34.5	2.2	N12 56.2	0.1	06 29
11 38	185 41.5	59.2	N22 15.6	0.4	22 Fri	263 26.6	2.2	N12 58.3	0.1	06 25
11 39	185 21.7	59.2	N22 05.6	0.4	23 Sat	264 18.8	2.2	N13 00.4	0.1	06 22
11 40	185 02.1	59.2	N21 55.0	0.5	24 SUN	265 11.2	2.2	N13 02.4	0.1	06 18
11 42	184 42.6	59.2	N21 43.7	0.5	25 Mon	266 03.8	2.2	N13 04.4	0.1	06 15
11 43	184 23.4	59.2	N21 31.8	0.5	26 Tu	266 56.5	2.2	N13 06.3	0.1	06 11
11 44	184 04.3	59.2	N21 19.2	0.5	27 Wed	267 49.3	2.2	N13 08.1	0.1	06 08
11 46	183 45.5	59.2	N21 06.1	0.6	28 Th	268 42.3	2.2	N13 10.0	0.1	06 04
11 47	183 26.9	59.2	N20 52.3	0.6	29 Fri	269 35.5	2.2	N13 11.7	0.1	06 01
11 48	183 08.5	59.2	N20 37.9	0.6	30 Sat	270 28.8	2.2	N13 13.4	0.1	05 57
11 49	182 50.4	59.3	N20 22.9	0.6	31 SUN	271 22.3	2.2	N13 15.0	0.1	05 54

VENUS, Av. Mag. −3.9
SHA July 5 269; 10 262; 15 256; 20 249; 25 242; 30 236

JUPITER, Av. Mag. −2.3
SHA July 5 326; 10 326; 15 325; 20 324; 25 324; 30 323

MARS Mer Pass h m	MARS GHA ° ′	MARS Mean Var/hr 15°+	MARS Dec ° ′	MARS Mean Var/hr ′	Day	SATURN GHA ° ′	SATURN Mean Var/hr 15°+	SATURN Dec ° ′	SATURN Mean Var/hr ′	SATURN Mer Pass h m
09 46	213 29.6	0.6	N21 23.6	0.3	1 Fri	87 53.1	2.4	S 1 56.8	0.0	18 06
09 45	213 44.0	0.6	N21 31.3	0.3	2 Sat	88 50.7	2.4	S 1 57.7	0.0	18 02
09 44	213 58.4	0.6	N21 38.8	0.3	3 SUN	89 48.2	2.4	S 1 58.7	0.0	17 58
09 43	214 12.8	0.6	N21 46.0	0.3	4 Mon	90 45.5	2.4	S 1 59.7	0.0	17 54
09 42	214 27.2	0.6	N21 53.0	0.3	5 Tu	91 42.8	2.4	S 2 00.7	0.0	17 50
09 41	214 41.6	0.6	N21 59.8	0.3	6 Wed	92 40.0	2.4	S 2 01.8	0.0	17 47
09 40	214 56.0	0.6	N22 06.5	0.3	7 Th	93 37.2	2.4	S 2 02.9	0.0	17 43
09 39	215 10.4	0.6	N22 12.9	0.3	8 Fri	94 34.2	2.4	S 2 04.1	0.1	17 39
09 38	215 24.8	0.6	N22 19.1	0.2	9 Sat	95 31.1	2.4	S 2 05.3	0.1	17 35
09 37	215 39.1	0.6	N22 25.0	0.2	10 SUN	96 28.0	2.4	S 2 06.5	0.1	17 31
09 36	215 53.5	0.6	N22 30.8	0.2	11 Mon	97 24.8	2.4	S 2 07.8	0.1	17 28
09 35	216 08.0	0.6	N22 36.4	0.2	12 Tu	98 21.5	2.4	S 2 09.1	0.1	17 24
09 34	216 22.4	0.6	N22 41.7	0.2	13 Wed	99 18.1	2.4	S 2 10.4	0.1	17 20
09 33	216 36.8	0.6	N22 46.8	0.2	14 Th	100 14.6	2.3	S 2 11.8	0.1	17 16
09 32	216 51.2	0.6	N22 51.8	0.2	15 Fri	101 11.0	2.3	S 2 13.1	0.1	17 13
09 31	217 05.7	0.6	N22 56.5	0.2	16 Sat	102 07.4	2.3	S 2 14.6	0.1	17 09
09 30	217 20.2	0.6	N23 01.0	0.2	17 SUN	103 03.7	2.3	S 2 16.0	0.1	17 05
09 29	217 34.7	0.6	N23 05.3	0.2	18 Mon	103 59.9	2.3	S 2 17.5	0.1	17 01
09 28	217 49.2	0.6	N23 09.4	0.2	19 Tu	104 56.0	2.3	S 2 19.0	0.1	16 58
09 27	218 03.7	0.6	N23 13.2	0.2	20 Wed	105 52.0	2.3	S 2 20.6	0.1	16 54
09 26	218 18.3	0.6	N23 16.9	0.1	21 Th	106 48.0	2.3	S 2 22.2	0.1	16 50
09 25	218 32.8	0.6	N23 20.3	0.1	22 Fri	107 43.8	2.3	S 2 23.8	0.1	16 46
09 24	218 47.5	0.6	N23 23.5	0.1	23 Sat	108 39.7	2.3	S 2 25.5	0.1	16 43
09 23	219 02.1	0.6	N23 26.6	0.1	24 SUN	109 35.4	2.3	S 2 27.1	0.1	16 39
09 22	219 16.8	0.6	N23 29.4	0.1	25 Mon	110 31.0	2.3	S 2 28.8	0.1	16 35
09 22	219 31.5	0.6	N23 32.0	0.1	26 Tu	111 26.6	2.3	S 2 30.6	0.1	16 32
09 21	219 46.3	0.6	N23 34.4	0.1	27 Wed	112 22.1	2.3	S 2 32.3	0.1	16 28
09 20	220 01.1	0.6	N23 36.6	0.1	28 Th	113 17.5	2.3	S 2 34.1	0.1	16 24
09 19	220 15.9	0.6	N23 38.5	0.1	29 Fri	114 12.9	2.3	S 2 36.0	0.1	16 21
09 18	220 30.9	0.6	N23 40.3	0.1	30 Sat	115 08.2	2.3	S 2 37.8	0.1	16 17
09 17	220 45.8	0.6	N23 41.9	0.1	31 SUN	116 03.4	2.3	S 2 39.7	0.1	16 13

MARS, Av. Mag. +1.4
SHA July 5 292; 10 288; 15 284; 20 281; 25 277; 30 273

SATURN, Av. Mag. +0.9
SHA July 5 169; 10 169; 15 169; 20 168; 25 168; 30 168

JULY 2011 — STARS

No.	Name	Mag	Transit h m	Dec ° '	SHA ° '
	0h GMT July I				
ψ	ARIES	–	5 24	–	357 44.8
1	Alpheratz	2.1	5 33	N29 09.3	353 16.9
2	Ankaa	2.4	5 51	S42 14.2	349 42.0
3	Schedar	2.2	6 05	N56 35.9	348 57.2
4	Diphda	2.0	6 08	S17 55.1	335 27.7
5	Achernar	0.5	7 02	S57 10.3	318 35.3
6	POLARIS	2.0	8 09	N89 18.6	328 02.4
7	Hamal	2.0	7 32	N23 31.0	315 19.5
8	Acamar	3.2	8 23	S40 15.2	314 16.6
9	Menkar	2.5	8 27	N 4 08.2	308 42.6
10	Mirfak	1.8	8 49	N49 53.9	290 51.2
11	Aldebaran	0.9	10 00	N16 31.9	281 13.6
12	Rigel	0.1	10 39	S 8 11.3	280 36.9
13	Capella	0.1	10 41	N46 00.4	278 33.7
14	Bellatrix	1.6	10 49	N 6 21.6	278 14.7
15	Elnath	1.7	10 50	N28 36.9	275 48.0
16	Alnilam	1.7	11 00	S 1 11.7	271 03.1
17	Betelgeuse	0.1–1.2	11 19	N 7 24.5	263 57.3
18	Canopus	-0.7	11 47	S52 42.1	258 35.2
19	Sirius	-1.5	12 09	S16 43.9	255 14.0
20	Adhara	1.5	12 22	S28 59.3	246 10.1
21	Castor	1.6	12 58	N31 51.7	245 01.5
22	Procyon	0.4	13 03	N 5 11.7	243 29.7
23	Pollux	1.1	13 09	N27 59.8	234 19.2
24	Avior	1.9	13 46	S59 33.0	222 53.9
25	Suhail	2.2	14 31	S43 28.9	221 40.9
26	Miaplacidus	1.7	14 36	S69 46.1	217 57.7
27	Alphard	2.0	14 51	S 8 42.6	207 45.2
28	Regulus	1.4	15 32	N11 53.4	193 53.7
29	Dubhe	1.8	16 27	N61 41.5	182 35.2
30	Denebola	2.1	17 12	N14 30.5	175 53.8
31	Gienah	2.6	17 39	S17 36.5	173 11.0
32	Acrux	1.3	17 50	S63 10.2	172 02.6
33	Gacrux	1.6	17 54	S57 11.0	167 53.7
34	Mimosa	1.3	18 11	S59 45.5	166 21.9
35	Alioth	1.8	18 17	N55 54.0	158 32.7
36	Spica	1.0	18 48	S11 13.4	152 59.8
37	Alkaid	1.9	19 10	N49 15.5	148 49.7
38	Hadar	0.6	19 27	S60 26.0	148 09.1
39	Menkent	2.1	19 29	S36 25.8	148 00.1
40	Arcturus	0.0	19 38	N19 07.4	145 56.9
41	Rigil Kent	-0.3	20 02	S60 53.2	139 53.4
42	Zuben'ubi	2.8	20 13	S16 05.5	137 06.8
43	Kochab	2.1	20 13	N74 06.7	137 19.4
44	Alphecca	2.2	20 57	N26 40.7	126 11.9
45	Antares	1.0	21 52	S26 27.5	112 27.6
46	Atria	1.9	22 12	S69 03.0	107 30.2
47	Sabik	2.4	22 33	S15 44.3	102 13.8
48	Shaula	1.6	22 56	S37 06.7	96 23.4
49	Rasalhague	2.1	22 57	N12 33.2	96 07.4
50	Eltanin	2.2	23 18	N51 29.4	90 46.3
51	Kaus Aust	1.9	23 46	S34 22.7	83 45.2
52	Vega	0.0	0 03	N38 47.8	80 39.5
53	Nunki	2.0	0 21	S26 16.8	75 59.6
54	Altair	0.8	1 16	N 8 54.1	62 09.2
55	Peacock	1.9	1 51	S56 41.6	53 20.7
56	Deneb	1.3	2 07	N45 19.4	49 32.0
57	Enif	2.4	3 09	N 9 55.8	33 48.2
58	Al Na'ir	1.7	3 34	S46 54.0	27 45.0
59	Fomalhaut	1.2	4 23	S29 33.4	15 25.2
60	Markab	2.5	4 30	N15 16.2	13 39.5

SUN AND MOON

SUN

Yr	Day of Mth	Week	Transit h m	Semi-Diam	Twilight h m	Sunrise h m (Lat 52°N)	Sunset h m	Twilight h m
182	1	Fri	12 04	15.8	02 56	03 44	20 23	21 11
183	2	Sat	12 04	15.8	02 57	03 45	20 23	21 11
184	3	Sun	12 04	15.8	02 57	03 46	20 22	21 10
185	4	Mon	12 04	15.8	02 58	03 46	20 22	21 10
186	5	Tu	12 05	15.8	02 59	03 47	20 22	21 09
187	6	Wed	12 05	15.8	03 00	03 48	20 21	21 08
188	7	Th	12 05	15.8	03 01	03 49	20 20	21 08
189	8	Fri	12 05	15.8	03 03	03 50	20 20	21 07
190	9	Sat	12 05	15.8	03 04	03 51	20 19	21 06
191	10	Sun	12 05	15.8	03 05	03 52	20 18	21 05
192	11	Mon	12 05	15.8	03 06	03 53	20 17	21 04
193	12	Tu	12 06	15.8	03 08	03 54	20 17	21 03
194	13	Wed	12 06	15.8	03 09	03 55	20 16	21 02
195	14	Th	12 06	15.8	03 10	03 56	20 15	21 00
196	15	Fri	12 06	15.8	03 12	03 57	20 14	20 59
197	16	Sat	12 06	15.8	03 13	03 59	20 13	20 58
198	17	Sun	12 06	15.8	03 15	04 00	20 12	20 57
199	18	Mon	12 06	15.8	03 16	04 01	20 10	20 55
200	19	Tu	12 06	15.8	03 18	04 02	20 09	20 54
201	20	Wed	12 06	15.8	03 19	04 04	20 08	20 52
202	21	Th	12 06	15.8	03 21	04 05	20 07	20 51
203	22	Fri	12 06	15.8	03 22	04 06	20 06	20 49
204	23	Sat	12 06	15.8	03 24	04 08	20 04	20 48
205	24	Sun	12 07	15.8	03 26	04 09	20 03	20 46
206	25	Mon	12 07	15.8	03 27	04 11	20 01	20 44
207	26	Tu	12 07	15.8	03 29	04 12	20 00	20 43
208	27	Wed	12 07	15.8	03 31	04 14	19 59	20 41
209	28	Th	12 07	15.8	03 33	04 15	19 57	20 39
210	29	Fri	12 06	15.8	03 34	04 17	19 55	20 37
211	30	Sat	12 06	15.8	03 36	04 18	19 54	20 36
212	31	Sun	12 06	15.8	03 38	04 20	19 52	20 34

MOON

Yr	Day of Mth	Week	Age days	Transit (Upper) h m	Diff m	Semi-diam	Hor Par	Moonrise h m (Lat 52°N)	Moonset h m
182	1	Fri	00	12 11	54	15.7	57.5	03 48	20 22
183	2	Sat	01	13 05	53	15.8	58.0	05 01	20 54
184	3	Sun	02	13 58	52	15.9	58.5	06 19	21 20
185	4	Mon	03	14 50	50	16.0	58.9	07 40	21 43
186	5	Tu	04	15 40	50	16.1	59.1	09 01	22 03
187	6	Wed	05	16 30	52	16.2	59.3	10 22	22 22
188	7	Th	06	17 20	54	16.2	59.3	11 43	22 43
189	8	Fri	07	18 12	54	16.2	59.3	13 05	23 06
190	9	Sat	08	19 06	56	16.1	59.2	14 27	23 35
191	10	Sun	09	20 02	58	16.1	59.0	15 47	—
192	11	Mon	10	21 00	58	16.0	58.7	17 01	00 10
193	12	Tu	11	21 58	57	15.9	58.3	18 05	00 56
194	13	Wed	12	22 55	55	15.8	57.9	18 56	01 52
195	14	Th	13	23 50	51	15.6	57.4	19 35	02 59
196	15	Fri	14	24 41	—	15.5	56.8	20 06	04 11
197	16	Sat	15	00 41	48	15.3	56.2	20 30	05 25
198	17	Sun	16	01 29	45	15.2	55.6	20 50	06 38
199	18	Mon	17	02 14	42	15.0	55.1	21 07	07 49
200	19	Tu	18	02 56	41	14.9	54.7	21 23	08 58
201	20	Wed	19	03 38	42	14.8	54.4	21 39	10 06
202	21	Th	20	04 19	42	14.8	54.3	21 56	11 13
203	22	Fri	21	05 01	44	14.8	54.4	22 16	12 20
204	23	Sat	22	05 45	45	14.8	54.4	22 39	13 28
205	24	Sun	23	06 30	49	14.9	54.8	23 08	14 35
206	25	Mon	24	07 19	51	15.1	55.3	23 44	15 39
207	26	Tu	25	08 10	53	15.2	55.9	—	16 40
208	27	Wed	26	09 03	55	15.4	56.7	00 32	17 33
209	28	Th	27	09 58	55	15.6	57.4	01 30	18 17
210	29	Fri	28	10 53	55	15.9	58.2	02 40	18 53
211	30	Sat	29	11 48	53	16.0	58.9	03 57	19 22
212	31	Sun	01	12 41	52	16.2	59.4	05 18	19 47

Lat Corr to Sunrise, Sunset etc.

Lat °	Twilight h m	Sunrise h m	Sunset h m	Twilight h m
N70	SAH	SAH	SAH	SAH
68	SAH	SAH	SAH	SAH
66	SAH	SAH	+2 19	SAH
64	Mon	TAN	+1 40	TAN
62	+1 14	-1 42	+1 40	+2 21
N60	+0 54	-1 14	+1 14	+1 27
58	+0 37	-0 54	+0 54	+0 56
56	+0 23	-0 37	+0 37	+0 34
54	+0 11	-0 22	+0 23	+0 15
50	-0 10	-0 10	+0 11	-0 13
N45	-0 30	+0 10	-0 10	-0 39
40	-0 46	+0 30	-0 30	-0 59
35	-0 59	+0 46	-0 46	-1 16
30	-1 11	+1 00	-0 59	-1 30
20	-1 31	+1 12	-1 11	-1 52
N10	-1 48	+1 32	-1 31	-2 11
0	-2 04	+1 46	-1 48	-2 27
S10	-2 21	+2 05	-2 04	-2 43
20	-2 38	+2 21	-2 36	-2 59
30	-2 57	+2 38	-2 54	-3 15
S35	-3 08	+2 57	-3 06	-3 35
40	-3 21	+3 08	-3 18	-3 47
45	-3 36	+3 21	-3 33	-3 51
S50	-4 02	+3 55	-3 51	-4 02

NOTES

The corrections to sunrise etc. are for middle of July. SAH means Sun above Horizon. TAN means Twilight all night.

Phases of the Moon

		d	h	m
●	New Moon	1	08	54
☽	First Quarter	8	06	29
○	Full Moon	15	06	40
◑	Last Quarter	23	05	02
●	New Moon	30	18	40

	d	h
Perigee	7	14
Apogee	21	23

SUN AND ARIES AUGUST 2011 SUN AND ARIES

GMT	SUN GHA ° ′	SUN Dec ° ′	ARIES GHA ° ′
Monday, 1st August			
00	178 24.2	**N18 09.2**	309 15.7
02	208 24.2	**18 07.9**	339 20.6
04	238 24.3	**18 06.7**	9 25.5
06	268 24.4	**18 05.4**	39 30.4
08	298 24.5	**18 04.2**	69 35.4
10	328 24.5	**18 02.9**	99 40.3
12	358 24.6	**18 01.7**	129 45.2
14	28 24.7	**18 00.4**	159 50.1
16	58 24.7	**17 59.2**	189 55.1
18	88 24.8	**17 57.9**	220 00.0
20	118 24.9	**17 56.6**	250 04.9
22	148 25.0	**N17 55.4**	280 09.9
Tuesday, 2nd August			
00	178 25.1	**N17 54.1**	310 14.8
02	208 25.1	**17 52.8**	340 19.7
04	238 25.2	**17 51.6**	10 24.6
06	268 25.3	**17 50.3**	40 29.6
08	298 25.4	**17 49.0**	70 34.5
10	328 25.5	**17 47.7**	100 39.4
12	358 25.6	**17 46.4**	130 44.4
14	28 25.6	**17 45.2**	160 49.3
16	58 25.7	**17 43.9**	190 54.2
18	88 25.8	**17 42.6**	220 59.1
20	118 25.9	**17 41.3**	251 04.1
22	148 26.0	**N17 40.0**	281 09.0
Wednesday, 3rd August			
00	178 26.1	**N17 38.7**	311 13.9
02	208 26.2	**17 37.4**	341 18.9
04	238 26.3	**17 36.1**	11 23.8
06	268 26.4	**17 34.8**	41 28.7
08	298 26.5	**17 33.5**	71 33.6
10	328 26.6	**17 32.2**	101 38.6
12	358 26.7	**17 30.9**	131 43.5
14	28 26.8	**17 29.6**	161 48.4
16	58 26.9	**17 28.3**	191 53.4
18	88 27.0	**17 27.0**	221 58.3
20	118 27.1	**17 25.7**	252 03.2
22	148 27.2	**N17 24.4**	282 08.1
Thursday, 4th August			
00	178 27.3	**N17 23.1**	312 13.1
02	208 27.4	**17 21.7**	342 18.0
04	238 27.5	**17 20.4**	12 22.9
06	268 27.6	**17 19.1**	42 27.8
08	298 27.7	**17 17.8**	72 32.8
10	328 27.8	**17 16.5**	102 37.7
12	358 27.9	**17 15.1**	132 42.6
14	28 28.1	**17 13.8**	162 47.6
16	58 28.2	**17 12.5**	192 52.5
18	88 28.3	**17 11.1**	222 57.4
20	118 28.4	**17 09.8**	253 02.3
22	148 28.5	**N17 08.5**	283 07.3
Friday, 5th August			
00	178 28.6	**N17 07.1**	313 12.2
02	208 28.8	**17 05.8**	343 17.1
04	238 28.9	**17 04.4**	13 22.1
06	268 29.0	**17 03.1**	43 27.0
08	298 29.1	**17 01.7**	73 31.9
10	328 29.3	**17 00.4**	103 36.8
12	358 29.4	**16 59.0**	133 41.8
14	28 29.5	**16 57.7**	163 46.7
16	58 29.6	**16 56.3**	193 51.6
18	88 29.8	**16 55.0**	223 56.6
20	118 29.9	**16 53.6**	254 01.5
22	148 30.0	**N16 52.3**	284 06.4

GMT	SUN GHA ° ′	SUN Dec ° ′	ARIES GHA ° ′
Saturday, 6th August			
00	178 30.1	**N16 50.9**	314 11.3
02	208 30.3	**16 49.5**	344 16.3
04	238 30.4	**16 48.2**	14 21.2
06	268 30.5	**16 46.8**	44 26.1
08	298 30.7	**16 45.4**	74 31.1
10	328 30.8	**16 44.1**	104 36.0
12	358 31.0	**16 42.7**	134 40.9
14	28 31.1	**16 41.3**	164 45.8
16	58 31.2	**16 39.9**	194 50.8
18	88 31.4	**16 38.6**	224 55.7
20	118 31.5	**16 37.2**	255 00.6
22	148 31.7	**N16 35.8**	285 05.6
Sunday, 7th August			
00	178 31.8	**N16 34.4**	315 10.5
02	208 32.0	**16 33.0**	345 15.4
04	238 32.1	**16 31.6**	15 20.3
06	268 32.2	**16 30.2**	45 25.3
08	298 32.4	**16 28.9**	75 30.2
10	328 32.5	**16 27.5**	105 35.1
12	358 32.7	**16 26.1**	135 40.0
14	28 32.8	**16 24.7**	165 45.0
16	58 33.0	**16 23.3**	195 49.9
18	88 33.2	**16 21.9**	225 54.8
20	118 33.3	**16 20.5**	255 59.8
22	148 33.5	**N16 19.1**	286 04.7
Monday, 8th August			
00	178 33.6	**N16 17.7**	316 09.6
02	208 33.8	**16 16.3**	346 14.5
04	238 33.9	**16 14.8**	16 19.5
06	268 34.1	**16 13.4**	46 24.4
08	298 34.3	**16 12.0**	76 29.3
10	328 34.4	**16 10.6**	106 34.3
12	358 34.6	**16 09.2**	136 39.2
14	28 34.7	**16 07.8**	166 44.1
16	58 34.9	**16 06.3**	196 49.0
18	88 35.1	**16 04.9**	226 54.0
20	118 35.2	**16 03.5**	256 58.9
22	148 35.4	**N16 02.1**	287 03.8
Tuesday, 9th August			
00	178 35.6	**N16 00.6**	317 08.8
02	208 35.8	**15 59.2**	347 13.7
04	238 35.9	**15 57.8**	17 18.6
06	268 36.1	**15 56.4**	47 23.5
08	298 36.3	**15 54.9**	77 28.5
10	328 36.4	**15 53.5**	107 33.4
12	358 36.6	**15 52.0**	137 38.3
14	28 36.8	**15 50.6**	167 43.3
16	58 37.0	**15 49.2**	197 48.2
18	88 37.2	**15 47.7**	227 53.1
20	118 37.3	**15 46.3**	257 58.0
22	148 37.5	**N15 44.8**	288 03.0
Wednesday, 10th August			
00	178 37.7	**N15 43.4**	318 07.9
02	208 37.9	**15 41.9**	348 12.8
04	238 38.1	**15 40.5**	18 17.8
06	268 38.2	**15 39.0**	48 22.7
08	298 38.4	**15 37.6**	78 27.6
10	328 38.6	**15 36.1**	108 32.5
12	358 38.8	**15 34.6**	138 37.5
14	28 39.0	**15 33.2**	168 42.4
16	58 39.2	**15 31.7**	198 47.3
18	88 39.4	**15 30.3**	228 52.3
20	118 39.6	**15 28.8**	258 57.2
22	148 39.8	**N15 27.3**	289 02.1

GMT	SUN GHA ° ′	SUN Dec ° ′	ARIES GHA ° ′
Thursday, 11th August			
00	178 40.0	**N15 25.9**	319 07.0
02	208 40.2	**15 24.4**	349 12.0
04	238 40.4	**15 22.9**	19 16.9
06	268 40.5	**15 21.4**	49 21.8
08	298 40.7	**15 20.0**	79 26.8
10	328 40.9	**15 18.5**	109 31.7
12	358 41.1	**15 17.0**	139 36.6
14	28 41.3	**15 15.5**	169 41.5
16	58 41.5	**15 14.0**	199 46.5
18	88 41.8	**15 12.6**	229 51.4
20	118 42.0	**15 11.1**	259 56.3
22	148 42.2	**N15 09.6**	290 01.3
Friday, 12th August			
00	178 42.4	**N15 08.1**	320 06.2
02	208 42.6	**15 06.6**	350 11.1
04	238 42.8	**15 05.1**	20 16.0
06	268 43.0	**15 03.6**	50 21.0
08	298 43.2	**15 02.1**	80 25.9
10	328 43.4	**15 00.6**	110 30.8
12	358 43.6	**14 59.1**	140 35.7
14	28 43.8	**14 57.6**	170 40.7
16	58 44.0	**14 56.1**	200 45.6
18	88 44.3	**14 54.6**	230 50.5
20	118 44.5	**14 53.1**	260 55.5
22	148 44.7	**N14 51.6**	291 00.4
Saturday, 13th August			
00	178 44.9	**N14 50.1**	321 05.3
02	208 45.1	**14 48.6**	351 10.2
04	238 45.4	**14 47.1**	21 15.2
06	268 45.6	**14 45.6**	51 20.1
08	298 45.8	**14 44.0**	81 25.0
10	328 46.0	**14 42.5**	111 30.0
12	358 46.2	**14 41.0**	141 34.9
14	28 46.5	**14 39.5**	171 39.8
16	58 46.7	**14 38.0**	201 44.7
18	88 46.9	**14 36.4**	231 49.7
20	118 47.1	**14 34.9**	261 54.6
22	148 47.4	**N14 33.4**	291 59.5
Sunday, 14th August			
00	178 47.6	**N14 31.8**	322 04.5
02	208 47.8	**14 30.3**	352 09.4
04	238 48.1	**14 28.8**	22 14.3
06	268 48.3	**14 27.3**	52 19.2
08	298 48.5	**14 25.7**	82 24.2
10	328 48.8	**14 24.2**	112 29.1
12	358 49.0	**14 22.6**	142 34.0
14	28 49.2	**14 21.1**	172 39.0
16	58 49.5	**14 19.6**	202 43.9
18	88 49.7	**14 18.0**	232 48.8
20	118 49.9	**14 16.5**	262 53.7
22	148 50.2	**N14 14.9**	292 58.7
Monday, 15th August			
00	178 50.4	**N14 13.4**	323 03.6
02	208 50.7	**14 11.8**	353 08.5
04	238 50.9	**14 10.3**	23 13.5
06	268 51.2	**14 08.7**	53 18.4
08	298 51.4	**14 07.2**	83 23.3
10	328 51.6	**14 05.6**	113 28.2
12	358 51.9	**14 04.1**	143 33.2
14	28 52.1	**14 02.5**	173 38.1
16	58 52.4	**14 00.9**	203 43.0
18	88 52.6	**13 59.4**	233 47.9
20	118 52.9	**13 57.8**	263 52.9
22	148 53.1	**N13 56.2**	293 57.8

GMT	SUN GHA ° ′	SUN Dec ° ′	ARIES GHA ° ′
Tuesday, 16th August			
00	178 53.4	**N13 54.7**	324 02.7
02	208 53.6	**13 53.1**	354 07.7
04	238 53.9	**13 51.5**	24 12.6
06	268 54.1	**13 50.0**	54 17.5
08	298 54.4	**13 48.4**	84 22.4
10	328 54.7	**13 46.8**	114 27.4
12	358 54.9	**13 45.2**	144 32.3
14	28 55.2	**13 43.7**	174 37.2
16	58 55.4	**13 42.1**	204 42.2
18	88 55.7	**13 40.5**	234 47.1
20	118 56.0	**13 38.9**	264 52.0
22	148 56.2	**N13 37.3**	294 56.9
Wednesday, 17th August			
00	178 56.5	**N13 35.8**	325 01.9
02	208 56.7	**13 34.2**	355 06.8
04	238 57.0	**13 32.6**	25 11.7
06	268 57.3	**13 31.0**	55 16.7
08	298 57.5	**13 29.4**	85 21.6
10	328 57.8	**13 27.8**	115 26.5
12	358 58.1	**13 26.2**	145 31.4
14	28 58.3	**13 24.6**	175 36.4
16	58 58.6	**13 23.0**	205 41.3
18	88 58.9	**13 21.4**	235 46.2
20	118 59.1	**13 19.8**	265 51.1
22	148 59.4	**N13 18.2**	295 56.1
Thursday, 18th August			
00	178 59.7	**N13 16.6**	326 01.0
02	209 00.0	**13 15.0**	356 05.9
04	239 00.2	**13 13.4**	26 10.9
06	269 00.5	**13 11.8**	56 15.8
08	299 00.8	**13 10.2**	86 20.7
10	329 01.1	**13 08.6**	116 25.6
12	359 01.3	**13 07.0**	146 30.6
14	29 01.6	**13 05.4**	176 35.5
16	59 01.9	**13 03.7**	206 40.4
18	89 02.2	**13 02.1**	236 45.4
20	119 02.5	**13 00.5**	266 50.3
22	149 02.7	**N12 58.9**	296 55.2
Friday, 19th August			
00	179 03.0	**N12 57.3**	327 00.1
02	209 03.3	**12 55.6**	357 05.1
04	239 03.6	**12 54.0**	27 10.0
06	269 03.9	**12 52.4**	57 14.9
08	299 04.2	**12 50.8**	87 19.9
10	329 04.5	**12 49.1**	117 24.8
12	359 04.7	**12 47.5**	147 29.7
14	29 05.0	**12 45.9**	177 34.6
16	59 05.3	**12 44.3**	207 39.6
18	89 05.6	**12 42.6**	237 44.5
20	119 05.9	**12 41.0**	267 49.4
22	149 06.2	**N12 39.3**	297 54.4
Saturday, 20th August			
00	179 06.5	**N12 37.7**	327 59.3
02	209 06.8	**12 36.1**	358 04.2
04	239 07.1	**12 34.4**	28 09.1
06	269 07.4	**12 32.8**	58 14.1
08	299 07.7	**12 31.1**	88 19.0
10	329 08.0	**12 29.5**	118 23.9
12	359 08.3	**12 27.9**	148 28.8
14	29 08.6	**12 26.2**	178 33.8
16	59 08.9	**12 24.6**	208 38.7
18	89 09.2	**12 22.9**	238 43.6
20	119 09.5	**12 21.3**	268 48.6
22	149 09.8	**N12 19.6**	298 53.5

GMT	SUN GHA ° ′	SUN Dec ° ′	ARIES GHA ° ′
Sunday, 21st August			
00	179 10.1	**N12 18.0**	328 58.4
02	209 10.4	**12 16.3**	359 03.3
04	239 10.7	**12 14.6**	29 08.3
06	269 11.0	**12 13.0**	59 13.2
08	299 11.3	**12 11.3**	89 18.1
10	329 11.6	**12 09.7**	119 23.1
12	359 11.9	**12 08.0**	149 28.0
14	29 12.2	**12 06.3**	179 32.9
16	59 12.5	**12 04.7**	209 37.8
18	89 12.8	**12 03.0**	239 42.8
20	119 13.1	**12 01.3**	269 47.7
22	149 13.4	**N11 59.7**	299 52.6
Monday, 22nd August			
00	179 13.8	**N11 58.0**	329 57.6
02	209 14.1	**11 56.3**	0 02.5
04	239 14.4	**11 54.7**	30 07.4
06	269 14.7	**11 53.0**	60 12.3
08	299 15.0	**11 51.3**	90 17.3
10	329 15.3	**11 49.6**	120 22.2
12	359 15.6	**11 47.9**	150 27.1
14	29 16.0	**11 46.3**	180 32.1
16	59 16.3	**11 44.6**	210 37.0
18	89 16.6	**11 42.9**	240 41.9
20	119 16.9	**11 41.2**	270 46.8
22	149 17.2	**N11 39.5**	300 51.8
Tuesday, 23rd August			
00	179 17.6	**N11 37.9**	330 56.7
02	209 17.9	**11 36.2**	1 01.6
04	239 18.2	**11 34.5**	31 06.6
06	269 18.5	**11 32.8**	61 11.5
08	299 18.8	**11 31.1**	91 16.4
10	329 19.2	**11 29.4**	121 21.3
12	359 19.5	**11 27.7**	151 26.3
14	29 19.8	**11 26.0**	181 31.2
16	59 20.1	**11 24.3**	211 36.1
18	89 20.5	**11 22.6**	241 41.1
20	119 20.8	**11 20.9**	271 46.0
22	149 21.1	**N11 19.2**	301 50.9
Wednesday, 24th August			
00	179 21.5	**N11 17.5**	331 55.8
02	209 21.8	**11 15.8**	2 00.8
04	239 22.1	**11 14.1**	32 05.7
06	269 22.5	**11 12.4**	62 10.6
08	299 22.8	**11 10.7**	92 15.6
10	329 23.1	**11 09.0**	122 20.5
12	359 23.5	**11 07.3**	152 25.4
14	29 23.8	**11 05.6**	182 30.3
16	59 24.1	**11 03.9**	212 35.3
18	89 24.5	**11 02.2**	242 40.2
20	119 24.8	**11 00.4**	272 45.1
22	149 25.1	**N10 58.7**	302 50.0
Thursday, 25th August			
00	179 25.5	**N10 57.0**	332 55.0
02	209 25.8	**10 55.3**	2 59.9
04	239 26.1	**10 53.6**	33 04.8
06	269 26.5	**10 51.9**	63 09.8
08	299 26.8	**10 50.1**	93 14.7
10	329 27.2	**10 48.4**	123 19.6
12	359 27.5	**10 46.7**	153 24.5
14	29 27.9	**10 45.0**	183 29.5
16	59 28.2	**10 43.2**	213 34.4
18	89 28.5	**10 41.5**	243 39.3
20	119 28.9	**10 39.8**	273 44.3
22	149 29.2	**N10 38.1**	303 49.2

GMT	SUN GHA ° ′	SUN Dec ° ′	ARIES GHA ° ′
Friday, 26th August			
00	179 29.6	**N10 36.3**	333 54.1
02	209 29.9	**10 34.6**	3 59.0
04	239 30.3	**10 32.9**	34 04.0
06	269 30.6	**10 31.1**	64 08.9
08	299 31.0	**10 29.4**	94 13.8
10	329 31.3	**10 27.7**	124 18.8
12	359 31.7	**10 25.9**	154 23.7
14	29 32.0	**10 24.2**	184 28.6
16	59 32.4	**10 22.4**	214 33.5
18	89 32.7	**10 20.7**	244 38.5
20	119 33.1	**10 19.0**	274 43.4
22	149 33.4	**N10 17.2**	304 48.3
Saturday, 27th August			
00	179 33.8	**N10 15.5**	334 53.3
02	209 34.1	**10 13.7**	4 58.2
04	239 34.5	**10 12.0**	35 03.1
06	269 34.9	**10 10.2**	65 08.0
08	299 35.2	**10 08.5**	95 13.0
10	329 35.6	**10 06.7**	125 17.9
12	359 35.9	**10 05.0**	155 22.8
14	29 36.3	**10 03.2**	185 27.8
16	59 36.6	**10 01.5**	215 32.7
18	89 37.0	**9 59.7**	245 37.6
20	119 37.4	**9 58.0**	275 42.5
22	149 37.7	**N 9 56.2**	305 47.5
Sunday, 28th August			
00	179 38.1	**N 9 54.5**	335 52.4
02	209 38.4	**9 52.7**	5 57.3
04	239 38.8	**9 50.9**	36 02.3
06	269 39.2	**9 49.2**	66 07.2
08	299 39.5	**9 47.4**	96 12.1
10	329 39.9	**9 45.6**	126 17.0
12	359 40.3	**9 43.9**	156 22.0
14	29 40.6	**9 42.1**	186 26.9
16	59 41.0	**9 40.4**	216 31.8
18	89 41.4	**9 38.6**	246 36.7
20	119 41.7	**9 36.8**	276 41.7
22	149 42.1	**N 9 35.0**	306 46.6
Monday, 29th August			
00	179 42.5	**N 9 33.3**	336 51.5
02	209 42.9	**9 31.5**	6 56.5
04	239 43.2	**9 29.7**	37 01.4
06	269 43.6	**9 28.0**	67 06.3
08	299 44.0	**9 26.2**	97 11.2
10	329 44.3	**9 24.4**	127 16.2
12	359 44.7	**9 22.6**	157 21.1
14	29 45.1	**9 20.9**	187 26.0
16	59 45.5	**9 19.1**	217 31.0
18	89 45.8	**9 17.3**	247 35.9
20	119 46.2	**9 15.5**	277 40.8
22	149 46.6	**N 9 13.7**	307 45.7
Tuesday, 30th August			
00	179 47.0	**N 9 12.0**	337 50.7
02	209 47.3	**9 10.2**	7 55.6
04	239 47.7	**9 08.4**	38 00.5
06	269 48.1	**9 06.6**	68 05.5
08	299 48.5	**9 04.8**	98 10.4
10	329 48.9	**9 03.0**	128 15.3
12	359 49.2	**9 01.2**	158 20.2
14	29 49.6	**8 59.4**	188 25.2
16	59 50.0	**8 57.7**	218 30.1
18	89 50.4	**8 55.9**	248 35.0
20	119 50.8	**8 54.1**	278 39.9
22	149 51.1	**N 8 52.3**	308 44.9

GMT	SUN GHA ° ′	SUN Dec ° ′	ARIES GHA ° ′
Wednesday, 31st August			
00	179 51.5	**N 8 50.5**	338 49.8
02	209 51.9	**8 48.7**	8 54.7
04	239 52.3	**8 46.9**	38 59.7
06	269 52.7	**N 8 45.1**	69 04.6
08	299 53.1	**N 8 43.3**	99 09.5
10	329 53.5	**8 41.5**	129 14.4
12	359 53.8	**8 39.7**	159 19.4
14	29 54.2	**N 8 37.9**	189 24.3
16	59 54.6	**N 8 36.1**	219 29.2
18	89 55.0	**8 34.3**	249 34.2
20	119 55.4	**8 32.5**	279 39.1
22	149 55.8	**N 8 30.7**	309 44.0

AUGUST 2011

MOON

Day	GMT hr	GHA ° '	Mean Var/hr 14°+	Dec ° '	Mean Var/hr '
1 Mon	0	163 44.8	28.6	N 8 59.2	13.5
	6	250 36.5	28.8	N 7 38.7	13.7
	12	337 28.9	28.9	N 6 16.3	14.0
	18	64 22.1	29.0	N 4 52.6	14.2
2 Tu	0	151 15.7	29.0	N 3 27.7	14.3
	6	238 09.8	29.1	N 2 02.1	14.3
	12	325 04.1	29.0	N 0 35.9	14.4
	18	51 58.4	29.0	S 0 50.3	14.3
3 Wed	0	138 52.7	29.0	S 2 16.4	14.2
	6	225 46.6	28.9	S 3 41.9	14.1
	12	312 40.2	28.8	S 5 06.6	13.9
	18	39 33.1	28.7	S 6 30.1	13.6
4 Th	0	126 25.2	28.5	S 7 52.1	13.3
	6	213 16.3	28.4	S 9 12.3	13.0
	12	300 06.4	28.1	S10 30.5	12.6
	18	26 55.3	27.9	S11 46.2	12.2
5 Fri	0	113 42.9	27.7	S12 59.3	11.6
	6	200 29.1	27.4	S14 09.4	11.1
	12	287 13.9	27.2	S15 16.3	10.6
	18	13 57.1	26.9	S16 19.7	9.9
6 Sat	0	100 38.9	26.6	S17 19.3	9.3
	6	187 19.1	26.4	S18 15.0	8.5
	12	273 57.9	26.2	S19 06.4	7.8
	18	0 35.3	26.0	S19 53.5	7.0
7 Sun	0	87 11.5	25.8	S20 35.9	6.2
	6	173 46.6	25.7	S21 13.6	5.4
	12	260 20.8	25.5	S21 46.5	4.6
	18	346 54.3	25.5	S22 14.3	3.7
8 Mon	0	73 27.3	25.5	S22 37.0	2.9
	6	160 00.1	25.5	S22 54.6	2.0
	12	246 33.0	25.5	S23 07.0	1.1
	18	333 06.1	25.7	S23 14.1	0.2
9 Tu	0	59 39.9	25.7	S23 16.1	0.6
	6	146 14.6	26.0	S23 13.0	1.4
	12	232 50.5	26.3	S23 04.8	2.3
	18	319 27.8	26.5	S22 51.7	3.0
10 Wed	0	46 06.7	26.9	S22 33.8	3.9
	6	132 47.6	27.2	S22 11.2	4.6
	12	219 30.6	27.6	S21 44.1	5.3
	18	306 15.8	28.0	S21 12.7	6.0
11 Th	0	33 03.5	28.4	S20 37.1	6.6
	6	119 53.6	28.8	S19 57.7	7.2
	12	206 46.4	29.2	S19 14.6	7.8
	18	293 41.8	29.7	S18 28.0	8.3
12 Fri	0	20 39.9	30.1	S17 38.2	8.8
	6	107 40.6	30.6	S16 45.5	9.3
	12	194 43.9	31.0	S15 50.0	9.8
	18	281 49.8	31.4	S14 51.9	10.1
13 Sat	0	8 58.2	31.8	S13 51.6	10.4
	6	96 08.9	32.2	S12 49.3	10.7
	12	183 22.0	32.5	S11 45.1	11.0
	18	270 37.2	32.9	S10 39.3	11.2
14 Sun	0	357 54.5	33.2	S 9 32.1	11.4
	6	85 13.6	33.5	S 8 23.7	11.6
	12	172 34.5	33.8	S 7 14.3	11.7
	18	259 56.9	34.0	S 6 04.1	11.8
15 Mon	0	347 20.8	34.2	S 4 53.3	11.9
	6	74 45.9	34.4	S 3 42.1	11.9
	12	162 12.0	34.5	S 2 30.6	12.0
	18	249 39.1	34.7	S 1 19.0	11.9
16 Tu	0	337 06.9	34.7	S 0 07.5	11.9
	6	64 35.3	34.8	N 1 03.8	11.8
	12	152 04.1	34.9	N 2 14.7	11.7
	18	239 33.0	34.8	N 3 25.1	11.6

Day	GMT hr	GHA ° '	Mean Var/hr 14°+	Dec ° '	Mean Var/hr '
17 Wed	0	327 02.1	34.8	N 4 34.8	11.5
	6	54 30.9	34.8	N 5 43.7	11.3
	12	141 59.5	34.7	N 6 51.6	11.1
	18	229 27.7	34.6	N 7 58.4	11.0
18 Th	0	316 55.2	34.4	N 9 04.0	10.6
	6	44 21.9	34.2	N10 08.2	10.5
	12	131 47.7	34.1	N11 10.9	10.2
	18	219 12.4	33.9	N12 11.9	9.9
19 Fri	0	306 35.9	33.6	N13 11.2	9.6
	6	33 58.1	33.4	N14 08.6	9.2
	12	121 18.7	33.2	N15 03.9	8.8
	18	208 37.7	32.8	N15 57.0	8.5
20 Sat	0	295 55.0	32.6	N16 47.7	8.0
	6	23 10.4	32.2	N17 36.0	7.6
	12	110 24.0	31.9	N18 21.6	7.1
	18	197 35.5	31.6	N19 04.5	6.6
21 Sun	0	284 45.0	31.2	N19 44.4	6.1
	6	11 52.4	30.9	N20 21.2	5.5
	12	98 57.7	30.5	N20 54.7	4.9
	18	186 00.9	30.2	N21 24.9	4.4
22 Mon	0	273 01.9	29.8	N21 51.6	3.8
	6	0 00.9	29.4	N22 14.5	3.2
	12	86 57.8	29.2	N22 33.6	2.4
	18	173 52.8	28.8	N22 48.7	1.8
23 Tu	0	260 45.9	28.6	N22 59.8	1.1
	6	347 37.3	28.3	N23 06.6	0.4
	12	74 27.1	28.1	N23 09.0	0.4
	18	161 15.4	27.8	N23 07.0	1.1
24 Wed	0	248 02.4	27.7	N23 00.5	1.9
	6	334 48.4	27.4	N22 49.4	2.7
	12	61 33.3	27.4	N22 33.6	3.4
	18	148 17.6	27.2	N22 13.1	4.2
25 Th	0	235 01.2	27.2	N21 48.0	5.0
	6	321 44.5	27.1	N21 18.1	5.8
	12	48 27.6	27.2	N20 43.6	6.6
	18	135 10.6	27.2	N20 04.4	7.3
26 Fri	0	221 53.8	27.2	N19 20.7	8.0
	6	308 37.1	27.3	N18 32.6	8.8
	12	35 20.9	27.4	N17 40.2	9.4
	18	122 05.1	27.4	N16 43.7	10.2
27 Sat	0	208 49.8	27.5	N15 43.1	10.8
	6	295 35.1	27.6	N14 38.8	11.3
	12	22 21.0	27.8	N13 30.8	11.9
	18	109 07.5	27.9	N12 19.6	12.4
28 Sun	0	195 54.6	27.9	N11 05.3	12.9
	6	282 42.2	28.1	N 9 48.1	13.3
	12	9 30.4	28.1	N 8 28.5	13.6
	18	96 18.9	28.1	N 7 06.7	13.9
29 Mon	0	183 07.7	28.2	N 5 43.0	14.2
	6	269 56.7	28.2	N 4 17.7	14.4
	12	356 45.8	28.2	N 2 51.3	14.5
	18	83 34.9	28.1	N 1 24.0	14.6
30 Tu	0	170 23.8	28.1	S 0 03.7	14.6
	6	257 12.3	28.0	S 1 31.6	14.6
	12	344 00.4	27.9	S 2 59.1	14.5
	18	70 47.8	27.7	S 4 26.1	14.3
31 Wed	0	157 34.5	27.7	S 5 52.0	14.1
	6	244 20.3	27.5	S 7 16.6	13.8
	12	331 05.2	27.3	S 8 39.4	13.5
	18	57 48.9	27.0	S10 00.1	13.0

PLANETS

VENUS Mer Pass h m	GHA ° '	Mean Var/hr 14°+	Dec ° '	Mean Var/hr '	Day	JUPITER GHA ° '	Mean Var/hr 15°+	Dec ° '	Mean Var/hr '	Mer Pass h m
11 50	182 32.5	59.3	N20 07.4	0.7	1 Mon	272 15.9	2.2	N13 16.6	0.1	05 50
11 52	182 14.8	59.3	N19 51.2	0.7	2 Tu	273 09.7	2.3	N13 18.1	0.1	05 46
11 53	181 57.4	59.3	N19 34.5	0.7	3 Wed	274 03.7	2.3	N13 19.6	0.1	05 43
11 54	181 40.2	59.3	N19 17.2	0.7	4 Th	274 57.8	2.3	N13 21.0	0.1	05 39
11 55	181 23.4	59.3	N18 59.4	0.8	5 Fri	275 52.1	2.3	N13 22.4	0.1	05 36
11 56	181 06.7	59.3	N18 41.1	0.8	6 Sat	276 46.6	2.3	N13 23.7	0.0	05 32
11 57	180 50.4	59.3	N18 22.2	0.8	7 SUN	277 41.2	2.3	N13 24.9	0.1	05 28
11 58	180 34.3	59.3	N18 02.8	0.8	8 Mon	278 36.0	2.3	N13 26.1	0.0	05 25
11 59	180 18.5	59.4	N17 42.9	0.9	9 Tu	279 31.0	2.3	N13 27.2	0.0	05 21
12 00	180 03.0	59.4	N17 22.5	0.9	10 Wed	280 26.1	2.3	N13 28.3	0.0	05 17
12 01	179 47.7	59.4	N17 01.6	0.9	11 Th	281 21.4	2.3	N13 29.3	0.0	05 14
12 02	179 32.7	59.4	N16 40.2	0.9	12 Fri	282 16.9	2.3	N13 30.2	0.0	05 10
12 03	179 18.0	59.4	N16 18.4	0.9	13 Sat	283 12.6	2.3	N13 31.1	0.0	05 06
12 04	179 03.6	59.4	N15 56.1	0.9	14 SUN	284 08.5	2.3	N13 32.0	0.0	05 03
12 05	178 49.4	59.4	N15 33.4	1.0	15 Mon	285 04.5	2.3	N13 32.7	0.0	04 59
12 06	178 35.4	59.4	N15 10.2	1.0	16 Tu	286 00.7	2.3	N13 33.4	0.0	04 55
12 07	178 21.8	59.4	N14 46.7	1.0	17 Wed	286 57.1	2.4	N13 34.1	0.0	04 51
12 08	178 08.4	59.5	N14 22.7	1.0	18 Th	287 53.7	2.4	N13 34.7	0.0	04 48
12 09	177 55.2	59.5	N13 58.3	1.0	19 Fri	288 50.5	2.4	N13 35.2	0.0	04 44
12 10	177 42.3	59.5	N13 33.6	1.1	20 Sat	289 47.4	2.4	N13 35.7	0.0	04 40
12 10	177 29.7	59.5	N13 08.4	1.1	21 SUN	290 44.6	2.4	N13 36.1	0.0	04 36
12 11	177 17.3	59.5	N12 43.0	1.1	22 Mon	291 41.9	2.4	N13 36.4	0.0	04 32
12 12	177 05.1	59.5	N12 17.1	1.1	23 Tu	292 39.4	2.4	N13 36.7	0.0	04 29
12 13	176 53.1	59.5	N11 51.0	1.1	24 Wed	293 37.2	2.4	N13 36.9	0.0	04 25
12 14	176 41.4	59.5	N11 24.5	1.1	25 Th	294 35.1	2.4	N13 37.1	0.0	04 21
12 14	176 29.9	59.5	N10 57.7	1.1	26 Fri	295 33.2	2.4	N13 37.2	0.0	04 17
12 15	176 18.6	59.5	N10 30.6	1.1	27 Sat	296 31.5	2.4	N13 37.3	0.0	04 13
12 16	176 07.5	59.5	N10 03.2	1.2	28 SUN	297 30.0	2.4	N13 37.2	0.0	04 09
12 17	175 56.6	59.6	N 9 35.5	1.2	29 Mon	298 28.7	2.5	N13 37.2	0.0	04 05
12 17	175 45.9	59.6	N 9 07.6	1.2	30 Tu	299 27.6	2.5	N13 37.0	0.0	04 02
12 18	175 35.3	59.6	N 8 39.5	1.2	31 Wed	300 26.7	2.5	N13 36.8	0.0	03 58

ENUS, Av. Mag. –4.0
SHA August 5 228; 10 222; 15 216; 20 210; 25 204; 30 198

JUPITER, Av. Mag. –2.5
SHA August 5 323; 10 322; 15 322; 20 322; 25 322; 30 322

MARS Mer Pass h m	GHA ° '	Mean Var/hr 15°+	Dec ° '	Mean Var/hr '	Day	SATURN GHA ° '	Mean Var/hr 15°+	Dec ° '	Mean Var/hr '	Mer Pass h m
09 16	221 00.8	0.6	N23 43.2	0.1	1 Mon	116 58.5	2.3	S 2 41.6	0.1	16 10
09 15	221 15.9	0.6	N23 44.4	0.0	2 Tu	117 53.6	2.3	S 2 43.5	0.1	16 06
09 14	221 31.1	0.6	N23 45.3	0.0	3 Wed	118 48.5	2.3	S 2 45.5	0.1	16 02
09 13	221 46.3	0.6	N23 46.0	0.0	4 Th	119 43.5	2.3	S 2 47.5	0.1	15 59
09 12	222 01.6	0.6	N23 46.6	0.0	5 Fri	120 38.3	2.3	S 2 49.5	0.1	15 55
09 10	222 16.9	0.6	N23 46.9	0.0	6 Sat	121 33.1	2.3	S 2 51.5	0.1	15 51
09 09	222 32.3	0.6	N23 47.1	0.0	7 SUN	122 27.8	2.3	S 2 53.6	0.1	15 48
09 08	222 47.8	0.7	N23 47.0	0.0	8 Mon	123 22.5	2.3	S 2 55.7	0.1	15 44
09 07	223 03.4	0.7	N23 46.7	0.0	9 Tu	124 17.1	2.3	S 2 57.8	0.1	15 40
09 06	223 19.1	0.7	N23 46.3	0.0	10 Wed	125 11.6	2.3	S 2 59.9	0.1	15 37
09 05	223 34.8	0.7	N23 45.6	0.0	11 Th	126 06.1	2.3	S 3 02.1	0.1	15 33
09 04	223 50.6	0.7	N23 44.8	0.0	12 Fri	127 00.5	2.3	S 3 04.2	0.1	15 30
09 03	224 06.5	0.7	N23 43.8	0.1	13 Sat	127 54.8	2.3	S 3 06.4	0.1	15 26
09 02	224 22.5	0.7	N23 42.5	0.1	14 SUN	128 49.1	2.3	S 3 08.7	0.1	15 22
09 01	224 38.6	0.7	N23 41.1	0.1	15 Mon	129 43.4	2.3	S 3 10.9	0.1	15 19
09 00	224 54.7	0.7	N23 39.5	0.1	16 Tu	130 37.5	2.3	S 3 13.2	0.1	15 15
08 59	225 11.0	0.7	N23 37.7	0.1	17 Wed	131 31.6	2.3	S 3 15.5	0.1	15 12
08 58	225 27.3	0.7	N23 35.7	0.1	18 Th	132 25.7	2.3	S 3 17.8	0.1	15 08
08 57	225 43.8	0.7	N23 33.6	0.1	19 Fri	133 19.7	2.2	S 3 20.1	0.1	15 04
08 56	226 00.3	0.7	N23 31.2	0.1	20 Sat	134 13.6	2.2	S 3 22.4	0.1	15 01
08 54	226 16.9	0.7	N23 28.7	0.1	21 SUN	135 07.5	2.2	S 3 24.8	0.1	14 57
08 53	226 33.7	0.7	N23 26.0	0.1	22 Mon	136 01.3	2.2	S 3 27.2	0.1	14 54
08 52	226 50.5	0.7	N23 23.1	0.1	23 Tu	136 55.1	2.2	S 3 29.6	0.1	14 50
08 51	227 07.5	0.7	N23 20.0	0.1	24 Wed	137 48.9	2.2	S 3 32.0	0.1	14 47
08 50	227 24.6	0.7	N23 16.8	0.1	25 Th	138 42.5	2.2	S 3 34.4	0.1	14 43
08 49	227 41.7	0.7	N23 13.4	0.2	26 Fri	139 36.2	2.2	S 3 36.9	0.1	14 39
08 48	227 59.0	0.7	N23 09.8	0.2	27 Sat	140 29.7	2.2	S 3 39.3	0.1	14 36
08 46	228 16.4	0.7	N23 06.0	0.2	28 SUN	141 23.3	2.2	S 3 41.8	0.1	14 32
08 45	228 34.0	0.7	N23 02.1	0.2	29 Mon	142 16.7	2.2	S 3 44.3	0.1	14 29
08 44	228 51.6	0.7	N22 58.0	0.2	30 Tu	143 10.2	2.2	S 3 46.8	0.1	14 25
08 43	229 09.4	0.7	N22 53.8	0.2	31 Wed	144 03.6	2.2	S 3 49.4	0.1	14 22

MARS, Av. Mag. +1.4
SHA August 5 269; 10 265; 15 262; 20 258; 25 254; 30 251

SATURN, Av. Mag. +0.9
SHA August 5 167; 10 167; 15 167; 20 166; 25 166; 30 165

AUGUST 2011

STARS

No.	Name	Mag	Transit h m	Dec ° '	SHA ° '
ψ	0h GMT August l — ARIES	–	3 22		357 44.6
1	Alpheratz	2.1	3 31	N29 09.4	353 16.6
2	Ankaa	2.4	3 49	S42 14.2	349 41.7
3	Schedar	2.2	4 04	N56 36.0	348 56.9
4	Diphda	2.0	4 06	S17 55.1	335 27.4
5	Achernar	0.5	5 00	S57 10.3	318 20.6
6	POLARIS	2.0	6 09	N89 18.6	328 02.1
7	Hamal	2.0	5 30	N23 31.1	315 19.2
8	Acamar	3.2	6 21	S40 15.2	314 16.4
9	Menkar	2.5	6 25	N 4 08.2	308 42.2
10	Mirfak	1.8	6 47	N49 54.0	290 51.0
11	Aldebaran	0.9	7 58	N16 31.9	281 13.4
12	Rigel	0.1	8 37	S 8 11.2	280 36.6
13	Capella	0.1	8 39	N46 00.4	278 33.6
14	Bellatrix	1.6	8 47	N 6 21.6	278 14.4
15	Elnath	1.7	8 49	N28 36.9	275 47.8
16	Alnilam	1.7	8 58	S 1 11.6	271 02.9
17	Betelgeuse	0.1–1.2	9 17	N 7 24.6	263 57.1
18	Canopus	-0.7	9 46	S52 41.9	258 35.1
19	Sirius	-1.5	10 07	S16 43.9	255 13.8
20	Adhara	1.5	10 20	S28 59.2	246 09.9
21	Castor	1.6	10 56	N31 51.7	245 01.4
22	Procyon	0.4	11 01	N 5 11.7	243 29.6
23	Pollux	1.1	11 07	N27 59.8	234 19.2
24	Avior	1.9	11 44	S59 32.8	222 53.9
25	Suhail	2.2	12 29	S43 28.8	221 41.0
26	Miaplacidus	1.7	12 34	S69 46.0	217 57.7
27	Alphard	2.0	12 49	S 8 42.6	207 45.2
28	Regulus	1.4	13 30	N11 54.6	193 53.8
29	Dubhe	1.8	14 25	N61 41.3	182 35.2
30	Denebola	2.1	15 10	N14 30.5	175 53.9
31	Gienah	2.6	15 37	S17 36.5	173 11.3
32	Acrux	1.3	15 48	S63 10.1	172 02.8
33	Gacrux	1.6	15 52	S57 10.9	167 53.9
34	Mimosa	1.3	16 09	S59 45.4	166 22.1
35	Alioth	1.8	16 15	N55 54.0	158 32.8
36	Spica	1.0	16 46	S11 13.4	153 00.0
37	Alkaid	1.9	17 08	N49 15.5	148 50.0
38	Hadar	0.6	17 25	S60 26.0	148 09.2
39	Menkent	2.1	17 27	S36 25.8	148 09.2
40	Arcturus	0.0	17 36	N19 07.5	145 57.0
41	Rigil Kent	-0.3	18 00	S60 53.2	139 53.7
42	Zuben'ubi	2.8	18 12	S16 05.4	137 06.9
43	Kochab	2.1	18 11	N74 06.7	137 09.0
44	Alphecca	2.2	18 55	N26 40.8	126 12.1
45	Antares	1.0	19 50	S26 27.5	112 27.7
46	Atria	1.9	20 10	S69 03.1	107 30.5
47	Sabik	2.4	20 31	S15 44.3	102 13.9
48	Shaula	1.6	20 54	S37 06.7	96 23.4
49	Rasalhague	2.1	20 55	N12 33.3	96 07.5
50	Eltanin	2.2	21 16	N51 29.6	90 46.4
51	Kaus Aust	1.9	21 44	S34 22.7	83 45.2
52	Vega	0.0	21 57	N38 48.0	80 39.5
53	Nunki	2.0	22 15	S26 16.8	75 59.6
54	Altair	0.8	23 11	N 8 54.2	62 09.2
55	Peacock	1.9	23 46	S56 41.7	53 20.6
56	Deneb	1.3	0 05	N45 19.5	49 31.9
57	Enif	2.4	1 08	N 9 55.9	33 48.1
58	Al Na'ir	1.7	1 32	S46 54.0	27 44.8
59	Fomalhaut	1.2	2 21	S29 33.3	15 25.0
60	Markab	2.5	2 28	N15 16.3	13 39.3

SUN AND MOON

SUN

Yr	Day	Mth Week	Transit h m	Semi-Diam	Twilight h m	Sunrise h m	Sunset h m	Twilight h m
213	1	Mon	12 06	15.8	03 40	04 21	19 51	20 32
214	2	Tu	12 06	15.8	03 41	04 23	19 49	20 30
215	3	Wed	12 06	15.8	03 43	04 24	19 47	20 28
216	4	Th	12 06	15.8	03 45	04 26	19 45	20 26
217	5	Fri	12 06	15.8	03 47	04 27	19 44	20 24
218	6	Sat	12 06	15.8	03 49	04 29	19 42	20 22
219	7	Sun	12 06	15.8	03 50	04 30	19 40	20 20
220	8	Mon	12 06	15.8	03 52	04 32	19 38	20 18
221	9	Tu	12 06	15.8	03 54	04 34	19 36	20 16
222	10	Wed	12 05	15.8	03 56	04 35	19 34	20 14
223	11	Th	12 05	15.8	03 58	04 37	19 33	20 11
224	12	Fri	12 05	15.8	04 00	04 38	19 31	20 09
225	13	Sat	12 05	15.8	04 03	04 40	19 29	20 07
226	14	Sun	12 05	15.8	04 05	04 42	19 27	20 05
227	15	Mon	12 04	15.8	04 07	04 43	19 25	20 03
228	16	Tu	12 04	15.8	04 09	04 45	19 23	20 00
229	17	Wed	12 04	15.8	04 11	04 47	19 21	19 58
230	18	Th	12 04	15.8	04 12	04 48	19 19	19 56
231	19	Fri	12 03	15.8	04 14	04 50	19 16	19 54
232	20	Sat	12 03	15.8	04 16	04 51	19 14	19 51
233	21	Sun	12 03	15.8	04 18	04 53	19 12	19 49
234	22	Mon	12 03	15.9	04 20	04 55	19 10	19 47
235	23	Tu	12 02	15.9	04 21	04 56	19 08	19 45
236	24	Wed	12 02	15.9	04 23	04 58	19 06	19 42
237	25	Th	12 02	15.9	04 25	05 00	19 04	19 40
238	26	Fri	12 02	15.9	04 27	05 01	19 01	19 38
239	27	Sat	12 01	15.9	04 29	05 03	18 59	19 35
240	28	Sun	12 01	15.9	04 30	05 04	18 57	19 33
241	29	Mon	12 01	15.9	04 32	05 06	18 55	19 30
242	30	Tu	12 01	15.9	04 32	05 08	18 53	19 28
243	31	Wed	12 00	15.9	04 34	05 09	18 50	19 26

Lat Corr to Sunrise, Sunset etc.

Lat °	Twilight h m	Sunrise h m	Sunset h m	Twilight h m
N70	TAN	-1 50	+1 46	TAN
68	-2 28	-1 26	+1 23	+2 26
66	-1 48	-1 08	+1 06	+1 47
64	-1 21	-0 53	+0 51	+1 20
62	-1 01	-0 41	+0 39	+1 00
N60	-0 45	-0 30	+0 29	+0 44
58	-0 32	-0 21	+0 21	+0 30
56	-0 20	-0 13	+0 13	+0 18
54	-0 10	-0 06	+0 06	+0 09
50	+0 06	+0 06	-0 06	-0 04
N45	+0 17	+0 18	-0 18	-0 27
40	+0 23	+0 28	-0 28	-0 38
35	+0 36	+0 37	-0 37	-0 48
30	+0 47	+0 44	-0 45	-0 58
20	+1 12	+0 57	-0 57	-1 13
N10	+1 24	+1 09	-1 08	-1 25
0	+1 35	+1 19	-1 18	-1 35
S10	+1 44	+1 29	-1 28	-1 46
20	+1 54	+1 40	-1 39	-1 56
30	+2 05	+1 52	-1 51	-2 06
S35	+2 10	+2 00	-1 57	-2 11
40	+2 16	+2 07	-2 05	-2 17
45	+2 23	+2 17	-2 15	-2 24
S50	+2 30	+2 27	-2 26	-2 32

NOTES
The corrections to sunrise, etc. are for middle of August. TAN means Twilight all night.

MOON

Yr	Day	Mth Week	Age days	Transit (Upper) h m	Diff m	Semi-diam	Hor Par	Moonrise (Lat 52°N) h m	Moonset (Lat 52°N) h m
213	1	Mon	02	13 33	52	16.3	59.8	06 41	20 08
214	2	Tu	03	14 25	51	16.3	59.9	08 05	20 29
215	3	Wed	04	15 16	52	16.3	59.9	09 28	20 50
216	4	Th	05	16 08	54	16.3	59.7	10 52	21 13
217	5	Fri	06	17 02	56	16.2	59.4	12 15	21 39
218	6	Sat	07	17 58	56	16.1	59.0	13 36	22 12
219	7	Sun	08	18 54	58	16.0	58.6	14 51	22 54
220	8	Mon	09	19 52	56	15.8	58.1	15 57	23 46
221	9	Tu	10	20 48	55	15.7	57.6	16 52	–
222	10	Wed	11	21 43	51	15.6	57.1	17 34	00 48
223	11	Th	12	22 34	49	15.4	56.6	18 07	01 57
224	12	Fri	13	23 23	46	15.3	56.1	18 33	03 10
225	13	Sat	14	24 09	–	15.1	55.6	18 55	04 22
226	14	Sun	15	00 09	43	15.0	55.1	19 13	05 34
227	15	Mon	16	00 52	42	14.9	54.8	19 30	06 43
228	16	Tu	17	01 34	42	14.8	54.4	19 46	07 51
229	17	Wed	18	02 16	41	14.8	54.2	20 03	08 59
230	18	Th	19	02 57	43	14.7	54.1	20 21	10 06
231	19	Fri	20	03 40	44	14.8	54.2	20 43	11 13
232	20	Sat	21	04 24	47	14.8	54.4	21 09	12 19
233	21	Sun	22	05 11	49	14.9	54.7	21 42	13 24
234	22	Mon	23	06 00	51	15.1	55.3	22 23	14 26
235	23	Tu	24	06 51	54	15.3	56.0	23 15	15 21
236	24	Wed	25	07 45	54	15.5	56.8	–	16 09
237	25	Th	26	08 39	54	15.7	57.7	00 19	16 48
238	26	Fri	27	09 33	54	16.0	58.6	01 31	17 20
239	27	Sat	28	10 27	54	16.2	59.4	02 50	17 47
240	28	Sun	29	11 21	52	16.4	60.1	04 13	18 11
241	29	Mon	00	12 13	53	16.5	60.6	05 38	18 32
242	30	Tu	01	13 06	54	16.6	60.8	07 04	18 54
243	31	Wed	02	14 00	55	16.5	60.7	08 30	19 17

Phases of the Moon

	d	h	m
First Quarter	6	11	08
Full Moon	13	18	57
Last Quarter	21	21	54
New Moon	29	03	04

	d	h
Perigee	2	21
Apogee	18	16
Perigee	30	18

SEPTEMBER 2011

GMT	SUN GHA ° ′	SUN Dec ° ′	ARIES GHA ° ′
Thursday, 1st September			
00	179 56.2	**N 8 28.9**	339 48.9
02	209 56.6	**8 27.1**	9 53.9
04	239 57.0	**8 25.3**	39 58.8
06	269 57.4	**8 23.4**	70 03.7
08	299 57.8	**8 21.6**	100 08.7
10	329 58.1	**8 19.8**	130 13.6
12	359 58.5	**8 18.0**	160 18.5
14	29 58.9	**8 16.2**	190 23.4
16	59 59.3	**8 14.4**	220 28.4
18	89 59.7	**8 12.6**	250 33.3
20	120 00.1	**8 10.8**	280 38.2
22	150 00.5	**N 8 08.9**	310 43.2
Friday, 2nd September			
00	180 00.9	**N 8 07.1**	340 48.1
02	210 01.3	**8 05.3**	10 53.0
04	240 01.7	**8 03.5**	40 57.9
06	270 02.1	**8 01.7**	71 02.9
08	300 02.5	**7 59.8**	101 07.8
10	330 02.9	**7 58.0**	131 12.7
12	0 03.3	**7 56.2**	161 17.6
14	30 03.7	**7 54.4**	191 22.6
16	60 04.1	**7 52.5**	221 27.5
18	90 04.5	**7 50.7**	251 32.4
20	120 04.9	**7 48.9**	281 37.4
22	150 05.3	**N 7 47.1**	311 42.3
Saturday, 3rd September			
00	180 05.7	**N 7 45.2**	341 47.2
02	210 06.1	**7 43.4**	11 52.1
04	240 06.5	**7 41.6**	41 57.1
06	270 06.9	**7 39.8**	72 02.0
08	300 07.3	**7 37.9**	102 06.9
10	330 07.7	**7 36.1**	132 11.9
12	0 08.2	**7 34.3**	162 16.8
14	30 08.6	**7 32.4**	192 21.7
16	60 09.0	**7 30.6**	222 26.6
18	90 09.4	**7 28.8**	252 31.6
20	120 09.8	**7 26.9**	282 36.5
22	150 10.2	**N 7 25.1**	312 41.4
Sunday, 4th September			
00	180 10.6	**N 7 23.2**	342 46.4
02	210 11.0	**7 21.4**	12 51.3
04	240 11.4	**7 19.6**	42 56.2
06	270 11.8	**7 17.7**	73 01.1
08	300 12.2	**7 15.9**	103 06.1
10	330 12.7	**7 14.0**	133 11.0
12	0 13.1	**7 12.2**	163 15.9
14	30 13.5	**7 10.4**	193 20.9
16	60 13.9	**7 08.5**	223 25.8
18	90 14.3	**7 06.7**	253 30.7
20	120 14.7	**7 04.8**	283 35.6
22	150 15.1	**N 7 03.0**	313 40.6
Monday, 5th September			
00	180 15.6	**N 7 01.1**	343 45.5
02	210 16.0	**6 59.3**	13 50.4
04	240 16.4	**6 57.4**	43 55.4
06	270 16.8	**6 55.6**	74 00.3
08	300 17.2	**6 53.7**	104 05.2
10	330 17.6	**6 51.9**	134 10.1
12	0 18.0	**6 50.0**	164 15.1
14	30 18.5	**6 48.2**	194 20.0
16	60 18.9	**6 46.3**	224 24.9
18	90 19.3	**6 44.5**	254 29.9
20	120 19.7	**6 42.6**	284 34.8
22	150 20.1	**N 6 40.8**	314 39.7

GMT	SUN GHA ° ′	SUN Dec ° ′	ARIES GHA ° ′
Tuesday, 6th September			
00	180 20.6	**N 6 38.9**	344 44.6
02	210 21.0	**6 37.0**	14 49.6
04	240 21.4	**6 35.2**	44 54.5
06	270 21.8	**6 33.3**	74 59.4
08	300 22.2	**6 31.5**	105 04.4
10	330 22.7	**6 29.6**	135 09.3
12	0 23.1	**6 27.7**	165 14.2
14	30 23.5	**6 25.9**	195 19.1
16	60 23.9	**6 24.0**	225 24.1
18	90 24.4	**6 22.2**	255 29.0
20	120 24.8	**6 20.3**	285 33.9
22	150 25.2	**N 6 18.4**	315 38.8
Wednesday, 7th September			
00	180 25.6	**N 6 16.6**	345 43.8
02	210 26.1	**6 14.7**	15 48.7
04	240 26.5	**6 12.8**	45 53.6
06	270 26.9	**6 11.0**	75 58.6
08	300 27.3	**6 09.1**	106 03.5
10	330 27.8	**6 07.2**	136 08.4
12	0 28.2	**6 05.4**	166 13.3
14	30 28.6	**6 03.5**	196 18.3
16	60 29.0	**6 01.6**	226 23.2
18	90 29.5	**5 59.7**	256 28.1
20	120 29.9	**5 57.9**	286 33.1
22	150 30.3	**N 5 56.0**	316 38.0
Thursday, 8th September			
00	180 30.8	**N 5 54.1**	346 42.9
02	210 31.2	**5 52.3**	16 47.8
04	240 31.6	**5 50.4**	46 52.8
06	270 32.1	**5 48.5**	76 57.7
08	300 32.5	**5 46.6**	107 02.6
10	330 32.9	**5 44.7**	137 07.6
12	0 33.3	**5 42.9**	167 12.5
14	30 33.8	**5 41.0**	197 17.4
16	60 34.2	**5 39.1**	227 22.3
18	90 34.6	**5 37.2**	257 27.3
20	120 35.1	**5 35.4**	287 32.2
22	150 35.5	**N 5 33.5**	317 37.1
Friday, 9th September			
00	180 35.9	**N 5 31.6**	347 42.1
02	210 36.4	**5 29.7**	17 47.0
04	240 36.8	**5 27.8**	47 51.9
06	270 37.2	**5 25.9**	77 56.8
08	300 37.7	**5 24.1**	108 01.8
10	330 38.1	**5 22.2**	138 06.7
12	0 38.5	**5 20.3**	168 11.6
14	30 39.0	**5 18.4**	198 16.6
16	60 39.4	**5 16.5**	228 21.5
18	90 39.9	**5 14.6**	258 26.4
20	120 40.3	**5 12.7**	288 31.3
22	150 40.7	**N 5 10.9**	318 36.3
Saturday, 10th September			
00	180 41.2	**N 5 09.0**	348 41.2
02	210 41.6	**5 07.1**	18 46.1
04	240 42.0	**5 05.2**	48 51.0
06	270 42.5	**5 03.3**	78 56.0
08	300 42.9	**5 01.4**	109 00.9
10	330 43.3	**4 59.5**	139 05.8
12	0 43.8	**4 57.6**	169 10.8
14	30 44.2	**4 55.7**	199 15.7
16	60 44.7	**4 53.8**	229 20.6
18	90 45.1	**4 51.9**	259 25.5
20	120 45.5	**4 50.0**	289 30.5
22	150 46.0	**N 4 48.2**	319 35.4

GMT	SUN GHA ° ′	SUN Dec ° ′	ARIES GHA ° ′
Sunday, 11th September			
00	180 46.4	**N 4 46.3**	349 40.3
02	210 46.9	**4 44.4**	19 45.3
04	240 47.3	**4 42.5**	49 50.2
06	270 47.7	**4 40.6**	79 55.1
08	300 48.2	**4 38.7**	110 00.0
10	330 48.6	**4 36.8**	140 05.0
12	0 49.1	**4 34.9**	170 09.9
14	30 49.5	**4 33.0**	200 14.8
16	60 49.9	**4 31.1**	230 19.8
18	90 50.4	**4 29.2**	260 24.7
20	120 50.8	**4 27.3**	290 29.6
22	150 51.3	**N 4 25.4**	320 34.5
Monday, 12th September			
00	180 51.7	**N 4 23.5**	350 39.5
02	210 52.2	**4 21.6**	20 44.4
04	240 52.6	**4 19.7**	50 49.3
06	270 53.0	**4 17.8**	80 54.3
08	300 53.5	**4 15.8**	110 59.2
10	330 53.9	**4 13.9**	141 04.1
12	0 54.4	**4 12.0**	171 09.0
14	30 54.8	**4 10.1**	201 14.0
16	60 55.3	**4 08.2**	231 18.9
18	90 55.7	**4 06.3**	261 23.8
20	120 56.1	**4 04.4**	291 28.7
22	150 56.6	**N 4 02.5**	321 33.7
Tuesday, 13th September			
00	180 57.0	**N 4 00.6**	351 38.6
02	210 57.5	**3 58.7**	21 43.5
04	240 57.9	**3 56.8**	51 48.5
06	270 58.4	**3 54.9**	81 53.4
08	300 58.8	**3 53.0**	111 58.3
10	330 59.2	**3 51.0**	142 03.2
12	0 59.7	**3 49.1**	172 08.2
14	31 00.1	**3 47.2**	202 13.1
16	61 00.6	**3 45.3**	232 18.0
18	91 01.0	**3 43.4**	262 23.0
20	121 01.5	**3 41.5**	292 27.9
22	151 01.9	**N 3 39.6**	322 32.8
Wednesday, 14th September			
00	181 02.4	**N 3 37.6**	352 37.7
02	211 02.8	**3 35.7**	22 42.7
04	241 03.3	**3 33.8**	52 47.6
06	271 03.7	**3 31.9**	82 52.5
08	301 04.2	**3 30.0**	112 57.5
10	331 04.6	**3 28.1**	143 02.4
12	1 05.0	**3 26.2**	173 07.3
14	31 05.5	**3 24.2**	203 12.2
16	61 05.9	**3 22.3**	233 17.2
18	91 06.4	**3 20.4**	263 22.1
20	121 06.8	**3 18.5**	293 27.0
22	151 07.3	**N 3 16.6**	323 31.9
Thursday, 15th September			
00	181 07.7	**N 3 14.6**	353 36.9
02	211 08.2	**3 12.7**	23 41.8
04	241 08.6	**3 10.8**	53 46.7
06	271 09.1	**3 08.9**	83 51.7
08	301 09.5	**3 07.0**	113 56.6
10	331 10.0	**3 05.0**	144 01.5
12	1 10.4	**3 03.1**	174 06.4
14	31 10.9	**3 01.2**	204 11.4
16	61 11.3	**2 59.3**	234 16.3
18	91 11.7	**2 57.3**	264 21.2
20	121 12.2	**2 55.4**	294 26.2
22	151 12.6	**N 2 53.5**	324 31.1

GMT	SUN GHA ° ′	SUN Dec ° ′	ARIES GHA ° ′
Friday, 16th September			
00	181 13.1	**N 2 51.6**	354 36.0
02	211 13.5	**2 49.6**	24 40.9
04	241 14.0	**2 47.7**	54 45.9
06	271 14.4	**2 45.8**	84 50.8
08	301 14.9	**2 43.9**	114 55.7
10	331 15.3	**2 41.9**	145 00.7
12	1 15.8	**2 40.0**	175 05.6
14	31 16.2	**2 38.1**	205 10.5
16	61 16.7	**2 36.2**	235 15.4
18	91 17.1	**2 34.2**	265 20.4
20	121 17.6	**2 32.3**	295 25.3
22	151 18.0	**N 2 30.4**	325 30.2
Saturday, 17th September			
00	181 18.5	**N 2 28.4**	355 35.2
02	211 18.9	**2 26.5**	25 40.1
04	241 19.4	**2 24.6**	55 45.0
06	271 19.8	**2 22.7**	85 49.9
08	301 20.2	**2 20.7**	115 54.9
10	331 20.7	**2 18.8**	145 59.8
12	1 21.1	**2 16.9**	176 04.7
14	31 21.6	**2 14.9**	206 09.6
16	61 22.0	**2 13.0**	236 14.6
18	91 22.5	**2 11.1**	266 19.5
20	121 22.9	**2 09.1**	296 24.4
22	151 23.4	**N 2 07.2**	326 29.4
Sunday, 18th September			
00	181 23.8	**N 2 05.3**	356 34.3
02	211 24.3	**2 03.3**	26 39.2
04	241 24.7	**2 01.4**	56 44.1
06	271 25.2	**1 59.5**	86 49.1
08	301 25.6	**1 57.5**	116 54.0
10	331 26.1	**1 55.6**	146 58.9
12	1 26.5	**1 53.7**	177 03.9
14	31 27.0	**1 51.7**	207 08.8
16	61 27.4	**1 49.8**	237 13.7
18	91 27.9	**1 47.9**	267 18.6
20	121 28.3	**1 45.9**	297 23.6
22	151 28.7	**N 1 44.0**	327 28.5
Monday, 19th September			
00	181 29.2	**N 1 42.1**	357 33.4
02	211 29.6	**1 40.1**	27 38.4
04	241 30.1	**1 38.2**	57 43.3
06	271 30.5	**1 36.2**	87 48.2
08	301 31.0	**1 34.3**	117 53.1
10	331 31.4	**1 32.4**	147 58.1
12	1 31.9	**1 30.4**	178 03.0
14	31 32.3	**1 28.5**	208 07.9
16	61 32.8	**1 26.5**	238 12.9
18	91 33.2	**1 24.6**	268 17.8
20	121 33.7	**1 22.7**	298 22.7
22	151 34.1	**N 1 20.7**	328 27.6
Tuesday, 20th September			
00	181 34.5	**N 1 18.8**	358 32.6
02	211 35.0	**1 16.8**	28 37.5
04	241 35.4	**1 14.9**	58 42.4
06	271 35.9	**1 13.0**	88 47.4
08	301 36.3	**1 11.0**	118 52.3
10	331 36.8	**1 09.1**	148 57.2
12	1 37.2	**1 07.1**	179 02.1
14	31 37.7	**1 05.2**	209 07.1
16	61 38.1	**1 03.3**	239 12.0
18	91 38.5	**1 01.3**	269 16.9
20	121 39.0	**0 59.4**	299 21.9
22	151 39.4	**N 0 57.4**	329 26.8

GMT	SUN GHA ° ′	SUN Dec ° ′	ARIES GHA ° ′
Wednesday, 21st September			
00	181 39.9	**N 0 55.5**	359 31.7
02	211 40.3	**0 53.6**	29 36.6
04	241 40.8	**0 51.6**	59 41.6
06	271 41.2	**0 49.7**	89 46.5
08	301 41.7	**0 47.7**	119 51.4
10	331 42.1	**0 45.8**	149 56.3
12	1 42.5	**0 43.8**	180 01.3
14	31 43.0	**0 41.9**	210 06.2
16	61 43.4	**0 40.0**	240 11.1
18	91 43.9	**0 38.0**	270 16.1
20	121 44.3	**0 36.1**	300 21.0
22	151 44.7	**N 0 34.1**	330 25.9
Thursday, 22nd September			
00	181 45.2	**N 0 32.2**	0 30.8
02	211 45.6	**0 30.2**	30 35.8
04	241 46.1	**0 28.3**	60 40.7
06	271 46.5	**0 26.3**	90 45.6
08	301 47.0	**0 24.4**	120 50.6
10	331 47.4	**0 22.4**	150 55.5
12	1 47.8	**0 20.5**	181 00.4
14	31 48.3	**0 18.6**	211 05.3
16	61 48.7	**0 16.6**	241 10.3
18	91 49.2	**0 14.7**	271 15.2
20	121 49.6	**0 12.7**	301 20.1
22	151 50.0	**N 0 10.8**	331 25.1
Friday, 23rd September			
00	181 50.5	**N 0 08.8**	1 30.0
02	211 50.9	**0 06.9**	31 34.9
04	241 51.4	**0 04.9**	61 39.8
06	271 51.8	**0 03.0**	91 44.8
08	301 52.2	**0 01.0**	121 49.7
10	331 52.7	**S 0 00.9**	151 54.6
12	1 53.1	**0 02.9**	181 59.6
14	31 53.5	**0 04.8**	212 04.5
16	61 54.0	**0 06.7**	242 09.4
18	91 54.4	**0 08.7**	272 14.3
20	121 54.9	**0 10.6**	302 19.3
22	151 55.3	**S 0 12.6**	332 24.2
Saturday, 24th September			
00	181 55.7	**S 0 14.5**	2 29.1
02	211 56.2	**0 16.5**	32 34.1
04	241 56.6	**0 18.4**	62 39.0
06	271 57.0	**0 20.4**	92 43.9
08	301 57.5	**0 22.3**	122 48.8
10	331 57.9	**0 24.3**	152 53.8
12	1 58.3	**0 26.2**	182 58.7
14	31 58.8	**0 28.2**	213 03.6
16	61 59.2	**0 30.1**	243 08.6
18	91 59.6	**0 32.1**	273 13.5
20	122 00.1	**0 34.0**	303 18.4
22	152 00.5	**S 0 36.0**	333 23.3
Sunday, 25th September			
00	182 00.9	**S 0 37.9**	3 28.3
02	212 01.4	**0 39.9**	33 33.2
04	242 01.8	**0 41.8**	63 38.1
06	272 02.2	**0 43.7**	93 43.0
08	302 02.7	**0 45.7**	123 48.0
10	332 03.1	**0 47.6**	153 52.9
12	2 03.5	**0 49.6**	183 57.8
14	32 04.0	**0 51.5**	214 02.8
16	62 04.4	**0 53.5**	244 07.7
18	92 04.8	**0 55.4**	274 12.6
20	122 05.3	**0 57.4**	304 17.5
22	152 05.7	**S 0 59.3**	334 22.5

GMT	SUN GHA ° ′	SUN Dec ° ′	ARIES GHA ° ′
Monday, 26th September			
00	182 06.1	**S 1 01.3**	4 27.4
02	212 06.5	**1 03.2**	34 32.3
04	242 07.0	**1 05.2**	64 37.3
06	272 07.4	**1 07.1**	94 42.2
08	302 07.8	**1 09.1**	124 47.1
10	332 08.3	**1 11.0**	154 52.0
12	2 08.7	**1 13.0**	184 57.0
14	32 09.1	**1 14.9**	215 01.9
16	62 09.5	**1 16.9**	245 06.8
18	92 10.0	**1 18.8**	275 11.8
20	122 10.4	**1 20.8**	305 16.7
22	152 10.8	**S 1 22.7**	335 21.6
Tuesday, 27th September			
00	182 11.2	**S 1 24.7**	5 26.5
02	212 11.7	**1 26.6**	35 31.5
04	242 12.1	**1 28.5**	65 36.4
06	272 12.5	**1 30.5**	95 41.3
08	302 12.9	**1 32.4**	125 46.2
10	332 13.4	**1 34.4**	155 51.2
12	2 13.8	**1 36.3**	185 56.1
14	32 14.2	**1 38.3**	216 01.0
16	62 14.6	**1 40.2**	246 06.0
18	92 15.1	**1 42.2**	276 10.9
20	122 15.5	**1 44.1**	306 15.8
22	152 15.9	**S 1 46.1**	336 20.7
Wednesday, 28th September			
00	182 16.3	**S 1 48.0**	6 25.7
02	212 16.7	**1 50.0**	36 30.6
04	242 17.2	**1 51.9**	66 35.5
06	272 17.6	**1 53.9**	96 40.5
08	302 18.0	**1 55.8**	126 45.4
10	332 18.4	**1 57.8**	156 50.3
12	2 18.8	**1 59.7**	186 55.2
14	32 19.3	**2 01.6**	217 00.2
16	62 19.7	**2 03.6**	247 05.1
18	92 20.1	**2 05.5**	277 10.0
20	122 20.5	**2 07.5**	307 15.0
22	152 20.9	**S 2 09.4**	337 19.9
Thursday, 29th September			
00	182 21.3	**S 2 11.4**	7 24.8
02	212 21.8	**2 13.3**	37 29.7
04	242 22.2	**2 15.3**	67 34.7
06	272 22.6	**2 17.2**	97 39.6
08	302 23.0	**2 19.2**	127 44.5
10	332 23.4	**2 21.1**	157 49.5
12	2 23.8	**2 23.0**	187 54.4
14	32 24.3	**2 25.0**	217 59.3
16	62 24.7	**2 26.9**	248 04.2
18	92 25.1	**2 28.9**	278 09.2
20	122 25.5	**2 30.8**	308 14.1
22	152 25.9	**S 2 32.8**	338 19.0
Friday, 30th September			
00	182 26.3	**S 2 34.7**	8 23.9
02	212 26.7	**2 36.6**	38 28.9
04	242 27.1	**2 38.6**	68 33.8
06	272 27.5	**2 40.5**	98 38.7
08	302 28.0	**2 42.5**	128 43.7
10	332 28.4	**2 44.4**	158 48.6
12	2 28.8	**2 46.4**	188 53.5
14	32 29.2	**2 48.3**	218 58.4
16	62 29.6	**2 50.2**	249 03.4
18	92 30.0	**2 52.2**	279 08.3
20	122 30.4	**2 54.1**	309 13.2
22	152 30.8	**S 2 56.1**	339 18.2

MOON

Day	GMT hr	GHA ° ′	Mean Var/hr 14°+	Dec ° ′	Mean Var/hr ′
1 Th	0	144 31.5	26.9	S11 18.4	12.5
	6	231 12.8	26.7	S12 33.9	12.0
	12	317 52.8	26.4	S13 46.3	11.5
	18	44 31.4	26.2	S14 55.3	10.8
2 Fri	0	131 08.8	26.0	S16 00.5	10.2
	6	217 44.9	25.8	S17 01.8	9.5
	12	304 19.7	25.6	S17 58.9	8.7
	18	30 53.4	25.4	S18 51.6	7.9
3 Sat	0	117 26.2	25.3	S19 39.6	7.1
	6	203 58.0	25.2	S20 22.7	6.3
	12	290 29.3	25.1	S21 00.9	5.5
	18	17 00.0	25.1	S21 34.0	4.6
4 Sun	0	103 30.6	25.1	S22 01.9	3.8
	6	190 01.2	25.2	S22 24.6	2.8
	12	276 32.0	25.2	S22 42.0	1.9
	18	3 03.5	25.4	S22 54.2	1.0
5 Mon	0	89 35.8	25.6	S23 01.1	0.2
	6	176 09.2	25.8	S23 02.8	0.6
	12	262 43.9	26.0	S22 59.4	1.4
	18	349 20.2	26.4	S22 51.0	2.3
6 Tu	0	75 58.4	26.7	S22 37.8	3.1
	6	162 38.6	27.1	S22 19.8	3.9
	12	249 21.1	27.5	S21 57.4	4.6
	18	336 05.8	27.9	S21 30.5	5.2
7 Wed	0	62 53.1	28.3	S20 59.5	5.9
	6	149 42.9	28.8	S20 24.5	6.5
	12	236 35.4	29.2	S19 45.8	7.1
	18	323 30.5	29.6	S19 03.6	7.7
8 Th	0	50 28.3	30.1	S18 18.0	8.1
	6	137 28.7	30.6	S17 29.3	8.7
	12	224 31.8	31.0	S16 37.8	9.1
	18	311 37.4	31.4	S15 43.5	9.5
9 Fri	0	38 45.5	31.8	S14 46.9	9.8
	6	125 55.9	32.2	S13 48.0	10.2
	12	213 08.7	32.5	S12 47.1	10.5
	18	300 23.5	32.9	S11 44.4	10.7
10 Sat	0	27 40.4	33.2	S10 40.1	11.0
	6	114 59.1	33.4	S 9 34.3	11.2
	12	202 19.6	33.7	S 8 27.4	11.4
	18	289 41.6	33.9	S 7 19.3	11.5
11 Sun	0	17 05.1	34.1	S 6 10.5	11.6
	6	104 29.8	34.3	S 5 00.9	11.6
	12	191 55.6	34.5	S 3 50.9	11.8
	18	279 22.3	34.6	S 2 40.5	11.8
12 Mon	0	6 49.8	34.7	S 1 29.9	11.8
	6	94 17.9	34.7	S 0 19.3	11.7
	12	181 46.5	34.8	N 0 51.2	11.7
	18	269 15.3	34.8	N 2 01.3	11.6
13 Tu	0	356 44.3	34.8	N 3 11.0	11.6
	6	84 13.2	34.8	N 4 20.1	11.4
	12	171 42.0	34.8	N 5 28.4	11.2
	18	259 10.4	34.7	N 6 35.9	11.1
14 Wed	0	346 38.4	34.5	N 7 42.3	10.9
	6	74 05.7	34.4	N 8 47.5	10.6
	12	161 32.3	34.3	N 9 51.3	10.4
	18	248 57.9	34.1	N10 53.7	10.1
15 Th	0	336 22.6	33.9	N11 54.5	9.8
	6	63 46.1	33.7	N12 53.4	9.5
	12	151 08.4	33.5	N13 50.5	9.1
	18	238 29.3	33.3	N14 45.5	8.8
16 Fri	0	325 48.8	32.9	N15 38.4	8.4
	6	53 06.7	32.7	N16 28.9	8.0
	12	140 23.0	32.4	N17 16.9	7.5
	18	227 37.7	32.1	N18 02.3	7.1
17 Sat	0	314 50.6	31.8	N18 44.9	6.6
	6	42 01.8	31.5	N19 24.6	6.0
	12	129 11.1	31.3	N20 01.3	5.5
	18	216 18.7	31.0	N20 34.8	5.0
18 Sun	0	303 24.5	30.7	N21 05.0	4.4
	6	30 28.5	30.3	N21 31.7	3.8
	12	117 30.8	30.1	N21 54.9	3.2
	18	204 31.4	29.8	N22 14.4	2.5
19 Mon	0	291 30.4	29.5	N22 30.0	1.9
	6	18 27.9	29.4	N22 41.8	1.2
	12	105 24.0	29.1	N22 49.5	0.5
	18	192 18.8	28.9	N22 53.1	0.2
20 Tu	0	279 12.4	28.8	N22 52.6	0.8
	6	6 04.9	28.6	N22 47.8	1.5
	12	92 56.5	28.4	N22 38.6	2.3
	18	179 47.3	28.4	N22 25.1	3.0
21 Wed	0	266 37.4	28.3	N22 07.3	3.8
	6	353 27.1	28.2	N21 45.0	4.5
	12	80 16.3	28.1	N21 18.3	5.3
	18	167 05.3	28.1	N20 47.2	5.9
22 Th	0	253 54.2	28.1	N20 11.8	6.6
	6	340 43.0	28.1	N19 32.0	7.4
	12	67 31.8	28.1	N18 48.1	8.0
	18	154 20.8	28.2	N18 00.0	8.8
23 Fri	0	241 09.9	28.3	N17 07.8	9.4
	6	327 59.2	28.2	N16 11.7	10.0
	12	54 48.8	28.3	N15 11.9	10.7
	18	141 38.5	28.3	N14 08.4	11.2
24 Sat	0	228 28.5	28.4	N13 01.4	11.7
	6	315 18.6	28.4	N11 51.3	12.2
	12	42 08.8	28.4	N10 38.1	12.7
	18	128 59.0	28.4	N 9 22.1	13.1
25 Sun	0	215 49.2	28.4	N 8 03.5	13.5
	6	302 39.2	28.3	N 6 42.7	13.8
	12	29 28.9	28.2	N 5 20.0	14.1
	18	116 18.2	28.1	N 3 55.5	14.3
26 Mon	0	203 07.0	28.0	N 2 29.8	14.5
	6	289 55.1	27.9	N 1 03.0	14.6
	12	16 42.4	27.7	S 0 24.4	14.6
	18	103 28.7	27.5	S 1 52.0	14.5
27 Tu	0	190 13.9	27.3	S 3 19.6	14.5
	6	276 57.8	27.1	S 4 46.6	14.4
	12	3 40.5	26.9	S 6 12.8	14.1
	18	90 21.6	26.6	S 7 37.6	13.8
28 Wed	0	177 01.2	26.3	S 9 00.7	13.5
	6	263 39.1	26.0	S10 21.7	13.1
	12	350 15.3	25.7	S11 40.2	12.6
	18	76 49.8	25.4	S12 55.8	12.0
29 Th	0	163 22.6	25.2	S14 08.2	11.4
	6	249 53.8	24.9	S15 16.8	10.8
	12	336 23.3	24.7	S16 21.6	10.0
	18	62 51.3	24.5	S17 22.0	9.3
30 Fri	0	149 17.9	24.2	S18 17.9	8.5
	6	235 43.4	24.1	S19 08.9	7.6
	12	322 07.9	23.9	S19 55.0	6.7
	18	48 31.7	23.9	S20 35.7	5.8

PLANETS

VENUS						JUPITER				
Mer Pass h m	GHA ° ′	Mean Var/hr 14°+	Dec ° ′	Mean Var/hr ′	Day	GHA ° ′	Mean Var/hr 15°+	Dec ° ′	Mean Var/hr ′	Mer Pass h m
12 19	175 25.0	59.6	N 8 11.1	1.2	1 Th	301 26.0	2.5	N13 36.5	0.0	03 54
12 19	175 14.8	59.6	N 7 42.4	1.2	2 Fri	302 25.4	2.5	N13 36.2	0.0	03 50
12 20	175 04.7	59.6	N 7 13.6	1.2	3 Sat	303 25.1	2.5	N13 35.8	0.0	03 46
12 21	174 54.8	59.6	N 6 44.5	1.2	4 SUN	304 25.0	2.5	N13 35.4	0.0	03 42
12 21	174 45.1	59.6	N 6 15.3	1.2	5 Mon	305 25.1	2.5	N13 34.8	0.0	03 38
12 22	174 35.5	59.6	N 5 45.9	1.2	6 Tu	306 25.4	2.5	N13 34.3	0.0	03 34
12 23	174 26.0	59.6	N 5 16.4	1.2	7 Wed	307 25.8	2.5	N13 33.6	0.0	03 30
12 23	174 16.6	59.6	N 4 46.6	1.2	8 Th	308 26.5	2.5	N13 32.9	0.0	03 26
12 24	174 07.3	59.6	N 4 16.8	1.3	9 Fri	309 27.3	2.5	N13 32.2	0.0	03 22
12 24	173 58.1	59.6	N 3 46.8	1.3	10 Sat	310 28.4	2.5	N13 31.4	0.0	03 18
12 25	173 49.0	59.6	N 3 16.7	1.3	11 SUN	311 29.6	2.6	N13 30.5	0.0	03 13
12 26	173 40.0	59.6	N 2 46.5	1.3	12 Mon	312 31.1	2.6	N13 29.6	0.0	03 09
12 26	173 31.0	59.6	N 2 16.2	1.3	13 Tu	313 32.7	2.6	N13 28.6	0.0	03 05
12 27	173 22.1	59.6	N 1 45.8	1.3	14 Wed	314 34.5	2.6	N13 27.5	0.0	03 01
12 27	173 13.2	59.6	N 1 15.4	1.3	15 Th	315 36.5	2.6	N13 26.4	0.0	02 57
12 28	173 04.4	59.6	N 0 44.9	1.3	16 Fri	316 38.7	2.6	N13 25.3	0.1	02 53
12 29	172 55.5	59.6	N 0 14.3	1.3	17 Sat	317 41.1	2.6	N13 24.0	0.1	02 49
12 29	172 46.7	59.6	S 0 16.2	1.3	18 SUN	318 43.6	2.6	N13 22.7	0.1	02 45
12 30	172 37.9	59.6	S 0 46.8	1.3	19 Mon	319 46.4	2.6	N13 21.4	0.1	02 40
12 30	172 29.0	59.6	S 1 17.4	1.3	20 Tu	320 49.3	2.6	N13 20.0	0.1	02 36
12 31	172 20.1	59.6	S 1 48.0	1.3	21 Wed	321 52.4	2.6	N13 18.6	0.1	02 32
12 32	172 11.2	59.6	S 2 18.6	1.3	22 Th	322 55.7	2.6	N13 17.1	0.1	02 28
12 32	172 02.3	59.6	S 2 49.2	1.3	23 Fri	323 59.1	2.7	N13 15.5	0.1	02 24
12 33	171 53.2	59.6	S 3 19.7	1.3	24 Sat	325 02.8	2.7	N13 13.9	0.1	02 19
12 33	171 44.2	59.6	S 3 50.1	1.3	25 SUN	326 06.6	2.7	N13 12.2	0.1	02 15
12 34	171 35.0	59.6	S 4 20.5	1.3	26 Mon	327 10.5	2.7	N13 10.5	0.1	02 11
12 35	171 25.7	59.6	S 4 50.8	1.3	27 Tu	328 14.7	2.7	N13 08.7	0.1	02 07
12 35	171 16.4	59.6	S 5 21.0	1.3	28 Wed	329 19.0	2.7	N13 06.9	0.1	02 02
12 36	171 06.9	59.6	S 5 51.2	1.3	29 Th	330 23.4	2.7	N13 05.1	0.1	01 58
12 37	170 57.3	59.6	S 6 21.2	1.2	30 Fri	331 28.0	2.7	N13 03.2	0.1	01 54

VENUS, Av. Mag. –3.9
SHA September 5 191; 10 185; 15 180; 20 174; 25 168; 30 163

JUPITER, Av. Mag. –2.7
SHA September 5 322; 10 322; 15 322; 20 322; 25 323; 30 323

MARS						SATURN				
Mer Pass h m	GHA ° ′	Mean Var/hr 15°+	Dec ° ′	Mean Var/hr ′	Day	GHA ° ′	Mean Var/hr 15°+	Dec ° ′	Mean Var/hr ′	Mer Pass h m
08 42	229 27.3	0.8	N22 49.4	0.2	1 Th	144 56.9	2.2	S 3 51.9	0.1	14 18
08 41	229 45.4	0.8	N22 44.8	0.2	2 Fri	145 50.2	2.2	S 3 54.5	0.1	14 15
08 39	230 03.5	0.8	N22 40.1	0.2	3 Sat	146 43.5	2.2	S 3 57.0	0.1	14 11
08 38	230 21.8	0.8	N22 35.2	0.2	4 SUN	147 36.7	2.2	S 3 59.6	0.1	14 07
08 37	230 40.2	0.8	N22 30.2	0.2	5 Mon	148 29.9	2.2	S 4 02.2	0.1	14 04
08 36	230 58.8	0.8	N22 25.0	0.2	6 Tu	149 23.0	2.2	S 4 04.8	0.1	14 00
08 34	231 17.5	0.8	N22 19.7	0.2	7 Wed	150 16.1	2.2	S 4 07.4	0.1	13 57
08 33	231 36.3	0.8	N22 14.2	0.2	8 Th	151 09.1	2.2	S 4 10.1	0.1	13 53
08 32	231 55.3	0.8	N22 08.6	0.2	9 Fri	152 02.2	2.2	S 4 12.7	0.1	13 50
08 31	232 14.4	0.8	N22 02.9	0.2	10 Sat	152 55.1	2.2	S 4 15.4	0.1	13 46
08 29	232 33.6	0.8	N21 57.0	0.3	11 SUN	153 48.1	2.2	S 4 18.0	0.1	13 43
08 28	232 53.0	0.8	N21 50.9	0.3	12 Mon	154 41.0	2.2	S 4 20.7	0.1	13 39
08 27	233 12.5	0.8	N21 44.8	0.3	13 Tu	155 33.9	2.2	S 4 23.4	0.1	13 36
08 25	233 32.2	0.8	N21 38.5	0.3	14 Wed	156 26.7	2.2	S 4 26.1	0.1	13 32
08 24	233 52.0	0.8	N21 32.0	0.3	15 Th	157 19.6	2.2	S 4 28.8	0.1	13 29
08 23	234 11.9	0.8	N21 25.5	0.3	16 Fri	158 12.3	2.2	S 4 31.5	0.1	13 25
08 21	234 32.0	0.8	N21 18.8	0.3	17 Sat	159 05.1	2.2	S 4 34.2	0.1	13 22
08 20	234 52.3	0.8	N21 12.0	0.3	18 SUN	159 57.8	2.2	S 4 36.9	0.1	13 18
08 19	235 12.6	0.9	N21 05.0	0.3	19 Mon	160 50.5	2.2	S 4 39.6	0.1	13 15
08 17	235 33.2	0.9	N20 58.0	0.3	20 Tu	161 43.2	2.2	S 4 42.3	0.1	13 11
08 16	235 53.8	0.9	N20 50.8	0.3	21 Wed	162 35.8	2.2	S 4 45.1	0.1	13 08
08 15	236 14.7	0.9	N20 43.5	0.3	22 Th	163 28.5	2.2	S 4 47.8	0.1	13 04
08 13	236 35.6	0.9	N20 36.1	0.3	23 Fri	164 21.1	2.2	S 4 50.6	0.1	13 01
08 12	236 56.8	0.9	N20 28.6	0.3	24 Sat	165 13.6	2.2	S 4 53.3	0.1	12 57
08 10	237 18.1	0.9	N20 21.0	0.3	25 SUN	166 06.2	2.2	S 4 56.1	0.1	12 54
08 09	237 39.5	0.9	N20 13.3	0.3	26 Mon	166 58.7	2.2	S 4 58.8	0.1	12 50
08 07	238 01.1	0.9	N20 05.5	0.3	27 Tu	167 51.2	2.2	S 5 01.6	0.1	12 47
08 06	238 22.9	0.9	N19 57.6	0.3	28 Wed	168 43.7	2.2	S 5 04.3	0.1	12 43
08 05	238 44.8	0.9	N19 49.6	0.3	29 Th	169 36.2	2.2	S 5 07.1	0.1	12 40
08 03	239 06.9	0.9	N19 41.5	0.3	30 Fri	170 28.6	2.2	S 5 09.9	0.1	12 36

MARS, Av. Mag. +1.4
SHA September 5 247; 10 244; 15 240; 20 237; 25 234; 30 231

SATURN, Av. Mag. +0.8
SHA September 5 165; 10 164; 15 164; 20 163; 25 163; 30 162

33

SEPTEMBER 2011

STARS

No.	Name	Mag	Transit h m	Dec ° '	SHA ° '
	0h GMT September 1				
ϒ	ARIES	—	1 21	—	—
1	Alpheratz	2.1	1 30	N29 09.5	357 44.5
2	Ankaa	2.4	1 47	S42 14.3	353 16.5
3	Schedar	2.2	2 02	N56 36.2	349 41.5
4	Diphda	2.0	2 05	S17 55.1	348 56.8
5	Achernar	0.5	2 58	S57 10.3	335 27.1
6	POLARIS	2.0	4 08	N89 18.7	318 07.2
7	Hamal	2.0	3 28	N23 31.2	328 01.9
8	Acamar	3.2	4 19	S40 15.2	315 19.0
9	Menkar	2.5	4 23	N 4 08.3	314 16.2
10	Mirfak	1.8	4 45	N49 54.1	308 41.9
11	Aldebaran	0.9	5 56	N16 32.0	290 50.7
12	Rigel	0.1	6 35	S 8 11.2	281 13.2
13	Capella	0.1	6 37	N46 00.4	280 36.3
14	Bellatrix	1.6	6 45	N 6 21.7	278 33.3
15	Elnath	1.7	6 47	N28 36.9	278 14.2
16	Alnilam	1.7	6 56	S 1 11.6	275 47.6
17	Betelgeuse	0.1–1.2	7 15	N 7 24.6	271 02.7
18	Canopus	−0.7	7 44	S52 41.9	263 56.8
19	Sirius	−1.5	8 05	S16 43.8	258 34.9
20	Adhara	1.5	8 18	S28 59.1	255 13.6
21	Castor	1.6	8 55	N31 51.6	246 09.7
22	Procyon	0.4	8 59	N 5 11.7	245 01.2
23	Pollux	1.1	9 05	N27 59.8	243 29.4
24	Avior	1.9	9 42	S59 32.7	234 19.0
25	Suhail	2.2	10 27	S43 28.7	222 53.7
26	Miaplacidus	1.7	10 32	S69 45.8	221 40.8
27	Alphard	2.0	10 47	S 8 42.5	217 57.6
28	Regulus	1.4	11 28	N11 54.6	207 45.1
29	Dubhe	1.8	12 23	N61 41.2	193 53.8
30	Denebola	2.1	13 08	N14 30.4	182 35.2
31	Gienah	2.6	13 35	S17 36.4	175 53.9
32	Acrux	1.3	13 46	S63 10.0	173 11.4
33	Gacrux	1.6	13 50	S57 10.8	172 02.9
34	Mimosa	1.3	14 07	S59 45.3	167 54.1
35	Alioth	1.8	14 13	N55 53.9	166 22.2
36	Spica	1.0	14 44	S11 13.3	158 32.8
37	Alkaid	1.9	15 06	N49 15.4	153 00.2
38	Hadar	0.6	15 23	S60 25.9	148 50.2
39	Menkent	2.1	15 26	S36 25.7	148 09.4
40	Arcturus	0.0	15 34	N19 07.4	145 57.1
41	Rigil Kent	−0.3	15 59	S60 53.1	139 53.9
42	Zuben'ubi	2.8	16 10	S16 05.4	137 07.0
43	Kochab	2.1	16 09	N74 06.7	137 20.5
44	Alphecca	2.2	16 53	N26 40.7	126 12.2
45	Antares	1.0	17 48	S26 27.5	112 27.9
46	Atria	1.9	18 08	S69 03.1	107 30.9
47	Sabik	2.4	18 29	S15 44.3	102 14.0
48	Shaula	1.6	18 52	S37 06.8	96 23.6
49	Rasalhague	2.1	18 53	N12 33.3	96 07.6
50	Eltanin	2.2	19 14	N51 29.6	90 46.6
51	Kaus Aust	1.9	19 42	S34 22.7	83 45.3
52	Vega	0.0	19 55	N38 48.0	80 39.7
53	Nunki	2.0	20 13	S26 16.8	75 59.7
54	Altair	0.8	21 09	N 8 54.2	62 09.3
55	Peacock	1.9	21 44	S56 41.8	53 20.7
56	Deneb	1.3	21 59	N45 19.7	49 32.0
57	Enif	2.4	23 02	N 9 56.0	33 48.1
58	Al Na'ir	1.7	23 26	S46 54.1	27 44.8
59	Fomalhaut	1.2	0 19	S29 33.4	15 25.0
60	Markab	2.5	0 26	N15 16.4	13 39.3

SUN AND MOON

SUN — Lat 52°N

Yr	Day of Mth/Week	Transit h m	Semi-Diam	Twilight h m	Sunrise h m	Sunset h m	Twilight h m
244	1 Th	12 00	15.9	04 36	05 11	18 48	19 23
245	2 Fri	12 00	15.9	04 37	05 13	18 46	19 21
246	3 Sat	11 59	15.9	04 39	05 14	18 44	19 19
247	4 Sun	11 59	15.9	04 41	05 16	18 41	19 16
248	5 Mon	11 59	15.9	04 43	05 18	18 39	19 14
249	6 Tu	11 58	15.9	04 44	05 19	18 37	19 11
250	7 Wed	11 58	15.9	04 46	05 21	18 34	19 09
251	8 Th	11 58	15.9	04 48	05 22	18 32	19 07
252	9 Fri	11 57	15.9	04 49	05 24	18 30	19 04
253	10 Sat	11 57	15.9	04 51	05 26	18 27	19 02
254	11 Sun	11 56	15.9	04 53	05 27	18 25	18 59
255	12 Mon	11 56	15.9	04 55	05 29	18 23	18 57
256	13 Tu	11 56	15.9	04 56	05 31	18 20	18 55
257	14 Wed	11 55	15.9	04 58	05 32	18 18	18 52
258	15 Th	11 55	15.9	05 00	05 34	18 16	18 50
259	16 Fri	11 55	15.9	05 01	05 35	18 13	18 47
260	17 Sat	11 54	15.9	05 03	05 37	18 11	18 45
261	18 Sun	11 54	15.9	05 05	05 39	18 09	18 43
262	19 Mon	11 54	15.9	05 06	05 40	18 06	18 40
263	20 Tu	11 53	16.0	05 08	05 42	18 04	18 38
264	21 Wed	11 53	16.0	05 10	05 44	18 02	18 35
265	22 Th	11 52	16.0	05 11	05 45	17 59	18 33
266	23 Fri	11 52	16.0	05 13	05 47	17 57	18 31
267	24 Sat	11 52	16.0	05 15	05 49	17 55	18 28
268	25 Sun	11 51	16.0	05 16	05 50	17 52	18 26
269	26 Mon	11 51	16.0	05 18	05 52	17 50	18 24
270	27 Tu	11 51	16.0	05 20	05 53	17 48	18 21
271	28 Wed	11 50	16.0	05 21	05 55	17 45	18 19
272	29 Th	11 50	16.0	05 23	05 57	17 43	18 17
273	30 Fri	11 50	16.0	05 25	05 58	17 41	18 14

MOON — Lat 52°N

Yr	Day of Mth/Week	Age days	Transit (Upper) h m	Semi-diam	Diff m	Hor Par	Moonrise h m	Moonset h m
244	1 Th	03	14 55	16.4	57	60.4	09 56	19 43
245	2 Fri	04	15 52	16.3	57	59.9	11 20	20 15
246	3 Sat	05	16 49	16.1	58	59.2	12 39	20 54
247	4 Sun	06	17 47	15.9	57	58.5	13 49	21 44
248	5 Mon	07	18 44	15.8	55	57.8	14 48	22 43
249	6 Tu	08	19 39	15.6	52	57.1	15 34	23 49
250	7 Wed	09	20 31	15.4	49	56.5	16 10	— —
251	8 Th	10	21 20	15.2	46	56.0	16 37	01 00
252	9 Fri	11	22 06	15.1	44	55.5	17 00	02 11
253	10 Sat	12	22 50	15.0	42	55.0	17 19	03 22
254	11 Sun	13	23 32	14.9	41	54.7	17 37	04 32
255	12 Mon	14	24 13	14.8	—	54.4	17 53	05 40
256	13 Tu	15	00 13	14.8	42	54.2	18 10	06 47
257	14 Wed	16	00 55	14.7	42	54.0	18 28	07 54
258	15 Th	17	01 37	14.7	44	54.1	18 49	09 00
259	16 Fri	18	02 21	14.7	45	54.1	19 13	10 07
260	17 Sat	19	03 06	14.8	48	54.3	19 43	11 12
261	18 Sun	20	03 54	14.9	50	54.7	20 20	12 14
262	19 Mon	21	04 44	15.0	51	55.2	21 07	13 10
263	20 Tu	22	05 35	15.2	52	55.9	22 04	14 00
264	21 Wed	23	06 27	15.4	53	56.7	23 11	14 42
265	22 Th	24	07 20	15.7	53	57.6	— —	15 16
266	23 Fri	25	08 13	16.0	54	58.5	00 25	15 45
267	24 Sat	26	09 05	16.2	56	59.5	01 44	16 10
268	25 Sun	27	09 58	16.4	57	60.3	03 06	16 33
269	26 Mon	28	10 51	16.6	60	60.9	04 31	16 54
270	27 Tu	00	11 45	16.7	60	61.3	05 58	17 17
271	28 Wed	01	12 41	16.7	57	61.3	07 26	17 43
272	29 Th	02	13 38	16.6	60	61.0	08 53	18 13
273	30 Fri	03	14 38	16.5	60	60.4	10 18	18 51

Lat Corr to Sunrise, Sunset etc.

Lat °	Twilight h m	Sunrise h m	Sunset h m	Twilight h m
N70	−0 53	−0 21	+0 20	+0 51
68	−0 42	−0 17	+0 16	+0 41
66	−0 34	−0 14	+0 13	+0 32
64	−0 27	−0 11	+0 10	+0 25
62	−0 20	−0 08	+0 08	+0 19
N60	−0 15	−0 06	+0 05	+0 15
58	−0 11	−0 04	+0 03	+0 10
56	−0 07	−0 02	+0 02	+0 07
54	−0 03	−0 01	−0 01	+0 03
50	+0 03	+0 02	−0 03	−0 03
N45	+0 09	+0 04	−0 06	−0 09
40	+0 12	+0 06	−0 08	−0 17
35	+0 16	+0 08	−0 09	−0 20
30	+0 20	+0 10	−0 11	−0 25
20	+0 24	+0 13	−0 13	−0 28
N10	+0 28	+0 16	−0 16	−0 28
0	+0 31	+0 18	−0 18	−0 32
S10	+0 33	+0 19	−0 20	−0 32
20	+0 33	+0 21	−0 22	−0 33
30	+0 34	+0 24	−0 26	−0 33
S35	+0 34	+0 26	−0 28	−0 32
40	+0 33	+0 27	−0 31	−0 32
45	+0 32	+0 28	−0 33	−0 31
S50	+0 31	+0 30	−0 35	−0 30

NOTES
The corrections to sunrise etc. are for middle of September.

Phases of the Moon

		d	h	m
☽	First Quarter	4	17	39
○	Full Moon	12	09	27
☾	Last Quarter	20	11	39
●	New Moon	27	11	09

	d	h
Apogee	15	06
Perigee	28	01

SUN AND ARIES OCTOBER 2011 SUN AND ARIES

GMT	SUN GHA ° ′	SUN Dec ° ′	ARIES GHA ° ′
Saturday, 1st October			
00	182 31.2	**S 2 58.0**	9 23.1
02	212 31.6	**3 00.0**	39 28.0
04	242 32.0	**3 01.9**	69 32.9
06	272 32.4	**3 03.8**	99 37.9
08	302 32.8	**3 05.8**	129 42.8
10	332 33.2	**3 07.7**	159 47.7
12	2 33.6	**3 09.7**	189 52.7
14	32 34.1	**3 11.6**	219 57.6
16	62 34.5	**3 13.5**	250 02.5
18	92 34.9	**3 15.5**	280 07.4
20	122 35.3	**3 17.4**	310 12.4
22	152 35.7	**S 3 19.3**	340 17.3
Sunday, 2nd October			
00	182 36.1	**S 3 21.3**	10 22.2
02	212 36.5	**3 23.2**	40 27.2
04	242 36.9	**3 25.2**	70 32.1
06	272 37.3	**3 27.1**	100 37.0
08	302 37.7	**3 29.0**	130 41.9
10	332 38.1	**3 31.0**	160 46.9
12	2 38.5	**3 32.9**	190 51.8
14	32 38.9	**3 34.8**	220 56.7
16	62 39.2	**3 36.8**	251 01.7
18	92 39.6	**3 38.7**	281 06.6
20	122 40.0	**3 40.6**	311 11.5
22	152 40.4	**S 3 42.6**	341 16.4
Monday, 3rd October			
00	182 40.8	**S 3 44.5**	11 21.4
02	212 41.2	**3 46.5**	41 26.3
04	242 41.6	**3 48.4**	71 31.2
06	272 42.0	**3 50.3**	101 36.2
08	302 42.4	**3 52.3**	131 41.1
10	332 42.8	**3 54.2**	161 46.0
12	2 43.2	**3 56.1**	191 50.9
14	32 43.6	**3 58.1**	221 55.9
16	62 44.0	**4 00.0**	252 00.8
18	92 44.4	**4 01.9**	282 05.7
20	122 44.7	**4 03.8**	312 10.7
22	152 45.1	**S 4 05.8**	342 15.6
Tuesday, 4th October			
00	182 45.5	**S 4 07.7**	12 20.5
02	212 45.9	**4 09.6**	42 25.4
04	242 46.3	**4 11.6**	72 30.4
06	272 46.7	**4 13.5**	102 35.3
08	302 47.1	**4 15.4**	132 40.2
10	332 47.5	**4 17.4**	162 45.1
12	2 47.8	**4 19.3**	192 50.1
14	32 48.2	**4 21.2**	222 55.0
16	62 48.6	**4 23.1**	252 59.9
18	92 49.0	**4 25.1**	283 04.9
20	122 49.4	**4 27.0**	313 09.8
22	152 49.8	**S 4 28.9**	343 14.7
Wednesday, 5th October			
00	182 50.1	**S 4 30.8**	13 19.6
02	212 50.5	**4 32.8**	43 24.6
04	242 50.9	**4 34.7**	73 29.5
06	272 51.3	**4 36.6**	103 34.4
08	302 51.7	**4 38.5**	133 39.4
10	332 52.0	**4 40.5**	163 44.3
12	2 52.4	**4 42.4**	193 49.2
14	32 52.8	**4 44.3**	223 54.1
16	62 53.2	**4 46.2**	253 59.1
18	92 53.5	**4 48.2**	284 04.0
20	122 53.9	**4 50.1**	314 08.9
22	152 54.3	**S 4 52.0**	344 13.9

GMT	SUN GHA ° ′	SUN Dec ° ′	ARIES GHA ° ′
Thursday, 6th October			
00	182 54.7	**S 4 53.9**	14 18.8
02	212 55.0	**4 55.9**	44 23.7
04	242 55.4	**4 57.8**	74 28.6
06	272 55.8	**4 59.7**	104 33.6
08	302 56.2	**5 01.6**	134 38.5
10	332 56.5	**5 03.5**	164 43.4
12	2 56.9	**5 05.4**	194 48.4
14	32 57.3	**5 07.4**	224 53.3
16	62 57.6	**5 09.3**	254 58.2
18	92 58.0	**5 11.2**	285 03.1
20	122 58.4	**5 13.1**	315 08.1
22	152 58.7	**S 5 15.0**	345 13.0
Friday, 7th October			
00	182 59.1	**S 5 17.0**	15 17.9
02	212 59.5	**5 18.9**	45 22.9
04	242 59.8	**5 20.8**	75 27.8
06	273 00.2	**5 22.7**	105 32.7
08	303 00.6	**5 24.6**	135 37.6
10	333 00.9	**5 26.5**	165 42.6
12	3 01.3	**5 28.4**	195 47.5
14	33 01.7	**5 30.3**	225 52.4
16	63 02.0	**5 32.3**	255 57.3
18	93 02.4	**5 34.2**	286 02.3
20	123 02.7	**5 36.1**	316 07.2
22	153 03.1	**S 5 38.0**	346 12.1
Saturday, 8th October			
00	183 03.5	**S 5 39.9**	16 17.1
02	213 03.8	**5 41.8**	46 22.0
04	243 04.2	**5 43.7**	76 26.9
06	273 04.5	**5 45.6**	106 31.8
08	303 04.9	**5 47.5**	136 36.8
10	333 05.2	**5 49.4**	166 41.7
12	3 05.6	**5 51.4**	196 46.6
14	33 05.9	**5 53.3**	226 51.6
16	63 06.3	**5 55.2**	256 56.5
18	93 06.7	**5 57.1**	287 01.4
20	123 07.0	**5 59.0**	317 06.3
22	153 07.4	**S 6 00.9**	347 11.3
Sunday, 9th October			
00	183 07.7	**S 6 02.8**	17 16.2
02	213 08.1	**6 04.7**	47 21.1
04	243 08.4	**6 06.6**	77 26.1
06	273 08.7	**6 08.5**	107 31.0
08	303 09.1	**6 10.4**	137 35.9
10	333 09.4	**6 12.3**	167 40.8
12	3 09.8	**6 14.2**	197 45.8
14	33 10.1	**6 16.1**	227 50.7
16	63 10.5	**6 18.0**	257 55.6
18	93 10.8	**6 19.9**	288 00.5
20	123 11.2	**6 21.8**	318 05.5
22	153 11.5	**S 6 23.7**	348 10.4
Monday, 10th October			
00	183 11.8	**S 6 25.6**	18 15.3
02	213 12.2	**6 27.5**	48 20.3
04	243 12.5	**6 29.4**	78 25.2
06	273 12.9	**6 31.3**	108 30.1
08	303 13.2	**6 33.2**	138 35.0
10	333 13.5	**6 35.1**	168 40.0
12	3 13.9	**6 36.9**	198 44.9
14	33 14.2	**6 38.8**	228 49.8
16	63 14.5	**6 40.7**	258 54.8
18	93 14.9	**6 42.6**	288 59.7
20	123 15.2	**6 44.5**	319 04.6
22	153 15.5	**S 6 46.4**	349 09.5

GMT	SUN GHA ° ′	SUN Dec ° ′	ARIES GHA ° ′
Tuesday, 11th October			
00	183 15.9	**S 6 48.3**	19 14.5
02	213 16.2	**6 50.2**	49 19.4
04	243 16.5	**6 52.1**	79 24.3
06	273 16.9	**6 54.0**	109 29.3
08	303 17.2	**6 55.8**	139 34.2
10	333 17.5	**6 57.7**	169 39.1
12	3 17.8	**6 59.6**	199 44.0
14	33 18.2	**7 01.5**	229 49.0
16	63 18.5	**7 03.4**	259 53.9
18	93 18.8	**7 05.3**	289 58.8
20	123 19.1	**7 07.2**	320 03.8
22	153 19.5	**S 7 09.0**	350 08.7
Wednesday, 12th October			
00	183 19.8	**S 7 10.9**	20 13.6
02	213 20.1	**7 12.8**	50 18.5
04	243 20.4	**7 14.7**	80 23.5
06	273 20.7	**7 16.6**	110 28.4
08	303 21.0	**7 18.4**	140 33.3
10	333 21.4	**7 20.3**	170 38.2
12	3 21.7	**7 22.2**	200 43.2
14	33 22.0	**7 24.1**	230 48.1
16	63 22.3	**7 25.9**	260 53.0
18	93 22.6	**7 27.8**	290 58.0
20	123 22.9	**7 29.7**	321 02.9
22	153 23.2	**S 7 31.6**	351 07.8
Thursday, 13th October			
00	183 23.6	**S 7 33.4**	21 12.7
02	213 23.9	**7 35.3**	51 17.7
04	243 24.2	**7 37.2**	81 22.6
06	273 24.5	**7 39.1**	111 27.5
08	303 24.8	**7 40.9**	141 32.5
10	333 25.1	**7 42.8**	171 37.4
12	3 25.4	**7 44.7**	201 42.3
14	33 25.7	**7 46.5**	231 47.2
16	63 26.0	**7 48.4**	261 52.2
18	93 26.3	**7 50.3**	291 57.1
20	123 26.6	**7 52.1**	322 02.0
22	153 26.9	**S 7 54.0**	352 07.0
Friday, 14th October			
00	183 27.2	**S 7 55.9**	22 11.9
02	213 27.5	**7 57.7**	52 16.8
04	243 27.8	**7 59.6**	82 21.7
06	273 28.1	**8 01.5**	112 26.7
08	303 28.4	**8 03.3**	142 31.6
10	333 28.7	**8 05.2**	172 36.5
12	3 29.0	**8 07.0**	202 41.5
14	33 29.3	**8 08.9**	232 46.4
16	63 29.6	**8 10.8**	262 51.3
18	93 29.9	**8 12.6**	292 56.2
20	123 30.2	**8 14.5**	323 01.2
22	153 30.4	**S 8 16.3**	353 06.1
Saturday, 15th October			
00	183 30.7	**S 8 18.2**	23 11.0
02	213 31.0	**8 20.0**	53 15.9
04	243 31.3	**8 21.9**	83 20.9
06	273 31.6	**8 23.7**	113 25.8
08	303 31.9	**8 25.6**	143 30.7
10	333 32.2	**8 27.4**	173 35.7
12	3 32.4	**8 29.3**	203 40.6
14	33 32.7	**8 31.1**	233 45.5
16	63 33.0	**8 33.0**	263 50.4
18	93 33.3	**8 34.8**	293 55.4
20	123 33.6	**8 36.7**	324 00.3
22	153 33.8	**S 8 38.5**	354 05.2

GMT	SUN GHA ° ′	SUN Dec ° ′	ARIES GHA ° ′
Sunday, 16th October			
00	183 34.1	**S 8 40.4**	24 10.2
02	213 34.4	**8 42.2**	54 15.1
04	243 34.7	**8 44.1**	84 20.0
06	273 34.9	**8 45.9**	114 24.9
08	303 35.2	**8 47.8**	144 29.9
10	333 35.5	**8 49.6**	174 34.8
12	3 35.8	**8 51.4**	204 39.7
14	33 36.0	**8 53.3**	234 44.7
16	63 36.3	**8 55.1**	264 49.6
18	93 36.6	**8 57.0**	294 54.5
20	123 36.8	**8 58.8**	324 59.4
22	153 37.1	**S 9 00.6**	355 04.4
Monday, 17th October			
00	183 37.4	**S 9 02.5**	25 09.3
02	213 37.6	**9 04.3**	55 14.2
04	243 37.9	**9 06.1**	85 19.2
06	273 38.1	**9 08.0**	115 24.1
08	303 38.4	**9 09.8**	145 29.0
10	333 38.7	**9 11.6**	175 33.9
12	3 38.9	**9 13.4**	205 38.9
14	33 39.2	**9 15.3**	235 43.8
16	63 39.4	**9 17.1**	265 48.7
18	93 39.7	**9 18.9**	295 53.7
20	123 39.9	**9 20.8**	325 58.6
22	153 40.2	**S 9 22.6**	356 03.5
Tuesday, 18th October			
00	183 40.4	**S 9 24.4**	26 08.4
02	213 40.7	**9 26.2**	56 13.4
04	243 40.9	**9 28.1**	86 18.3
06	273 41.2	**9 29.9**	116 23.2
08	303 41.4	**9 31.7**	146 28.2
10	333 41.7	**9 33.5**	176 33.1
12	3 41.9	**9 35.3**	206 38.0
14	33 42.2	**9 37.2**	236 42.9
16	63 42.4	**9 39.0**	266 47.9
18	93 42.7	**9 40.8**	296 52.8
20	123 42.9	**9 42.6**	326 57.7
22	153 43.1	**S 9 44.4**	357 02.7
Wednesday, 19th October			
00	183 43.4	**S 9 46.2**	27 07.6
02	213 43.6	**9 48.0**	57 12.5
04	243 43.9	**9 49.8**	87 17.4
06	273 44.1	**9 51.7**	117 22.4
08	303 44.3	**9 53.5**	147 27.3
10	333 44.6	**9 55.3**	177 32.2
12	3 44.8	**9 57.1**	207 37.1
14	33 45.0	**9 58.9**	237 42.1
16	63 45.2	**10 00.7**	267 47.0
18	93 45.5	**10 02.5**	297 51.9
20	123 45.7	**10 04.3**	327 56.9
22	153 45.9	**S10 06.1**	358 01.8
Thursday, 20th October			
00	183 46.2	**S10 07.9**	28 06.7
02	213 46.4	**10 09.7**	58 11.6
04	243 46.6	**10 11.5**	88 16.6
06	273 46.8	**10 13.3**	118 21.5
08	303 47.0	**10 15.1**	148 26.4
10	333 47.3	**10 16.9**	178 31.4
12	3 47.5	**10 18.7**	208 36.3
14	33 47.7	**10 20.5**	238 41.2
16	63 47.9	**10 22.3**	268 46.1
18	93 48.1	**10 24.1**	298 51.1
20	123 48.3	**10 25.9**	328 56.0
22	153 48.6	**S10 27.6**	359 00.9

M GMT	SUN GHA ° ′	SUN Dec ° ′	ARIES GHA ° ′
Friday, 21st October			
00	183 48.8	**S10 29.4**	29 05.9
02	213 49.0	**10 31.2**	59 10.8
04	243 49.2	**10 33.0**	89 15.7
06	273 49.4	**10 34.8**	119 20.6
08	303 49.6	**10 36.6**	149 25.6
10	333 49.8	**10 38.4**	179 30.5
12	3 50.0	**10 40.1**	209 35.4
14	33 50.2	**10 41.9**	239 40.4
16	63 50.4	**10 43.7**	269 45.3
18	93 50.6	**10 45.5**	299 50.2
20	123 50.8	**10 47.3**	329 55.1
22	153 51.0	**S10 49.0**	0 00.1
Saturday, 22nd October			
00	183 51.2	**S10 50.8**	30 05.0
02	213 51.4	**10 52.6**	60 09.9
04	243 51.6	**10 54.4**	90 14.9
06	273 51.8	**10 56.1**	120 19.8
08	303 52.0	**10 57.9**	150 24.7
10	333 52.2	**10 59.7**	180 29.6
12	3 52.4	**11 01.4**	210 34.6
14	33 52.6	**11 03.2**	240 39.5
16	63 52.7	**11 05.0**	270 44.4
18	93 52.9	**11 06.7**	300 49.3
20	123 53.1	**11 08.5**	330 54.3
22	153 53.3	**S11 10.3**	0 59.2
Sunday, 23rd October			
00	183 53.5	**S11 12.0**	31 04.1
02	213 53.7	**11 13.8**	61 09.1
04	243 53.8	**11 15.5**	91 14.0
06	273 54.0	**11 17.3**	121 18.9
08	303 54.2	**11 19.1**	151 23.8
10	333 54.4	**11 20.8**	181 28.8
12	3 54.6	**11 22.6**	211 33.7
14	33 54.7	**11 24.3**	241 38.6
16	63 54.9	**11 26.1**	271 43.6
18	93 55.1	**11 27.8**	301 48.5
20	123 55.2	**11 29.6**	331 53.4
22	153 55.4	**S11 31.3**	1 58.3
Monday, 24th October			
00	183 55.6	**S11 33.1**	32 03.3
02	213 55.8	**11 34.8**	62 08.2
04	243 55.9	**11 36.6**	92 13.1
06	273 56.1	**11 38.3**	122 18.1
08	303 56.2	**11 40.0**	152 23.0
10	333 56.4	**11 41.8**	182 27.9
12	3 56.6	**11 43.5**	212 32.8
14	33 56.7	**11 45.3**	242 37.8
16	63 56.9	**11 47.0**	272 42.7
18	93 57.0	**11 48.7**	302 47.6
20	123 57.2	**11 50.5**	332 52.6
22	153 57.4	**S11 52.2**	2 57.5
Tuesday, 25th October			
00	183 57.5	**S11 53.9**	33 02.4
02	213 57.7	**11 55.7**	63 07.3
04	243 57.8	**11 57.4**	93 12.3
06	273 58.0	**11 59.1**	123 17.2
08	303 58.1	**12 00.9**	153 22.1
10	333 58.3	**12 02.6**	183 27.0
12	3 58.4	**12 04.3**	213 32.0
14	33 58.5	**12 06.0**	243 36.9
16	63 58.7	**12 07.8**	273 41.8
18	93 58.8	**12 09.5**	303 46.8
20	123 59.0	**12 11.2**	333 51.7
22	153 59.1	**S12 12.9**	3 56.6

GMT	SUN GHA ° ′	SUN Dec ° ′	ARIES GHA ° ′
Wednesday, 26th October			
00	183 59.2	**S12 14.6**	34 01.5
02	213 59.4	**12 16.4**	64 06.5
04	243 59.5	**12 18.1**	94 11.4
06	273 59.7	**12 19.8**	124 16.3
08	303 59.8	**12 21.5**	154 21.3
10	333 59.9	**12 23.2**	184 26.2
12	4 00.0	**12 24.9**	214 31.1
14	34 00.2	**12 26.6**	244 36.0
16	64 00.3	**12 28.3**	274 41.0
18	94 00.4	**12 30.0**	304 45.9
20	124 00.6	**12 31.7**	334 50.8
22	154 00.7	**S12 33.4**	4 55.8
Thursday, 27th October			
00	184 00.8	**S12 35.1**	35 00.7
02	214 00.9	**12 36.8**	65 05.6
04	244 01.0	**12 38.5**	95 10.5
06	274 01.2	**12 40.2**	125 15.5
08	304 01.3	**12 41.9**	155 20.4
10	334 01.4	**12 43.6**	185 25.3
12	4 01.5	**12 45.3**	215 30.3
14	34 01.6	**12 47.0**	245 35.2
16	64 01.7	**12 48.7**	275 40.1
18	94 01.9	**12 50.4**	305 45.0
20	124 02.0	**12 52.1**	335 50.0
22	154 02.1	**S12 53.8**	5 54.9
Friday, 28th October			
00	184 02.2	**S12 55.5**	35 59.8
02	214 02.3	**12 57.1**	66 04.7
04	244 02.4	**12 58.8**	96 09.7
06	274 02.5	**13 00.5**	126 14.6
08	304 02.6	**13 02.2**	156 19.5
10	334 02.7	**13 03.9**	186 24.5
12	4 02.8	**13 05.5**	216 29.4
14	34 02.9	**13 07.2**	246 34.3
16	64 03.0	**13 08.9**	276 39.2
18	94 03.1	**13 10.6**	306 44.2
20	124 03.2	**13 12.2**	336 49.1
22	154 03.3	**S13 13.9**	6 54.0
Saturday, 29th October			
00	184 03.4	**S13 15.6**	36 59.0
02	214 03.5	**13 17.2**	67 03.9
04	244 03.5	**13 18.9**	97 08.8
06	274 03.6	**13 20.6**	127 13.7
08	304 03.7	**13 22.2**	157 18.7
10	334 03.8	**13 23.9**	187 23.6
12	4 03.9	**13 25.5**	217 28.5
14	34 04.0	**13 27.2**	247 33.5
16	64 04.1	**13 28.9**	277 38.4
18	94 04.1	**13 30.5**	307 43.3
20	124 04.2	**13 32.2**	337 48.2
22	154 04.3	**S13 33.8**	7 53.2
Sunday, 30th October			
00	184 04.4	**S13 35.5**	37 58.1
02	214 04.4	**13 37.1**	68 03.0
04	244 04.5	**13 38.8**	98 08.0
06	274 04.6	**13 40.4**	128 12.9
08	304 04.7	**13 42.0**	158 17.8
10	334 04.7	**13 43.7**	188 22.7
12	4 04.8	**13 45.3**	218 27.7
14	34 04.9	**13 47.0**	248 32.6
16	64 04.9	**13 48.6**	278 37.5
18	94 05.0	**13 50.2**	308 42.5
20	124 05.1	**13 51.9**	338 47.4
22	154 05.1	**S13 53.5**	8 52.3

GMT	SUN GHA ° ′	SUN Dec ° ′	ARIES GHA ° ′
Monday, 31st October			
00	184 05.2	**S13 55.1**	38 57.2
02	214 05.2	**13 56.8**	69 02.2
04	244 05.3	**13 58.4**	99 07.1
06	274 05.4	**S14 00.0**	129 12.0
08	304 05.4	**S14 01.7**	159 17.0
10	334 05.5	**14 03.3**	189 21.9
12	4 05.5	**14 04.9**	219 26.8
14	34 05.6	**S14 06.5**	249 31.7
16	64 05.6	**S14 08.1**	279 36.7
18	94 05.7	**14 09.8**	309 41.6
20	124 05.7	**14 11.4**	339 46.5
22	154 05.8	**S14 13.0**	9 51.5

MOON

Day	GMT hr	GHA ° '	Mean Var/hr 14°+	Dec ° '	Mean Var/hr '
1 Sat	0	134 55.1	23.9	S21 11.1	4.9
	6	221 18.4	23.9	S21 41.0	4.0
	12	307 41.8	24.0	S22 05.4	3.1
	18	34 05.8	24.2	S22 24.2	2.1
2 Sun	0	120 30.7	24.4	S22 37.4	1.2
	6	206 56.7	24.6	S22 45.1	0.3
	12	293 24.3	24.9	S22 47.4	0.6
	18	19 53.6	25.3	S22 44.3	1.4
3 Mon	0	106 25.0	25.6	S22 36.1	2.3
	6	192 58.8	26.0	S22 22.8	3.1
	12	279 35.0	26.5	S22 04.7	3.8
	18	6 14.0	27.0	S21 42.0	4.6
4 Tu	0	92 55.9	27.5	S21 14.9	5.3
	6	179 40.7	28.0	S20 43.6	6.0
	12	266 28.5	28.5	S20 08.4	6.6
	18	353 19.4	29.0	S19 29.4	7.2
5 Wed	0	80 13.4	29.5	S18 47.0	7.7
	6	167 10.4	30.1	S18 01.4	8.1
	12	254 10.4	30.5	S17 12.8	8.6
	18	341 13.3	31.0	S16 21.3	9.0
6 Th	0	68 19.0	31.4	S15 27.4	9.4
	6	155 27.4	31.9	S14 31.1	9.8
	12	242 38.4	32.3	S13 32.7	10.0
	18	329 51.8	32.7	S12 32.5	10.4
7 Fri	0	57 07.4	33.0	S11 30.5	10.6
	6	144 25.2	33.3	S10 26.9	10.9
	12	231 44.9	33.6	S 9 22.1	11.0
	18	319 06.4	33.9	S 8 16.1	11.2
8 Sat	0	46 29.4	34.1	S 7 09.1	11.3
	6	133 53.9	34.3	S 6 01.3	11.4
	12	221 19.6	34.4	S 4 52.8	11.5
	18	308 46.4	34.6	S 3 43.9	11.6
9 Sun	0	36 14.0	34.7	S 2 34.6	11.6
	6	123 42.4	34.9	S 1 25.1	11.6
	12	211 11.2	34.9	S 0 15.6	11.6
	18	298 40.5	34.9	N 0 53.8	11.5
10 Mon	0	26 09.9	34.9	N 2 02.9	11.5
	6	113 39.4	34.9	N 3 11.6	11.3
	12	201 08.7	34.8	N 4 19.8	11.2
	18	288 37.7	34.8	N 5 27.2	11.1
11 Tu	0	16 06.4	34.6	N 6 33.8	10.9
	6	103 34.4	34.5	N 7 39.4	10.8
	12	191 01.7	34.4	N 8 43.9	10.5
	18	278 28.2	34.2	N 9 47.1	10.2
12 Wed	0	5 53.7	34.1	N10 48.9	10.0
	6	93 18.1	33.9	N11 49.0	9.7
	12	180 41.3	33.6	N12 47.5	9.4
	18	268 03.2	33.4	N13 44.0	9.0
13 Th	0	355 23.7	33.2	N14 38.6	8.7
	6	82 42.8	32.9	N15 30.9	8.3
	12	170 00.3	32.7	N16 21.0	7.9
	18	257 16.2	32.4	N17 08.5	7.5
14 Fri	0	344 30.5	32.1	N17 53.4	7.0
	6	71 43.1	31.8	N18 35.6	6.5
	12	158 54.1	31.5	N19 14.8	6.0
	18	246 03.4	31.2	N19 51.0	5.4
15 Sat	0	333 11.1	31.0	N20 24.1	4.9
	6	60 17.2	30.7	N20 53.8	4.4
	12	147 21.7	30.4	N21 20.0	3.8
	18	234 24.8	30.3	N21 42.7	3.2
16 Sun	0	321 26.5	30.0	N22 01.8	2.5
	6	48 26.8	29.9	N22 17.1	1.8
	12	135 26.0	29.6	N22 28.5	1.2
	18	222 24.1	29.6	N22 36.0	0.5
17 Mon	0	309 21.2	29.4	N22 39.5	0.2
	6	36 17.5	29.3	N22 39.0	0.9
	12	123 13.1	29.2	N22 34.3	1.5
	18	210 08.2	29.1	N22 25.5	2.2
18 Tu	0	297 02.9	29.1	N22 12.6	2.9
	6	23 57.3	29.0	N21 55.5	3.6
	12	110 51.5	29.0	N21 34.2	4.3
	18	197 45.7	29.0	N21 08.8	4.9
19 Wed	0	284 39.9	29.0	N20 39.3	5.7
	6	11 34.3	29.1	N20 05.7	6.3
	12	98 28.9	29.1	N19 28.1	7.0
	18	185 23.7	29.2	N18 46.7	7.6
20 Th	0	272 18.9	29.3	N18 01.3	8.2
	6	359 14.4	29.3	N17 12.2	8.8
	12	86 10.3	29.4	N16 19.5	9.4
	18	173 06.4	29.4	N15 23.3	10.0
21 Fri	0	260 02.9	29.4	N14 23.7	10.5
	6	346 59.6	29.4	N13 20.8	11.0
	12	73 56.4	29.5	N12 14.9	11.5
	18	160 53.3	29.5	N11 06.0	12.0
22 Sat	0	247 50.1	29.4	N 9 54.5	12.4
	6	334 46.8	29.4	N 8 40.3	12.8
	12	61 43.1	29.3	N 7 23.9	13.1
	18	148 39.1	29.2	N 6 05.4	13.4
23 Sun	0	235 34.4	29.1	N 4 45.1	13.7
	6	322 29.0	28.9	N 3 23.2	13.9
	12	49 22.6	28.7	N 2 00.0	14.1
	18	136 15.1	28.5	N 0 35.7	14.1
24 Mon	0	223 06.3	28.3	S 0 49.2	14.2
	6	309 56.1	28.0	S 2 14.5	14.2
	12	36 44.2	27.7	S 3 39.8	14.2
	18	123 30.6	27.4	S 5 04.7	14.1
25 Tu	0	210 15.0	27.0	S 6 28.9	13.8
	6	296 57.3	26.7	S 7 52.1	13.6
	12	23 37.4	26.2	S 9 13.7	13.3
	18	110 15.2	25.9	S10 33.4	12.9
26 Wed	0	196 50.6	25.5	S11 50.8	12.4
	6	283 23.6	25.1	S13 05.5	11.9
	12	9 54.2	24.6	S14 17.0	11.3
	18	96 22.5	24.3	S15 25.0	10.6
27 Th	0	182 48.5	23.9	S16 29.0	9.9
	6	269 12.4	23.6	S17 28.8	9.2
	12	355 34.3	23.3	S18 24.0	8.3
	18	81 54.5	23.1	S19 14.3	7.4
28 Fri	0	168 13.2	22.9	S19 59.4	6.5
	6	254 30.8	22.8	S20 39.0	5.6
	12	340 47.6	22.7	S21 13.1	4.7
	18	67 04.1	22.8	S21 41.4	3.7
29 Sat	0	153 20.6	22.8	S22 03.8	2.6
	6	239 37.5	23.0	S22 20.4	1.7
	12	325 55.4	23.3	S22 31.1	0.8
	18	52 14.5	23.5	S22 36.1	0.3
30 Sun	0	138 35.3	23.8	S22 35.3	1.1
	6	224 58.2	24.2	S22 28.9	2.0
	12	311 23.5	24.7	S22 17.1	2.9
	18	37 51.6	25.2	S22 00.1	3.8
31 Mon	0	124 22.6	25.8	S21 38.2	4.5
	6	210 56.7	26.3	S21 11.5	5.2
	12	297 34.2	26.9	S20 40.4	6.0
	18	24 15.2	27.5	S20 05.1	6.6

PLANETS

VENUS Mer Pass h m	VENUS GHA ° '	VENUS Mean Var/hr 14°+	VENUS Dec ° '	VENUS Mean Var/hr '	Day	JUPITER GHA ° '	JUPITER Mean Var/hr 15°+	JUPITER Dec ° '	JUPITER Mean Var/hr '	JUPITER Mer Pass h m
12 37	170 47.6	59.6	S 6 51.0	1.2	1 Sat	332 32.8	2.7	N13 01.2	0.1	01 49
12 38	170 37.8	59.6	S 7 20.7	1.2	2 SUN	333 37.7	2.7	N12 59.2	0.1	01 45
12 39	170 27.8	59.6	S 7 50.3	1.2	3 Mon	334 42.7	2.7	N12 57.1	0.1	01 41
12 39	170 17.6	59.6	S 8 19.7	1.2	4 Tu	335 47.9	2.7	N12 55.1	0.1	01 37
12 40	170 07.3	59.6	S 8 48.9	1.2	5 Wed	336 53.2	2.7	N12 52.9	0.1	01 32
12 41	169 56.8	59.6	S 9 17.9	1.2	6 Th	337 58.7	2.7	N12 50.8	0.1	01 28
12 41	169 46.2	59.5	S 9 46.8	1.2	7 Fri	339 04.3	2.7	N12 48.5	0.1	01 23
12 42	169 35.3	59.5	S10 15.3	1.2	8 Sat	340 09.9	2.7	N12 46.3	0.1	01 19
12 43	169 24.2	59.5	S10 43.7	1.2	9 SUN	341 15.7	2.8	N12 44.0	0.1	01 15
12 44	169 12.9	59.5	S11 11.8	1.2	10 Mon	342 21.7	2.8	N12 41.7	0.1	01 10
12 44	169 01.4	59.5	S11 39.7	1.1	11 Tu	343 27.7	2.8	N12 39.3	0.1	01 06
12 45	168 49.7	59.5	S12 07.3	1.1	12 Wed	344 33.8	2.8	N12 37.0	0.1	01 02
12 46	168 37.7	59.5	S12 34.6	1.1	13 Th	345 40.0	2.8	N12 34.6	0.1	00 57
12 47	168 25.5	59.5	S13 01.6	1.1	14 Fri	346 46.4	2.8	N12 32.1	0.1	00 53
12 48	168 13.0	59.5	S13 28.3	1.1	15 Sat	347 52.8	2.8	N12 29.6	0.1	00 48
12 48	168 00.3	59.5	S13 54.7	1.1	16 SUN	348 59.2	2.8	N12 27.2	0.1	00 44
12 49	167 47.3	59.4	S14 20.7	1.1	17 Mon	350 05.8	2.8	N12 24.6	0.1	00 39
12 50	167 34.0	59.4	S14 46.4	1.1	18 Tu	351 12.4	2.8	N12 22.1	0.1	00 35
12 51	167 20.5	59.4	S15 11.7	1.0	19 Wed	352 19.1	2.8	N12 19.6	0.1	00 31
12 52	167 06.7	59.4	S15 36.7	1.0	20 Th	353 25.9	2.8	N12 17.0	0.1	00 26
12 53	166 52.5	59.4	S16 01.2	1.0	21 Fri	354 32.7	2.8	N12 14.4	0.1	00 22
12 54	166 38.1	59.4	S16 25.4	1.0	22 Sat	355 39.5	2.8	N12 11.8	0.1	00 17
12 55	166 23.4	59.4	S16 49.1	1.0	23 SUN	356 46.4	2.8	N12 09.2	0.1	00 13
12 56	166 08.3	59.4	S17 12.4	1.0	24 Mon	357 53.4	2.8	N12 06.6	0.1	00 08
12 57	165 53.0	59.3	S17 35.3	0.9	25 Tu	359 00.3	2.8	N12 03.9	0.1	00 04
12 58	165 37.4	59.3	S17 57.7	0.9	26 Wed	0 07.3	2.8	N12 01.3	0.1	00 00
12 59	165 21.4	59.3	S18 19.6	0.9	27 Th	1 14.3	2.8	N11 58.7	0.1	23 51
13 00	165 05.1	59.3	S18 41.1	0.9	28 Fri	2 21.4	2.8	N11 56.0	0.1	23 46
13 01	164 48.5	59.3	S19 02.1	0.9	29 Sat	3 28.4	2.8	N11 53.4	0.1	23 42
13 03	164 31.7	59.3	S19 22.5	0.8	30 SUN	4 35.4	2.8	N11 50.7	0.1	23 37
13 04	164 14.5	59.3	S19 42.5	0.8	31 Mon	5 42.4	2.8	N11 48.1	0.1	23 33

VENUS, Av. Mag. –3.9
SHA October 5 157; 10 151; 15 145; 20 139; 25 133; 30 127

JUPITER, Av. Mag. –2.9
SHA October 5 324; 10 324; 15 325; 20 325; 25 326; 30 327

MARS Mer Pass h m	MARS GHA ° '	MARS Mean Var/hr 15°+	MARS Dec ° '	MARS Mean Var/hr '	Day	SATURN GHA ° '	SATURN Mean Var/hr 15°+	SATURN Dec ° '	SATURN Mean Var/hr '	SATURN Mer Pass h m
08 02	239 29.1	0.9	N19 33.3	0.3	1 Sat	171 21.0	2.2	S 5 12.6	0.1	12 33
08 00	239 51.5	0.9	N19 25.0	0.4	2 SUN	172 13.5	2.2	S 5 15.4	0.1	12 29
07 59	240 14.1	0.9	N19 16.6	0.3	3 Mon	173 05.9	2.2	S 5 18.2	0.1	12 26
07 57	240 36.8	1.0	N19 08.2	0.4	4 Tu	173 58.3	2.2	S 5 20.9	0.1	12 22
07 56	240 59.7	1.0	N18 59.7	0.4	5 Wed	174 50.6	2.2	S 5 23.7	0.1	12 19
07 54	241 22.8	1.0	N18 51.0	0.4	6 Th	175 43.0	2.2	S 5 26.5	0.1	12 15
07 52	241 46.0	1.0	N18 42.4	0.4	7 Fri	176 35.4	2.2	S 5 29.2	0.1	12 12
07 51	242 09.4	1.0	N18 33.6	0.4	8 Sat	177 27.7	2.2	S 5 32.0	0.1	12 08
07 49	242 33.0	1.0	N18 24.8	0.4	9 SUN	178 20.0	2.2	S 5 34.8	0.1	12 05
07 48	242 56.7	1.0	N18 15.9	0.4	10 Mon	179 12.4	2.2	S 5 37.5	0.1	12 01
07 46	243 20.6	1.0	N18 06.9	0.4	11 Tu	180 04.7	2.2	S 5 40.3	0.1	11 58
07 45	243 44.6	1.0	N17 57.9	0.4	12 Wed	180 57.0	2.2	S 5 43.0	0.1	11 54
07 43	244 08.8	1.0	N17 48.8	0.4	13 Th	181 49.3	2.2	S 5 45.8	0.1	11 51
07 41	244 33.2	1.0	N17 39.6	0.4	14 Fri	182 41.7	2.2	S 5 48.5	0.1	11 48
07 40	244 57.8	1.0	N17 30.4	0.4	15 Sat	183 34.0	2.2	S 5 51.2	0.1	11 44
07 38	245 22.5	1.0	N17 21.1	0.4	16 SUN	184 26.3	2.2	S 5 54.0	0.1	11 41
07 36	245 47.4	1.0	N17 11.8	0.4	17 Mon	185 18.6	2.2	S 5 56.7	0.1	11 37
07 35	246 12.5	1.1	N17 02.5	0.4	18 Tu	186 10.9	2.2	S 5 59.4	0.1	11 34
07 33	246 37.7	1.1	N16 53.1	0.4	19 Wed	187 03.2	2.2	S 6 02.1	0.1	11 30
07 31	247 03.1	1.1	N16 43.6	0.4	20 Th	187 55.5	2.2	S 6 04.8	0.1	11 27
07 30	247 28.7	1.1	N16 34.1	0.4	21 Fri	188 47.8	2.2	S 6 07.5	0.1	11 23
07 28	247 54.5	1.1	N16 24.6	0.4	22 Sat	189 40.2	2.2	S 6 10.2	0.1	11 20
07 26	248 20.4	1.1	N16 15.0	0.4	23 SUN	190 32.5	2.2	S 6 12.9	0.1	11 16
07 24	248 46.5	1.1	N16 05.4	0.4	24 Mon	191 24.8	2.2	S 6 15.6	0.1	11 13
07 23	249 12.8	1.1	N15 55.8	0.4	25 Tu	192 17.2	2.2	S 6 18.3	0.1	11 09
07 21	249 39.3	1.1	N15 46.2	0.4	26 Wed	193 09.5	2.2	S 6 20.9	0.1	11 06
07 19	250 06.0	1.1	N15 36.5	0.4	27 Th	194 01.9	2.2	S 6 23.6	0.1	11 02
07 17	250 32.8	1.1	N15 26.8	0.4	28 Fri	194 54.2	2.2	S 6 26.2	0.1	10 59
07 15	250 59.9	1.1	N15 17.1	0.4	29 Sat	195 46.6	2.2	S 6 28.8	0.1	10 55
07 14	251 27.1	1.1	N15 07.4	0.4	30 SUN	196 39.0	2.2	S 6 31.4	0.1	10 52
07 12	251 54.5	1.2	N14 57.7	0.4	31 Mon	197 31.4	2.2	S 6 34.0	0.1	10 48

MARS, Av. Mag. +1.2
SHA October 5 228; 10 225; 15 222; 20 219; 25 216; 30 213

SATURN, Av. Mag. +0.7
SHA October 5 162; 10 161; 15 160; 20 160; 25 159; 30 159

OCTOBER 2011

STARS

No.	Name	Mag	Transit h m	Dec ° '	SHA ° '
	0h GMT October 1				
ψ	ARIES	–	23 19	– –	357 44.4
1	Alpheratz	2.1	23 28	N29 09.6	353 16.4
2	Ankaa	2.4	23 45	S42 14.4	349 41.4
3	Schedar	2.2	0 04	N56 36.3	348 56.7
4	Diphda	2.0	0 07	S17 55.1	335 27.0
5	Achernar	0.5	1 00	S57 10.5	317 57.4
6	POLARIS	2.0	2 10	N89 18.8	328 01.8
7	Hamal	2.0	1 30	N23 31.2	315 18.8
8	Acamar	3.2	2 21	S40 15.3	314 16.0
9	Menkar	2.5	2 25	N 4 08.3	308 41.7
10	Mirfak	1.8	2 47	N49 54.1	290 50.5
11	Aldebaran	0.9	3 58	N16 32.0	281 13.0
12	Rigel	0.1	4 37	S 8 11.2	280 36.0
13	Capella	0.1	4 39	N46 00.4	278 33.1
14	Bellatrix	1.6	4 47	N 6 21.7	278 13.9
15	Elnath	1.7	4 49	N28 37.0	275 47.4
16	Alnilam	1.7	4 58	S 1 11.6	271 02.5
17	Betelgeuse	0.1–1.2	5 17	N 7 24.6	263 56.5
18	Canopus	-0.7	5 46	S52 41.9	258 34.7
19	Sirius	-1.5	6 07	S16 43.8	255 13.4
20	Adhara	1.5	6 21	S28 59.1	246 09.5
21	Castor	1.6	6 57	N31 51.6	245 01.0
22	Procyon	0.4	7 01	N 5 11.7	245 01.0
23	Pollux	1.1	7 07	N27 59.7	243 29.2
24	Avior	1.9	7 44	S59 32.6	234 18.7
25	Suhail	2.2	8 29	S43 28.6	222 53.5
26	Miaplacidus	1.7	8 34	S69 45.7	221 40.4
27	Alphard	2.0	8 49	S 8 42.5	217 57.4
28	Regulus	1.4	9 30	N11 54.5	207 45.0
29	Dubhe	1.8	10 25	N61 41.0	193 53.6
30	Denebola	2.1	11 10	N14 30.4	182 35.3
31	Gienah	2.6	11 37	S17 36.4	175 53.9
32	Acrux	1.3	11 48	S63 09.8	173 11.4
33	Gacrux	1.6	11 52	S57 10.7	172 02.9
34	Mimosa	1.3	12 09	S59 45.2	167 54.0
35	Alioth	1.8	12 15	N55 53.7	166 22.2
36	Spica	1.0	12 46	S11 13.3	158 32.8
37	Alkaid	1.9	13 08	N49 15.3	153 00.3
38	Hadar	0.6	13 25	S60 25.8	148 50.3
39	Menkent	2.1	13 28	S36 25.6	148 09.4
40	Arcturus	0.0	13 36	N19 07.4	145 57.1
41	Rigil Kent	-0.3	14 01	S60 53.0	139 54.4
42	Zuben'ubi	2.8	14 12	S16 05.4	137 07.1
43	Kochab	2.1	14 11	N74 06.5	137 20.9
44	Alphecca	2.2	14 55	N26 40.7	126 12.3
45	Antares	1.0	15 50	S26 27.4	112 28.8
46	Atria	1.9	16 10	S69 03.0	107 31.2
47	Sabik	2.4	16 31	S15 44.3	102 14.1
48	Shaula	1.6	16 54	S37 06.7	96 23.7
49	Rasalhague	2.1	16 55	N12 33.3	96 07.7
50	Eltanin	2.2	17 16	N51 29.6	90 46.9
51	Kaus Aust	1.9	17 45	S34 22.7	83 45.5
52	Vega	0.0	17 57	N38 48.0	80 39.9
53	Nunki	2.0	18 15	S26 16.8	75 59.8
54	Altair	0.8	19 11	N 8 54.3	62 09.4
55	Peacock	1.9	19 46	S56 41.9	53 20.9
56	Deneb	1.3	20 01	N45 19.8	49 32.2
57	Enif	2.4	21 04	N 9 56.0	33 48.1
58	Al Na'ir	1.7	21 28	S46 54.2	27 44.9
59	Fomalhaut	1.2	22 17	S29 33.5	15 25.0
60	Markab	2.5	22 24	N15 16.4	13 39.3

SUN AND MOON

SUN

Yr	Day of Mth	Week	Transit h m	Semi-Diam	Lat 52°N Twilight h m	Sunrise h m	Sunset h m	Twilight h m
274	1	Sat	11 50	16.0	05 26	06 00	17 38	18 12
275	2	Sun	11 49	16.0	05 28	06 02	17 36	18 10
276	3	Mon	11 49	16.0	05 30	06 03	17 34	18 07
277	4	Tu	11 49	16.0	05 31	06 05	17 32	18 05
278	5	Wed	11 49	16.0	05 33	06 07	17 29	18 03
279	6	Th	11 48	16.0	05 35	06 09	17 27	18 01
280	7	Fri	11 48	16.0	05 36	06 10	17 25	17 58
281	8	Sat	11 48	16.0	05 38	06 12	17 22	17 56
282	9	Sun	11 47	16.0	05 40	06 14	17 20	17 54
283	10	Mon	11 47	16.0	05 41	06 15	17 18	17 52
284	11	Tu	11 47	16.1	05 43	06 17	17 16	17 50
285	12	Wed	11 46	16.1	05 45	06 19	17 13	17 47
286	13	Th	11 46	16.1	05 47	06 20	17 11	17 45
287	14	Fri	11 46	16.1	05 48	06 22	17 09	17 43
288	15	Sat	11 45	16.1	05 50	06 24	17 07	17 41
289	16	Sun	11 45	16.1	05 52	06 26	17 05	17 39
290	17	Mon	11 45	16.1	05 53	06 27	17 03	17 37
291	18	Tu	11 45	16.1	05 55	06 29	17 00	17 35
292	19	Wed	11 45	16.1	05 57	06 31	16 58	17 33
293	20	Th	11 45	16.1	05 58	06 33	16 56	17 31
294	21	Fri	11 45	16.1	06 00	06 34	16 54	17 29
295	22	Sat	11 45	16.1	06 02	06 36	16 52	17 27
296	23	Sun	11 44	16.1	06 03	06 38	16 50	17 25
297	24	Mon	11 44	16.1	06 05	06 40	16 48	17 23
298	25	Tu	11 44	16.1	06 07	06 42	16 46	17 21
299	26	Wed	11 44	16.1	06 08	06 43	16 44	17 19
300	27	Th	11 44	16.1	06 10	06 45	16 42	17 17
301	28	Fri	11 44	16.1	06 12	06 47	16 40	17 15
302	29	Sat	11 44	16.1	06 13	06 49	16 38	17 13
303	30	Sun	11 44	16.1	06 15	06 50	16 36	17 11
304	31	Mon	11 44	16.1	06 17	06 52	16 34	17 10

MOON

Yr	Day of Mth	Week	Age days	Transit (Upper) h m	Diff	Semi-diam	Hor Par	Lat 52°N Moonrise	Moonset
274	1	Sat	04	15 38	59	16.3	59.6	11 34	19 38
275	2	Sun	05	16 37	57	16.0	58.8	12 39	20 36
276	3	Mon	06	17 34	54	15.8	57.9	13 31	21 42
277	4	Tu	07	18 28	50	15.5	57.0	14 10	22 52
278	5	Wed	08	19 18	46	15.3	56.3	14 41	– –
279	6	Th	09	20 04	44	15.2	55.6	15 05	00 03
280	7	Fri	10	20 48	43	15.0	55.1	15 25	01 14
281	8	Sat	11	21 31	41	14.9	54.7	15 43	02 23
282	9	Sun	12	22 12	42	14.8	54.3	16 00	03 31
283	10	Mon	13	22 54	42	14.7	54.1	16 17	04 38
284	11	Tu	14	23 36	43	14.7	54.0	16 35	05 44
285	12	Wed	15	24 19	–	14.7	54.0	16 55	06 51
286	13	Th	16	00 19	45	14.7	54.0	17 18	07 57
287	14	Fri	17	01 04	47	14.7	54.1	17 46	09 02
288	15	Sat	18	01 51	49	14.8	54.4	18 21	10 05
289	16	Sun	19	02 40	50	14.9	54.7	19 05	11 03
290	17	Mon	20	03 30	51	15.0	55.2	19 57	11 55
291	18	Tu	21	04 21	51	15.2	55.8	20 59	12 38
292	19	Wed	22	05 12	51	15.4	56.5	22 08	13 14
293	20	Th	23	06 03	51	15.6	57.3	23 22	13 44
294	21	Fri	24	06 54	50	15.8	58.2	– –	14 10
295	22	Sat	25	07 44	51	15.9	59.2	00 40	14 33
296	23	Sun	26	08 35	53	16.1	60.0	02 01	14 54
297	24	Mon	27	09 28	54	16.4	60.8	03 24	15 16
298	25	Tu	28	10 22	57	16.6	61.2	04 50	15 40
299	26	Wed	29	11 19	59	16.7	61.4	06 18	16 08
300	27	Th	01	12 18	62	16.7	61.2	07 46	16 43
301	28	Fri	02	13 20	62	16.6	60.8	09 09	17 27
302	29	Sat	03	14 22	60	16.4	60.0	10 22	18 22
303	30	Sun	04	15 22	57	16.1	59.1	11 21	19 27
304	31	Mon	05	16 19	53	15.8	58.2	12 07	20 38

Lat Corr to Sunrise, Sunset etc.

Lat °	Twilight h m	Sunrise h m	Sunset h m	Twilight h m
N70	+0 19	+0 49	-0 51	-0 20
68	+0 16	+0 40	-0 41	-0 16
66	+0 13	+0 32	-0 33	-0 14
64	+0 10	+0 25	-0 27	-0 11
62	+0 08	+0 20	-0 21	-0 09
N60	+0 06	+0 14	-0 16	-0 07
58	+0 05	+0 11	-0 11	-0 05
56	+0 03	+0 06	-0 07	-0 04
54	+0 01	+0 03	-0 04	-0 02
50	-0 02	-0 03	+0 03	+0 01
N45	-0 05	-0 10	+0 09	+0 05
40	-0 08	-0 15	+0 14	+0 07
35	-0 14	-0 19	+0 19	+0 10
30	-0 18	-0 23	+0 23	+0 13
20	-0 24	-0 30	+0 30	+0 18
N10	-0 29	-0 42	+0 37	+0 23
0	-0 36	-0 49	+0 43	+0 29
S10	-0 43	-0 55	+0 49	+0 36
20	-0 53	-1 03	+0 57	+0 44
30	-1 00	-1 08	+1 05	+0 54
S35	-1 00	-1 13	+1 10	+1 00
40	-1 16	-1 19	+1 15	+1 08
45	-1 16	-1 27	+1 21	+1 16
S50	-1 27	-1 27	+1 28	+1 26

NOTES
The corrections to sunrise etc. are for middle of October.

Phases of the Moon

		d	h	m
◐	First Quarter	4	03	15
○	Full Moon	12	02	06
◑	Last Quarter	20	03	30
●	New Moon	26	19	56

	d	h
Apogee	12	12
Perigee	26	12

NOVEMBER 2011

SUN AND ARIES

Tuesday, 1st November

GMT	SUN GHA ° ′	SUN Dec ° ′	ARIES GHA ° ′
00	184 05.8	S14 14.6	39 56.4
02	214 05.8	14 16.2	70 01.3
04	244 05.9	14 17.8	100 06.2
06	274 05.9	14 19.4	130 11.2
08	304 06.0	14 21.0	160 16.1
10	334 06.0	14 22.7	190 21.0
12	4 06.0	14 24.3	220 26.0
14	34 06.1	14 25.9	250 30.9
16	64 06.1	14 27.5	280 35.8
18	94 06.1	14 29.1	310 40.7
20	124 06.2	14 30.7	340 45.7
22	154 06.2	S14 32.2	10 50.6

Wednesday, 2nd November

GMT	SUN GHA ° ′	SUN Dec ° ′	ARIES GHA ° ′
00	184 06.2	S14 33.8	40 55.5
02	214 06.3	14 35.4	71 00.4
04	244 06.3	14 37.0	101 05.4
06	274 06.3	14 38.6	131 10.3
08	304 06.3	14 40.2	161 15.2
10	334 06.3	14 41.8	191 20.2
12	4 06.4	14 43.4	221 25.1
14	34 06.4	14 45.0	251 30.0
16	64 06.4	14 46.5	281 34.9
18	94 06.4	14 48.1	311 39.9
20	124 06.4	14 49.7	341 44.8
22	154 06.4	S14 51.3	11 49.7

Thursday, 3rd November

GMT	SUN GHA ° ′	SUN Dec ° ′	ARIES GHA ° ′
00	184 06.5	S14 52.8	41 54.7
02	214 06.5	14 54.4	71 59.6
04	244 06.5	14 56.0	102 04.5
06	274 06.5	14 57.5	132 09.4
08	304 06.5	14 59.1	162 14.4
10	334 06.5	15 00.7	192 19.3
12	4 06.5	15 02.2	222 24.2
14	34 06.5	15 03.8	252 29.2
16	64 06.5	15 05.4	282 34.1
18	94 06.5	15 06.9	312 39.0
20	124 06.5	15 08.5	342 43.9
22	154 06.5	S15 10.0	12 48.9

Friday, 4th November

GMT	SUN GHA ° ′	SUN Dec ° ′	ARIES GHA ° ′
00	184 06.5	S15 11.6	42 53.8
02	214 06.5	15 13.1	72 58.7
04	244 06.5	15 14.7	103 03.7
06	274 06.5	15 16.2	133 08.6
08	304 06.5	15 17.8	163 13.5
10	334 06.4	15 19.3	193 18.4
12	4 06.4	15 20.9	223 23.4
14	34 06.4	15 22.4	253 28.3
16	64 06.4	15 24.0	283 33.2
18	94 06.4	15 25.5	313 38.2
20	124 06.4	15 27.0	343 43.1
22	154 06.3	S15 28.6	13 48.0

Saturday, 5th November

GMT	SUN GHA ° ′	SUN Dec ° ′	ARIES GHA ° ′
00	184 06.3	S15 30.1	43 52.9
02	214 06.3	15 31.6	73 57.9
04	244 06.3	15 33.2	104 02.8
06	274 06.3	15 34.7	134 07.7
08	304 06.2	15 36.2	164 12.6
10	334 06.2	15 37.7	194 17.6
12	4 06.2	15 39.2	224 22.5
14	34 06.1	15 40.8	254 27.4
16	64 06.1	15 42.3	284 32.4
18	94 06.1	15 43.8	314 37.3
20	124 06.0	15 45.3	344 42.2
22	154 06.0	S15 46.8	14 47.1

Sunday, 6th November

GMT	SUN GHA ° ′	SUN Dec ° ′	ARIES GHA ° ′
00	184 06.0	S15 48.3	44 52.1
02	214 05.9	15 49.8	74 57.0
04	244 05.9	15 51.4	105 01.9
06	274 05.8	15 52.9	135 06.9
08	304 05.8	15 54.4	165 11.8
10	334 05.7	15 55.9	195 16.7
12	4 05.7	15 57.4	225 21.6
14	34 05.6	15 58.9	255 26.6
16	64 05.6	16 00.4	285 31.5
18	94 05.5	16 01.9	315 36.4
20	124 05.5	16 03.3	345 41.4
22	154 05.4	S16 04.8	15 46.3

Monday, 7th November

GMT	SUN GHA ° ′	SUN Dec ° ′	ARIES GHA ° ′
00	184 05.4	S16 06.3	45 51.2
02	214 05.3	16 07.8	75 56.1
04	244 05.3	16 09.3	106 01.1
06	274 05.2	16 10.8	136 06.0
08	304 05.1	16 12.3	166 10.9
10	334 05.1	16 13.7	196 15.9
12	4 05.0	16 15.2	226 20.8
14	34 05.0	16 16.7	256 25.7
16	64 04.9	16 18.2	286 30.6
18	94 04.8	16 19.6	316 35.6
20	124 04.7	16 21.1	346 40.5
22	154 04.7	S16 22.6	16 45.4

Tuesday, 8th November

GMT	SUN GHA ° ′	SUN Dec ° ′	ARIES GHA ° ′
00	184 04.6	S16 24.0	46 50.3
02	214 04.5	16 25.5	76 55.3
04	244 04.5	16 27.0	107 00.2
06	274 04.4	16 28.4	137 05.1
08	304 04.3	16 29.9	167 10.1
10	334 04.2	16 31.3	197 15.0
12	4 04.1	16 32.8	227 19.9
14	34 04.1	16 34.2	257 24.8
16	64 04.0	16 35.7	287 29.8
18	94 03.9	16 37.1	317 34.7
20	124 03.8	16 38.6	347 39.6
22	154 03.7	S16 40.0	17 44.6

Wednesday, 9th November

GMT	SUN GHA ° ′	SUN Dec ° ′	ARIES GHA ° ′
00	184 03.6	S16 41.5	47 49.5
02	214 03.5	16 42.9	77 54.4
04	244 03.4	16 44.4	107 59.3
06	274 03.3	16 45.8	138 04.3
08	304 03.2	16 47.2	168 09.2
10	334 03.1	16 48.7	198 14.1
12	4 03.0	16 50.1	228 19.1
14	34 02.9	16 51.5	258 24.0
16	64 02.8	16 53.0	288 28.9
18	94 02.7	16 54.4	318 33.8
20	124 02.6	16 55.8	348 38.8
22	154 02.5	S16 57.2	18 43.7

Thursday, 10th November

GMT	SUN GHA ° ′	SUN Dec ° ′	ARIES GHA ° ′
00	184 02.4	S16 58.6	48 48.6
02	214 02.3	17 00.1	78 53.6
04	244 02.2	17 01.5	108 58.5
06	274 02.1	17 02.9	139 03.4
08	304 02.0	17 04.3	169 08.3
10	334 01.9	17 05.7	199 13.3
12	4 01.7	17 07.1	229 18.2
14	34 01.6	17 08.5	259 23.1
16	64 01.5	17 09.9	289 28.0
18	94 01.4	17 11.3	319 33.0
20	124 01.3	17 12.7	349 37.9
22	154 01.1	S17 14.1	19 42.8

Friday, 11th November

GMT	SUN GHA ° ′	SUN Dec ° ′	ARIES GHA ° ′
00	184 01.0	S17 15.5	49 47.8
02	214 00.9	17 16.9	79 52.7
04	244 00.8	17 18.3	109 57.6
06	274 00.6	17 19.7	140 02.5
08	304 00.5	17 21.1	170 07.5
10	334 00.4	17 22.5	200 12.4
12	4 00.2	17 23.8	230 17.3
14	34 00.1	17 25.2	260 22.3
16	64 00.0	17 26.6	290 27.2
18	93 59.8	17 28.0	320 32.1
20	123 59.7	17 29.3	350 37.0
22	153 59.5	S17 30.7	20 42.0

Saturday, 12th November

GMT	SUN GHA ° ′	SUN Dec ° ′	ARIES GHA ° ′
00	183 59.4	S17 32.1	50 46.9
02	213 59.2	17 33.5	80 51.8
04	243 59.1	17 34.8	110 56.8
06	273 59.0	17 36.2	141 01.7
08	303 58.8	17 37.5	171 06.6
10	333 58.7	17 38.9	201 11.5
12	3 58.5	17 40.3	231 16.5
14	33 58.3	17 41.6	261 21.4
16	63 58.2	17 43.0	291 26.3
18	93 58.0	17 44.3	321 31.3
20	123 57.9	17 45.7	351 36.2
22	153 57.7	S17 47.0	21 41.1

Sunday, 13th November

GMT	SUN GHA ° ′	SUN Dec ° ′	ARIES GHA ° ′
00	183 57.6	S17 48.4	51 46.0
02	213 57.4	17 49.7	81 51.0
04	243 57.2	17 51.0	111 55.9
06	273 57.1	17 52.4	142 00.8
08	303 56.9	17 53.7	172 05.8
10	333 56.7	17 55.0	202 10.7
12	3 56.6	17 56.4	232 15.6
14	33 56.4	17 57.7	262 20.5
16	63 56.2	17 59.0	292 25.5
18	93 56.0	18 00.4	322 30.4
20	123 55.9	18 01.7	352 35.3
22	153 55.7	S18 03.0	22 40.3

Monday, 14th November

GMT	SUN GHA ° ′	SUN Dec ° ′	ARIES GHA ° ′
00	183 55.5	S18 04.3	52 45.2
02	213 55.3	18 05.6	82 50.1
04	243 55.2	18 07.0	112 55.0
06	273 55.0	18 08.3	143 00.0
08	303 54.8	18 09.6	173 04.9
10	333 54.6	18 10.9	203 09.8
12	3 54.4	18 12.2	233 14.8
14	33 54.2	18 13.5	263 19.7
16	63 54.0	18 14.8	293 24.6
18	93 53.8	18 16.1	323 29.5
20	123 53.6	18 17.4	353 34.5
22	153 53.4	S18 18.7	23 39.4

Tuesday, 15th November

GMT	SUN GHA ° ′	SUN Dec ° ′	ARIES GHA ° ′
00	183 53.3	S18 20.0	53 44.3
02	213 53.1	18 21.3	83 49.3
04	243 52.9	18 22.6	113 54.2
06	273 52.7	18 23.8	143 59.1
08	303 52.4	18 25.1	174 04.0
10	333 52.2	18 26.4	204 09.0
12	3 52.0	18 27.7	234 13.9
14	33 51.8	18 29.0	264 18.8
16	63 51.6	18 30.2	294 23.7
18	93 51.4	18 31.5	324 28.7
20	123 51.2	18 32.8	354 33.6
22	153 51.0	S18 34.0	24 38.5

Wednesday, 16th November

GMT	SUN GHA ° ′	SUN Dec ° ′	ARIES GHA ° ′
00	183 50.8	S18 35.3	54 43.5
02	213 50.6	18 36.6	84 48.4
04	243 50.3	18 37.8	114 53.3
06	273 50.1	18 39.1	144 58.2
08	303 49.9	18 40.3	175 03.2
10	333 49.7	18 41.6	205 08.1
12	3 49.5	18 42.9	235 13.0
14	33 49.2	18 44.1	265 18.0
16	63 49.0	18 45.3	295 22.9
18	93 48.8	18 46.6	325 27.8
20	123 48.5	18 47.8	355 32.7
22	153 48.3	S18 49.1	25 37.7

Thursday, 17th November

GMT	SUN GHA ° ′	SUN Dec ° ′	ARIES GHA ° ′
00	183 48.1	S18 50.3	55 42.6
02	213 47.8	18 51.5	85 47.5
04	243 47.6	18 52.8	115 52.5
06	273 47.4	18 54.0	145 57.4
08	303 47.1	18 55.2	176 02.3
10	333 46.9	18 56.5	206 07.2
12	3 46.7	18 57.7	236 12.2
14	33 46.4	18 58.9	266 17.1
16	63 46.2	19 00.1	296 22.0
18	93 45.9	19 01.3	326 27.0
20	123 45.7	19 02.6	356 31.9
22	153 45.4	S19 03.8	26 36.8

Friday, 18th November

GMT	SUN GHA ° ′	SUN Dec ° ′	ARIES GHA ° ′
00	183 45.2	S19 05.0	56 41.7
02	213 44.9	19 06.2	86 46.7
04	243 44.7	19 07.4	116 51.6
06	273 44.4	19 08.6	146 56.5
08	303 44.2	19 09.8	177 01.5
10	333 43.9	19 11.0	207 06.4
12	3 43.6	19 12.2	237 11.3
14	33 43.4	19 13.4	267 16.2
16	63 43.1	19 14.6	297 21.2
18	93 42.9	19 15.8	327 26.1
20	123 42.6	19 16.9	357 31.0
22	153 42.3	S19 18.1	27 36.0

Saturday, 19th November

GMT	SUN GHA ° ′	SUN Dec ° ′	ARIES GHA ° ′
00	183 42.1	S19 19.3	57 40.9
02	213 41.8	19 20.5	87 45.8
04	243 41.5	19 21.7	117 50.7
06	273 41.2	19 22.8	147 55.7
08	303 41.0	19 24.0	178 00.6
10	333 40.7	19 25.2	208 05.5
12	3 40.4	19 26.3	238 10.4
14	33 40.1	19 27.5	268 15.4
16	63 39.9	19 28.7	298 20.3
18	93 39.6	19 29.8	328 25.2
20	123 39.3	19 31.0	358 30.2
22	153 39.0	S19 32.1	28 35.1

Sunday, 20th November

GMT	SUN GHA ° ′	SUN Dec ° ′	ARIES GHA ° ′
00	183 38.7	S19 33.3	58 40.0
02	213 38.4	19 34.4	88 44.9
04	243 38.2	19 35.6	118 49.9
06	273 37.9	19 36.7	148 54.8
08	303 37.6	19 37.9	178 59.7
10	333 37.3	19 39.0	209 04.7
12	3 37.0	19 40.2	239 09.6
14	33 36.7	19 41.3	269 14.5
16	63 36.4	19 42.4	299 19.4
18	93 36.1	19 43.6	329 24.4
20	123 35.8	19 44.7	359 29.3
22	153 35.5	S19 45.8	29 34.2

Monday, 21st November

GMT	SUN GHA ° ′	SUN Dec ° ′	ARIES GHA ° ′
00	183 35.2	S19 46.9	59 39.2
02	213 34.9	19 48.0	89 44.1
04	243 34.6	19 49.2	119 49.0
06	273 34.3	19 50.3	149 53.9
08	303 34.0	19 51.4	179 58.9
10	333 33.7	19 52.5	210 03.8
12	3 33.3	19 53.6	240 08.7
14	33 33.0	19 54.7	270 13.7
16	63 32.7	19 55.8	300 18.6
18	93 32.4	19 56.9	330 23.5
20	123 32.1	19 58.0	0 28.4
22	153 31.8	S19 59.1	30 33.4

Tuesday, 22nd November

GMT	SUN GHA ° ′	SUN Dec ° ′	ARIES GHA ° ′
00	183 31.4	S20 00.2	60 38.3
02	213 31.1	20 01.3	90 43.2
04	243 30.8	20 02.4	120 48.1
06	273 30.5	20 03.5	150 53.1
08	303 30.1	20 04.5	180 58.0
10	333 29.8	20 05.6	211 02.9
12	3 29.5	20 06.7	241 07.9
14	33 29.2	20 07.8	271 12.8
16	63 28.8	20 08.8	301 17.7
18	93 28.5	20 09.9	331 22.6
20	123 28.2	20 11.0	1 27.6
22	153 27.8	S20 12.0	31 32.5

Wednesday, 23rd November

GMT	SUN GHA ° ′	SUN Dec ° ′	ARIES GHA ° ′
00	183 27.5	S20 13.1	61 37.4
02	213 27.1	20 14.2	91 42.4
04	243 26.8	20 15.2	121 47.3
06	273 26.5	20 16.3	151 52.2
08	303 26.1	20 17.3	181 57.1
10	333 25.8	20 18.4	212 02.1
12	3 25.4	20 19.4	242 07.0
14	33 25.1	20 20.5	272 11.9
16	63 24.7	20 21.5	302 16.9
18	93 24.4	20 22.5	332 21.8
20	123 24.0	20 23.6	2 26.7
22	153 23.7	S20 24.6	32 31.6

Thursday, 24th November

GMT	SUN GHA ° ′	SUN Dec ° ′	ARIES GHA ° ′
00	183 23.3	S20 25.6	62 36.6
02	213 23.0	20 26.7	92 41.5
04	243 22.6	20 27.7	122 46.4
06	273 22.3	20 28.7	152 51.4
08	303 21.9	20 29.7	182 56.3
10	333 21.5	20 30.7	213 01.2
12	3 21.2	20 31.8	243 06.1
14	33 20.8	20 32.8	273 11.1
16	63 20.5	20 33.8	303 16.0
18	93 20.1	20 34.8	333 20.9
20	123 19.7	20 35.8	3 25.9
22	153 19.4	S20 36.8	33 30.8

Friday, 25th November

GMT	SUN GHA ° ′	SUN Dec ° ′	ARIES GHA ° ′
00	183 19.0	S20 37.8	63 35.7
02	213 18.6	20 38.8	93 40.6
04	243 18.2	20 39.8	123 45.6
06	273 17.9	20 40.8	153 50.5
08	303 17.5	20 41.8	183 55.4
10	333 17.1	20 42.7	214 00.4
12	3 16.7	20 43.7	244 05.3
14	33 16.4	20 44.7	274 10.2
16	63 16.0	20 45.7	304 15.1
18	93 15.6	20 46.7	334 20.1
20	123 15.2	20 47.6	4 25.0
22	153 14.8	S20 48.6	34 29.9

Saturday, 26th November

GMT	SUN GHA ° ′	SUN Dec ° ′	ARIES GHA ° ′
00	183 14.4	S20 49.6	64 34.9
02	213 14.1	20 50.5	94 39.8
04	243 13.7	20 51.5	124 44.7
06	273 13.3	20 52.4	154 49.6
08	303 12.9	20 53.4	184 54.6
10	333 12.5	20 54.4	214 59.5
12	3 12.1	20 55.3	245 04.4
14	33 11.7	20 56.3	275 09.3
16	63 11.3	20 57.2	305 14.3
18	93 10.9	20 58.1	335 19.2
20	123 10.5	20 59.1	5 24.1
22	153 10.1	S21 00.0	35 29.1

Sunday, 27th November

GMT	SUN GHA ° ′	SUN Dec ° ′	ARIES GHA ° ′
00	183 09.7	S21 00.9	65 34.0
02	213 09.3	21 01.9	95 38.9
04	243 08.9	21 02.8	125 43.8
06	273 08.5	21 03.7	155 48.8
08	303 08.1	21 04.7	185 53.7
10	333 07.7	21 05.6	215 58.6
12	3 07.3	21 06.5	246 03.6
14	33 06.9	21 07.4	276 08.5
16	63 06.5	21 08.3	306 13.4
18	93 06.1	21 09.2	336 18.3
20	123 05.6	21 10.1	6 23.3
22	153 05.2	S21 11.0	36 28.2

Monday, 28th November

GMT	SUN GHA ° ′	SUN Dec ° ′	ARIES GHA ° ′
00	183 04.8	S21 11.9	66 33.1
02	213 04.4	21 12.8	96 38.1
04	243 04.0	21 13.7	126 43.0
06	273 03.6	21 14.6	156 47.9
08	303 03.1	21 15.5	186 52.8
10	333 02.7	21 16.4	216 57.8
12	3 02.3	21 17.3	247 02.7
14	33 01.9	21 18.2	277 07.6
16	63 01.4	21 19.0	307 12.6
18	93 01.0	21 19.9	337 17.5
20	123 00.6	21 20.8	7 22.4
22	153 00.2	S21 21.7	37 27.3

Tuesday, 29th November

GMT	SUN GHA ° ′	SUN Dec ° ′	ARIES GHA ° ′
00	182 59.7	S21 22.5	67 32.3
02	212 59.3	21 23.4	97 37.2
04	242 58.9	21 24.3	127 42.1
06	272 58.4	21 25.1	157 47.1
08	302 58.0	21 26.0	187 52.0
10	332 57.6	21 26.8	217 56.9
12	2 57.1	21 27.7	248 01.8
14	32 56.7	21 28.5	278 06.8
16	62 56.2	21 29.4	308 11.7
18	92 55.8	21 30.2	338 16.6
20	122 55.4	21 31.0	8 21.6
22	152 54.9	S21 31.9	38 26.5

Wednesday, 30th November

GMT	SUN GHA ° ′	SUN Dec ° ′	ARIES GHA ° ′
00	182 54.5	S21 32.7	68 31.4
02	212 54.0	21 33.5	98 36.3
04	242 53.6	21 34.4	128 41.3
06	272 53.1	21 35.2	158 46.2
08	302 52.7	21 36.0	188 51.1
10	332 52.2	21 36.8	218 56.1
12	2 51.8	21 37.7	249 01.0
14	32 51.3	21 38.5	279 05.9
16	62 50.9	21 39.3	309 10.8
18	92 50.4	21 40.1	339 15.8
20	122 50.0	21 40.9	9 20.7
22	152 49.5	S21 41.7	39 25.6

MOON

Day	GMT hr	GHA ° ′	Mean Var/hr 14°+	Dec ° ′	Mean Var/hr ′
1 Tu	0	110 59.7	28.0	S19 25.9	7.2
	6	197 47.7	28.7	S18 43.1	7.7
	12	284 39.3	29.3	S17 57.0	8.2
	18	11 34.3	29.7	S17 07.7	8.7
2 Wed	0	98 32.7	30.4	S16 15.7	9.1
	6	185 34.4	30.9	S15 21.1	9.6
	12	272 39.2	31.3	S14 24.2	9.8
	18	359 47.1	31.8	S13 25.3	10.1
3 Th	0	86 57.8	32.3	S12 24.5	10.5
	6	174 11.1	32.7	S11 22.1	10.7
	12	261 26.9	33.1	S10 18.2	10.9
	18	348 44.9	33.3	S 9 13.1	11.1
4 Fri	0	76 05.1	33.7	S 8 07.0	11.2
	6	163 27.1	33.9	S 7 00.0	11.3
	12	250 50.7	34.2	S 5 52.2	11.4
	18	338 15.8	34.4	S 4 43.9	11.4
5 Sat	0	65 42.2	34.6	S 3 35.2	11.5
	6	153 09.6	34.7	S 2 26.3	11.5
	12	240 37.8	34.8	S 1 17.3	11.6
	18	328 06.7	34.9	S 0 08.3	11.5
6 Sun	0	55 36.0	35.0	N 1 00.6	11.4
	6	143 05.7	35.0	N 2 09.1	11.4
	12	230 35.4	34.9	N 3 17.2	11.2
	18	318 05.0	34.9	N 4 24.7	11.2
7 Mon	0	45 34.3	34.8	N 5 31.6	10.9
	6	133 03.2	34.7	N 6 37.5	10.9
	12	220 31.5	34.5	N 7 42.5	10.6
	18	307 59.0	34.4	N 8 46.4	10.4
8 Tu	0	35 25.7	34.2	N 9 49.0	10.2
	6	122 51.2	34.1	N10 50.2	10.0
	12	210 15.7	33.8	N11 49.9	9.6
	18	297 38.8	33.6	N12 47.8	9.4
9 Wed	0	25 00.5	33.4	N13 44.0	9.0
	6	112 20.7	33.0	N14 38.1	8.6
	12	199 39.4	32.8	N15 30.1	8.2
	18	286 56.4	32.6	N16 19.8	7.9
10 Th	0	14 11.7	32.2	N17 07.1	7.4
	6	101 25.2	32.0	N17 51.7	6.9
	12	188 37.0	31.7	N18 33.6	6.4
	18	275 47.0	31.4	N19 12.6	6.0
11 Fri	0	2 55.3	31.1	N19 48.5	5.4
	6	90 01.9	30.8	N20 21.2	4.9
	12	177 06.8	30.5	N20 50.6	4.3
	18	264 10.1	30.2	N21 16.5	3.7
12 Sat	0	351 12.0	30.1	N21 38.9	3.0
	6	78 12.5	29.9	N21 57.5	2.5
	12	165 11.7	29.7	N22 12.4	1.7
	18	252 09.8	29.5	N22 23.4	1.1
13 Sun	0	339 06.9	29.4	N22 30.4	0.5
	6	66 03.2	29.3	N22 33.3	0.3
	12	152 58.9	29.2	N22 32.2	0.9
	18	239 54.1	29.2	N22 27.0	1.6
14 Mon	0	326 49.0	29.2	N22 17.7	2.3
	6	53 43.7	29.1	N22 04.2	3.0
	12	140 38.5	29.1	N21 46.6	3.7
	18	227 33.4	29.2	N21 24.9	4.4
15 Tu	0	314 28.6	29.2	N20 59.2	5.0
	6	41 24.2	29.4	N20 29.5	5.6
	12	128 20.4	29.4	N19 55.9	6.3
	18	215 17.1	29.5	N19 18.5	6.9
16 Wed	0	302 14.6	29.7	N18 37.3	7.5
	6	29 12.7	29.8	N17 52.5	8.1
	12	116 11.5	29.9	N17 04.2	8.7
	18	203 11.1	30.0	N16 12.5	9.2

Day	GMT hr	GHA ° ′	Mean Var/hr 14°+	Dec ° ′	Mean Var/hr ′
17 Th	0	290 11.4	30.2	N15 17.6	9.8
	6	17 12.4	30.2	N14 19.6	10.2
	12	104 13.9	30.3	N13 18.6	10.7
	18	191 16.0	30.4	N12 14.8	11.2
18 Fri	0	278 18.4	30.4	N11 08.3	11.5
	6	5 21.2	30.5	N 9 59.4	11.9
	12	92 24.1	30.5	N 8 48.3	12.3
	18	179 27.0	30.4	N 7 35.0	12.5
19 Sat	0	266 29.7	30.4	N 6 19.8	12.8
	6	353 32.1	30.3	N 5 03.0	13.0
	12	80 34.0	30.2	N 3 44.7	13.3
	18	167 35.2	30.1	N 2 25.2	13.4
20 Sun	0	254 35.4	29.9	N 1 04.7	13.5
	6	341 34.6	29.7	S 0 16.6	13.7
	12	68 32.4	29.4	S 1 38.3	13.7
	18	155 28.6	29.1	S 3 00.3	13.7
21 Mon	0	242 23.2	28.8	S 4 22.1	13.6
	6	329 15.8	28.4	S 5 43.5	13.4
	12	56 06.3	28.0	S 7 04.2	13.3
	18	142 54.4	27.6	S 8 23.9	13.0
22 Tu	0	229 40.1	27.2	S 9 42.1	12.7
	6	316 23.3	26.7	S10 58.5	12.4
	12	43 03.7	26.3	S12 12.8	11.9
	18	129 41.3	25.8	S13 24.5	11.4
23 Wed	0	216 16.1	25.3	S14 33.3	10.9
	6	302 48.0	24.8	S15 38.9	10.3
	12	29 17.2	24.4	S16 40.8	9.6
	18	115 43.6	23.9	S17 38.6	8.8
24 Th	0	202 07.5	23.6	S18 32.0	8.1
	6	288 29.0	23.2	S19 20.8	7.2
	12	14 48.5	22.9	S20 04.5	6.3
	18	101 06.1	22.6	S20 42.9	5.4
25 Fri	0	187 22.3	22.5	S21 15.8	4.4
	6	Eclipse of the Sun occurs today			
	12	359 52.1	22.4	S22 04.4	2.5
	18	86 06.5	22.5	S22 19.9	1.5
26 Sat	0	172 21.3	22.6	S22 29.4	0.6
	6	258 36.8	22.8	S22 33.0	0.5
	12	344 53.7	23.2	S22 30.7	1.4
	18	71 12.3	23.5	S22 22.6	2.3
27 Sun	0	157 33.0	23.9	S22 09.0	3.2
	6	243 56.2	24.4	S21 50.0	4.1
	12	330 22.3	24.9	S21 25.8	4.9
	18	56 51.5	25.5	S20 56.6	5.6
28 Mon	0	143 24.0	26.1	S20 22.9	6.4
	6	230 00.1	26.6	S19 44.9	7.1
	12	316 39.8	27.3	S19 02.8	7.6
	18	43 23.1	27.9	S18 17.0	8.2
29 Tu	0	130 10.2	28.5	S17 27.8	8.8
	6	217 01.0	29.1	S16 35.6	9.2
	12	303 55.3	29.7	S15 40.5	9.6
	18	30 53.2	30.2	S14 42.9	10.0
30 Wed	0	117 54.5	30.8	S13 43.1	10.3
	6	204 59.1	31.3	S12 41.3	10.6
	12	292 06.7	31.8	S11 37.8	10.8
	18	19 17.2	32.2	S10 32.9	11.0

PLANETS

VENUS						JUPITER				
Mer Pass h m	GHA ° ′	Mean Var/hr 14°+	Dec ° ′	Mean Var/hr ′	Day	GHA ° ′	Mean Var/hr 15°+	Dec ° ′	Mean Var/hr ′	Mer Pass h m
13 05	163 57.0	59.3	S20 01.9	0.8	1 Tu	6 49.4	2.8	N11 45.5	0.1	23 28
13 06	163 39.2	59.2	S20 20.7	0.8	2 Wed	7 56.4	2.8	N11 42.9	0.1	23 24
13 07	163 21.1	59.2	S20 39.0	0.7	3 Th	9 03.4	2.8	N11 40.3	0.1	23 19
13 09	163 02.7	59.2	S20 56.8	0.7	4 Fri	10 10.3	2.8	N11 37.7	0.1	23 15
13 10	162 44.0	59.2	S21 13.9	0.7	5 Sat	11 17.2	2.8	N11 35.1	0.1	23 11
13 11	162 25.0	59.2	S21 30.5	0.7	6 SUN	12 24.0	2.8	N11 32.6	0.1	23 06
13 12	162 05.7	59.2	S21 46.4	0.6	7 Mon	13 30.8	2.8	N11 30.0	0.1	23 02
13 14	161 46.2	59.2	S22 01.8	0.6	8 Tu	14 37.5	2.8	N11 27.5	0.1	22 57
13 15	161 26.4	59.2	S22 16.5	0.6	9 Wed	15 44.2	2.8	N11 25.0	0.1	22 53
13 16	161 06.3	59.2	S22 30.6	0.6	10 Th	16 50.8	2.8	N11 22.6	0.1	22 48
13 18	160 46.0	59.1	S22 44.0	0.5	11 Fri	17 57.3	2.8	N11 20.1	0.1	22 44
13 19	160 25.4	59.1	S22 56.8	0.5	12 Sat	19 03.7	2.8	N11 17.7	0.1	22 40
13 20	160 04.6	59.1	S23 08.9	0.5	13 SUN	20 10.1	2.8	N11 15.4	0.1	22 35
13 22	159 43.5	59.1	S23 20.3	0.4	14 Mon	21 16.4	2.8	N11 13.0	0.1	22 31
13 23	159 22.3	59.1	S23 31.1	0.4	15 Tu	22 22.5	2.8	N11 10.7	0.1	22 26
13 25	159 00.8	59.1	S23 41.1	0.4	16 Wed	23 28.6	2.8	N11 08.5	0.1	22 22
13 26	158 39.2	59.1	S23 50.5	0.4	17 Th	24 34.6	2.7	N11 06.2	0.1	22 18
13 28	158 17.3	59.1	S23 59.1	0.3	18 Fri	25 40.5	2.7	N11 04.1	0.1	22 13
13 29	157 55.3	59.1	S24 07.1	0.3	19 Sat	26 46.2	2.7	N11 01.9	0.1	22 09
13 31	157 33.2	59.1	S24 14.3	0.3	20 SUN	27 51.9	2.7	N10 59.8	0.1	22 05
13 32	157 10.9	59.1	S24 20.8	0.2	21 Mon	28 57.4	2.7	N10 57.8	0.1	22 00
13 34	156 48.5	59.1	S24 26.6	0.2	22 Tu	30 02.8	2.7	N10 55.8	0.1	21 56
13 35	156 26.0	59.1	S24 31.6	0.2	23 Wed	31 08.0	2.7	N10 53.8	0.1	21 52
13 37	156 03.3	59.1	S24 35.9	0.1	24 Th	32 13.2	2.7	N10 51.9	0.1	21 47
13 38	155 40.6	59.1	S24 39.4	0.1	25 Fri	33 18.1	2.7	N10 50.1	0.1	21 43
13 40	155 17.9	59.1	S24 42.2	0.1	26 Sat	34 23.0	2.7	N10 48.3	0.1	21 39
13 41	154 55.1	59.0	S24 44.2	0.1	27 SUN	35 27.7	2.7	N10 46.5	0.1	21 34
13 43	154 32.3	59.0	S24 45.5	0.0	28 Mon	36 32.2	2.7	N10 44.9	0.1	21 30
13 44	154 09.4	59.1	S24 46.1	0.0	29 Tu	37 36.6	2.7	N10 43.3	0.1	21 26
13 46	153 46.6	59.0	S24 45.9	0.0	30 Wed	38 40.8	2.7	N10 41.7	0.1	21 21

VENUS, Av. Mag. −3.9
SHA November 5 119; 10 112; 15 106; 20 99; 25 92; 30 85

JUPITER, Av. Mag. −2.9
SHA November 5 327; 10 328; 15 329; 20 329; 25 330; 30 330

MARS						SATURN				
Mer Pass h m	GHA ° ′	Mean Var/hr 15°+	Dec ° ′	Mean Var/hr ′	Day	GHA ° ′	Mean Var/hr 15°+	Dec ° ′	Mean Var/hr ′	Mer Pass h m
07 10	252 22.1	1.2	N14 47.9	0.4	1 Tu	198 23.8	2.2	S 6 36.6	0.1	10 45
07 08	252 49.9	1.2	N14 38.2	0.4	2 Wed	199 16.3	2.2	S 6 39.2	0.1	10 41
07 06	253 17.9	1.2	N14 28.4	0.4	3 Th	200 08.7	2.2	S 6 41.8	0.1	10 38
07 04	253 46.1	1.2	N14 18.7	0.4	4 Fri	201 01.2	2.2	S 6 44.3	0.1	10 34
07 02	254 14.5	1.2	N14 09.0	0.4	5 Sat	201 53.7	2.2	S 6 46.9	0.1	10 31
07 01	254 43.0	1.2	N13 59.2	0.4	6 SUN	202 46.2	2.2	S 6 49.4	0.1	10 27
06 59	255 11.8	1.2	N13 49.5	0.4	7 Mon	203 38.8	2.2	S 6 51.9	0.1	10 24
06 57	255 40.7	1.2	N13 39.8	0.4	8 Tu	204 31.4	2.2	S 6 54.4	0.1	10 20
06 55	256 09.9	1.2	N13 30.0	0.4	9 Wed	205 23.9	2.2	S 6 56.9	0.1	10 17
06 53	256 39.3	1.2	N13 20.3	0.4	10 Th	206 16.6	2.2	S 6 59.3	0.1	10 13
06 51	257 08.8	1.2	N13 10.7	0.4	11 Fri	207 09.2	2.2	S 7 01.8	0.1	10 10
06 49	257 38.6	1.2	N13 01.0	0.4	12 Sat	208 01.9	2.2	S 7 04.2	0.1	10 06
06 47	258 08.5	1.3	N12 51.4	0.4	13 SUN	208 54.6	2.2	S 7 06.6	0.1	10 03
06 45	258 38.7	1.3	N12 41.7	0.4	14 Mon	209 47.3	2.2	S 7 09.0	0.1	09 59
06 43	259 09.1	1.3	N12 32.2	0.4	15 Tu	210 40.0	2.2	S 7 11.4	0.1	09 56
06 41	259 39.7	1.3	N12 22.6	0.4	16 Wed	211 32.8	2.2	S 7 13.7	0.1	09 52
06 39	260 10.5	1.3	N12 13.1	0.4	17 Th	212 25.7	2.2	S 7 16.1	0.1	09 49
06 37	260 41.5	1.3	N12 03.6	0.4	18 Fri	213 18.5	2.2	S 7 18.4	0.1	09 45
06 35	261 12.7	1.3	N11 54.2	0.4	19 Sat	214 11.4	2.2	S 7 20.7	0.1	09 42
06 32	261 44.2	1.3	N11 44.8	0.4	20 SUN	215 04.3	2.2	S 7 23.0	0.1	09 38
06 30	262 15.8	1.3	N11 35.5	0.4	21 Mon	215 57.3	2.2	S 7 25.2	0.1	09 35
06 28	262 47.7	1.3	N11 26.2	0.4	22 Tu	216 50.3	2.2	S 7 27.5	0.1	09 31
06 26	263 19.9	1.3	N11 17.0	0.4	23 Wed	217 43.3	2.2	S 7 29.7	0.1	09 28
06 24	263 52.3	1.4	N11 07.8	0.4	24 Th	218 36.4	2.2	S 7 31.9	0.1	09 24
06 22	264 24.9	1.4	N10 58.7	0.4	25 Fri	219 29.5	2.2	S 7 34.1	0.1	09 21
06 20	264 57.8	1.4	N10 49.7	0.4	26 Sat	220 22.7	2.2	S 7 36.2	0.1	09 17
06 17	265 30.9	1.4	N10 40.7	0.4	27 SUN	221 15.9	2.2	S 7 38.4	0.1	09 14
06 15	266 04.2	1.4	N10 31.8	0.4	28 Mon	222 09.2	2.2	S 7 40.5	0.1	09 10
06 13	266 37.8	1.4	N10 23.0	0.4	29 Tu	223 02.5	2.2	S 7 42.6	0.1	09 06
06 11	267 11.7	1.4	N10 14.3	0.4	30 Wed	223 55.8	2.2	S 7 44.6	0.1	09 03

MARS, Av. Mag. +0.9
SHA November 5 210; 10 208; 15 205; 20 203; 25 201; 30 199

SATURN, Av. Mag. +0.8
SHA November 5 158; 10 157; 15 157; 20 156; 25 156; 30 155

NOVEMBER 2011

STARS

No.	Name	Mag	Transit h m	Dec ° '	SHA ° '
	0h GMT November 1				
Ψ	ARIES	-	21 17	-	-
1	Alpheratz	2.1	21 26	N29 09.7	357 44.5
2	Ankaa	2.4	21 44	S42 14.5	353 16.5
3	Schedar	2.2	21 58	N56 36.5	349 41.4
4	Diphda	2.0	22 01	S17 55.2	348 56.8
5	Achernar	0.5	22 55	S57 10.6	335 27.0
6	POLARIS	2.0	0 09	N89 19.0	317 53.2
7	Hamal	2.0	23 24	N23 31.3	328 01.7
8	Acamar	3.2	0 19	S40 15.4	315 18.7
9	Menkar	2.5	0 23	N 4 08.3	314 15.9
10	Mirfak	1.8	0 45	N49 54.3	308 41.5
11	Aldebaran	0.9	1 57	N16 32.0	290 50.3
12	Rigel	0.1	2 35	S 8 11.2	281 12.8
13	Capella	0.1	2 37	N46 00.5	280 35.7
14	Bellatrix	1.6	2 46	N 6 21.6	278 32.9
15	Elnath	1.7	2 47	N28 37.0	278 13.7
16	Alnilam	1.7	2 57	S 1 11.7	275 47.2
17	Betelgeuse	0.1 −1.2	3 16	N 7 24.5	271 02.2
18	Canopus	−0.7	3 44	S52 42.0	263 56.2
19	Sirius	−1.5	4 05	S16 43.9	258 34.5
20	Adhara	1.5	4 19	S28 59.2	255 13.2
21	Castor	1.6	4 55	N31 51.5	246 09.2
22	Procyon	0.4	4 59	N 5 11.6	245 00.7
23	Pollux	1.1	5 05	N27 59.7	243 28.9
24	Avior	1.9	5 42	S59 32.7	234 18.3
25	Suhail	2.2	6 28	S43 28.7	222 53.2
26	Miaplacidus	1.7	6 33	S69 45.7	221 39.9
27	Alphard	2.0	6 47	S 8 42.6	217 57.2
28	Regulus	1.4	7 28	N11 54.4	207 44.8
29	Dubhe	1.8	8 23	N61 40.9	193 53.3
30	Denebola	2.1	9 08	N14 30.2	182 35.0
31	Gienah	2.6	9 35	S17 36.4	175 53.7
32	Acrux	1.3	9 46	S63 09.7	173 11.1
33	Gacrux	1.6	9 50	S57 10.6	172 02.6
34	Mimosa	1.3	10 07	S59 45.1	167 53.8
35	Alioth	1.8	10 13	N55 57.1	166 22.1
36	Spica	1.0	10 44	S11 13.3	158 32.7
37	Alkaid	1.9	11 06	N49 15.1	153 00.2
38	Hadar	0.6	11 23	S60 25.7	148 50.2
39	Menkent	2.1	11 26	S36 25.6	148 09.3
40	Arcturus	0.0	11 35	N19 07.2	145 57.1
41	Rigil Kent	−0.3	11 59	S60 52.9	139 54.0
42	Zuben'ubi	2.8	12 10	S16 05.4	137 07.0
43	Kochab	2.1	12 09	N74 06.3	137 21.1
44	Alphecca	2.2	12 53	N26 40.6	126 12.3
45	Antares	1.0	13 48	S26 27.4	112 28.0
46	Atria	1.9	14 08	S69 02.9	107 31.4
47	Sabik	2.4	14 29	S15 44.3	102 14.2
48	Shaula	1.6	14 52	S37 06.7	96 23.8
49	Rasalhague	2.1	14 53	N12 33.3	96 07.8
50	Eltanin	2.2	15 15	N51 29.5	90 47.1
51	Kaus Aust	1.9	15 43	S34 22.7	83 45.6
52	Vega	0.0	15 55	N38 48.0	80 40.0
53	Nunki	2.0	16 14	S26 16.8	76 00.0
54	Altair	0.8	17 09	N 8 54.2	62 09.5
55	Peacock	1.9	17 44	S56 41.9	53 21.2
56	Deneb	1.3	17 59	N45 19.8	49 32.4
57	Enif	2.4	19 02	N 9 56.0	33 48.2
58	Al Na'ir	1.7	19 26	S46 54.3	27 45.1
59	Fomalhaut	1.2	20 15	S29 33.5	15 25.1
60	Markab	2.5	20 22	N15 16.4	13 39.3

SUN AND MOON

SUN

Yr	Mth	Day of Week	Transit h m	Semi-Diam	Twilight h m	Sunrise h m	Sunset h m (Lat 52°N)	Twilight h m
305	1	Tu	11 44	16.1	06 19	06 54	16 32	17 08
306	2	Wed	11 44	16.1	06 20	06 56	16 31	17 06
307	3	Th	11 44	16.1	06 22	06 58	16 29	17 05
308	4	Fri	11 44	16.2	06 24	06 59	16 27	17 03
309	5	Sat	11 44	16.2	06 25	07 01	16 25	17 01
310	6	Sun	11 44	16.2	06 27	07 03	16 23	17 00
311	7	Mon	11 44	16.2	06 29	07 05	16 22	16 58
312	8	Tu	11 44	16.2	06 30	07 07	16 20	16 57
313	9	Wed	11 44	16.2	06 32	07 08	16 18	16 55
314	10	Th	11 44	16.2	06 34	07 10	16 17	16 54
315	11	Fri	11 44	16.2	06 35	07 12	16 15	16 52
316	12	Sat	11 44	16.2	06 37	07 14	16 14	16 51
317	13	Sun	11 44	16.2	06 38	07 16	16 12	16 49
318	14	Mon	11 44	16.2	06 40	07 17	16 11	16 48
319	15	Tu	11 45	16.2	06 42	07 19	16 09	16 47
320	16	Wed	11 45	16.2	06 43	07 21	16 08	16 46
321	17	Th	11 45	16.2	06 45	07 23	16 07	16 44
322	18	Fri	11 45	16.2	06 46	07 24	16 05	16 43
323	19	Sat	11 45	16.2	06 48	07 26	16 04	16 42
324	20	Sun	11 46	16.2	06 50	07 28	16 03	16 41
325	21	Mon	11 46	16.2	06 51	07 29	16 02	16 40
326	22	Tu	11 46	16.2	06 53	07 31	16 01	16 39
327	23	Wed	11 46	16.2	06 54	07 33	15 59	16 38
328	24	Th	11 47	16.2	06 56	07 34	15 58	16 37
329	25	Fri	11 47	16.2	06 57	07 36	15 57	16 36
330	26	Sat	11 47	16.2	06 59	07 37	15 56	16 35
331	27	Sun	11 48	16.2	07 00	07 39	15 56	16 35
332	28	Mon	11 48	16.2	07 01	07 41	15 55	16 34
333	29	Tu	11 48	16.2	07 03	07 42	15 54	16 33
334	30	Wed	11 49	16.2	07 04	07 44	15 53	16 33

MOON

Yr	Mth	Day of Week	Age days	Transit (Upper) h m	Diff m	Semi-diam	Hor Par	Moonrise h m (Lat 52°N)	Moonset h m
305	1	Tu	06	17 12	49	15.6	57.2	12 41	21 51
306	2	Wed	07	18 01	45	15.4	56.3	13 08	23 03
307	3	Th	08	18 46	44	15.1	55.6	13 30	—
308	4	Fri	09	19 30	41	15.0	55.0	13 49	00 14
309	5	Sat	10	20 11	41	14.9	54.5	14 06	01 22
310	6	Sun	11	20 52	42	14.8	54.2	14 23	02 29
311	7	Mon	12	21 34	43	14.7	54.0	14 41	03 35
312	8	Tu	13	22 17	44	14.7	54.0	15 00	04 41
313	9	Wed	14	23 01	47	14.8	54.2	15 22	05 48
314	10	Th	15	23 48	48	14.8	54.2	15 49	06 53
315	11	Fri	16	24 36	—	14.8	54.4	16 22	07 57
316	12	Sat	17	00 36	50	14.9	54.7	17 03	08 58
317	13	Sun	18	01 26	51	15.0	55.0	17 53	09 51
318	14	Mon	19	02 17	52	15.1	55.5	18 52	10 37
319	15	Tu	20	03 09	50	15.2	56.0	19 59	11 15
320	16	Wed	21	03 59	50	15.4	56.6	21 10	11 47
321	17	Th	22	04 49	49	15.6	57.2	22 25	12 13
322	18	Fri	23	05 38	49	15.8	58.0	23 41	12 36
323	19	Sat	24	06 27	49	16.0	58.7	—	12 57
324	20	Sun	25	07 16	51	16.2	59.5	01 00	13 18
325	21	Mon	26	08 07	54	16.4	60.1	02 22	13 40
326	22	Tu	27	09 01	57	16.5	60.7	03 46	14 05
327	23	Wed	28	09 58	60	16.6	60.9	05 11	14 35
328	24	Th	29	10 58	63	16.6	60.9	06 36	15 13
329	25	Fri	00	12 01	62	16.5	60.6	07 55	16 03
330	26	Sat	01	13 03	60	16.4	60.0	09 03	17 04
331	27	Sun	02	14 03	57	16.1	59.3	09 57	18 15
332	28	Mon	03	15 00	52	15.9	58.3	10 37	19 30
333	29	Tu	04	15 52	48	15.6	57.4	11 08	20 45
334	30	Wed	05	16 40	45	15.4	56.5	11 33	21 58

Lat Corr to Sunrise, Sunset etc.

Lat °	Twilight h m	Sunrise h m	Sunset h m	Twilight h m
N70	+1 27	+2 27	−2 27	−1 27
68	+1 09	+1 52	−1 52	−1 10
66	+0 56	+1 27	−1 26	−0 57
64	+0 44	+1 07	−1 07	−0 45
62	+0 34	+0 51	−0 51	−0 35
N60	+0 26	+0 38	−0 38	−0 26
58	+0 19	+0 26	−0 26	−0 18
56	+0 12	+0 17	−0 16	−0 12
54	+0 06	+0 08	−0 08	−0 06
50	−0 06	−0 07	+0 07	+0 06
N45	−0 17	−0 22	+0 22	+0 16
40	−0 26	−0 34	+0 34	+0 26
35	−0 34	−0 45	+0 45	+0 34
30	−0 42	−0 54	+0 54	+0 42
20	−0 56	−1 10	+1 10	+0 57
N10	−1 09	−1 24	+1 25	+1 10
0	−1 21	−1 37	+1 38	+1 22
S10	−1 36	−1 50	+1 51	+1 35
20	−1 52	−2 05	+2 06	+1 51
30	−2 10	−2 21	+2 23	+2 10
35	−2 22	−2 31	+2 33	+2 22
40	−2 36	−2 42	+2 44	+2 36
45	−2 52	−2 55	+2 57	+2 53
S50	−3 14	−3 12	+3 13	+3 13

NOTES

The corrections to sunrise etc. are for middle of November.

Phases of the Moon

		d	h	m
☽	First Quarter	2	16	38
○	Full Moon	10	20	16
☾	Last Quarter	18	15	09
●	New Moon	25	06	10

	d	h
Apogee	8	13
Perigee	23	23

GMT	SUN GHA ° ′	SUN Dec ° ′	ARIES GHA ° ′
Thursday, 1st December			
00	182 49.1	**S21 42.5**	69 30.6
02	212 48.6	**21 43.3**	99 35.5
04	242 48.1	**21 44.1**	129 40.4
06	272 47.7	**21 44.9**	159 45.3
08	302 47.2	**21 45.7**	189 50.3
10	332 46.8	**21 46.4**	219 55.2
12	2 46.3	**21 47.2**	250 00.1
14	32 45.8	**21 48.0**	280 05.0
16	62 45.4	**21 48.8**	310 10.0
18	92 44.9	**21 49.6**	340 14.9
20	122 44.4	**21 50.3**	10 19.8
22	152 44.0	**S21 51.1**	40 24.8
Friday, 2nd December			
00	182 43.5	**S21 51.9**	70 29.7
02	212 43.0	**21 52.6**	100 34.6
04	242 42.5	**21 53.4**	130 39.5
06	272 42.1	**21 54.1**	160 44.5
08	302 41.6	**21 54.9**	190 49.4
10	332 41.1	**21 55.6**	220 54.3
12	2 40.6	**21 56.4**	250 59.3
14	32 40.2	**21 57.1**	281 04.2
16	62 39.7	**21 57.9**	311 09.1
18	92 39.2	**21 58.6**	341 14.0
20	122 38.7	**21 59.3**	11 19.0
22	152 38.2	**S22 00.1**	41 23.9
Saturday, 3rd December			
00	182 37.8	**S22 00.8**	71 28.8
02	212 37.3	**22 01.5**	101 33.8
04	242 36.8	**22 02.2**	131 38.7
06	272 36.3	**22 03.0**	161 43.6
08	302 35.8	**22 03.7**	191 48.5
10	332 35.3	**22 04.4**	221 53.5
12	2 34.8	**22 05.1**	251 58.4
14	32 34.3	**22 05.8**	282 03.3
16	62 33.9	**22 06.5**	312 08.3
18	92 33.4	**22 07.2**	342 13.2
20	122 32.9	**22 07.9**	12 18.1
22	152 32.4	**S22 08.6**	42 23.0
Sunday, 4th December			
00	182 31.9	**S22 09.3**	72 28.0
02	212 31.4	**22 10.0**	102 32.9
04	242 30.9	**22 10.7**	132 37.8
06	272 30.4	**22 11.4**	162 42.8
08	302 29.9	**22 12.1**	192 47.7
10	332 29.4	**22 12.7**	222 52.6
12	2 28.9	**22 13.4**	252 57.5
14	32 28.4	**22 14.1**	283 02.5
16	62 27.9	**22 14.8**	313 07.4
18	92 27.4	**22 15.4**	343 12.3
20	122 26.9	**22 16.1**	13 17.2
22	152 26.4	**S22 16.7**	43 22.2
Monday, 5th December			
00	182 25.9	**S22 17.4**	73 27.1
02	212 25.3	**22 18.1**	103 32.0
04	242 24.8	**22 18.7**	133 37.0
06	272 24.3	**22 19.4**	163 41.9
08	302 23.8	**22 20.0**	193 46.8
10	332 23.3	**22 20.6**	223 51.7
12	2 22.8	**22 21.3**	253 56.7
14	32 22.3	**22 21.9**	284 01.6
16	62 21.8	**22 22.6**	314 06.5
18	92 21.2	**22 23.2**	344 11.5
20	122 20.7	**22 23.8**	14 16.4
22	152 20.2	**S22 24.4**	44 21.3

GMT	SUN GHA ° ′	SUN Dec ° ′	ARIES GHA ° ′
Tuesday, 6th December			
00	182 19.7	**S22 25.1**	74 26.2
02	212 19.2	**22 25.7**	104 31.2
04	242 18.7	**22 26.3**	134 36.1
06	272 18.1	**22 26.9**	164 41.0
08	302 17.6	**22 27.5**	194 46.0
10	332 17.1	**22 28.1**	224 50.9
12	2 16.6	**22 28.7**	254 55.8
14	32 16.0	**22 29.3**	285 00.7
16	62 15.5	**22 29.9**	315 05.7
18	92 15.0	**22 30.5**	345 10.6
20	122 14.5	**22 31.1**	15 15.5
22	152 13.9	**S22 31.7**	45 20.5
Wednesday, 7th December			
00	182 13.4	**S22 32.3**	75 25.4
02	212 12.9	**22 32.9**	105 30.3
04	242 12.3	**22 33.4**	135 35.2
06	272 11.8	**22 34.0**	165 40.2
08	302 11.3	**22 34.6**	195 45.1
10	332 10.7	**22 35.2**	225 50.0
12	2 10.2	**22 35.7**	255 55.0
14	32 09.7	**22 36.3**	285 59.9
16	62 09.1	**22 36.9**	316 04.8
18	92 08.6	**22 37.4**	346 09.7
20	122 08.1	**22 38.0**	16 14.7
22	152 07.5	**S22 38.5**	46 19.6
Thursday, 8th December			
00	182 07.0	**S22 39.1**	76 24.5
02	212 06.4	**22 39.6**	106 29.4
04	242 05.9	**22 40.2**	136 34.4
06	272 05.4	**22 40.7**	166 39.3
08	302 04.8	**22 41.2**	196 44.2
10	332 04.3	**22 41.8**	226 49.2
12	2 03.7	**22 42.3**	256 54.1
14	32 03.2	**22 42.8**	286 59.0
16	62 02.6	**22 43.3**	317 03.9
18	92 02.1	**22 43.9**	347 08.9
20	122 01.5	**22 44.4**	17 13.8
22	152 01.0	**S22 44.9**	47 18.7
Friday, 9th December			
00	182 00.4	**S22 45.4**	77 23.7
02	211 59.9	**22 45.9**	107 28.6
04	241 59.3	**22 46.4**	137 33.5
06	271 58.8	**22 46.9**	167 38.4
08	301 58.2	**22 47.4**	197 43.4
10	331 57.7	**22 47.9**	227 48.3
12	1 57.1	**22 48.4**	257 53.2
14	31 56.6	**22 48.9**	287 58.2
16	61 56.0	**22 49.4**	318 03.1
18	91 55.5	**22 49.9**	348 08.0
20	121 54.9	**22 50.3**	18 12.9
22	151 54.4	**S22 50.8**	48 17.9
Saturday, 10th December			
00	181 53.8	**S22 51.3**	78 22.8
02	211 53.2	**22 51.8**	108 27.7
04	241 52.7	**22 52.2**	138 32.7
06	271 52.1	**22 52.7**	168 37.6
08	301 51.6	**22 53.2**	198 42.5
10	331 51.0	**22 53.6**	228 47.4
12	1 50.4	**22 54.1**	258 52.4
14	31 49.9	**22 54.5**	288 57.3
16	61 49.3	**22 55.0**	319 02.2
18	91 48.7	**22 55.4**	349 07.2
20	121 48.2	**22 55.9**	19 12.1
22	151 47.6	**S22 56.3**	49 17.0

GMT	SUN GHA ° ′	SUN Dec ° ′	ARIES GHA ° ′
Sunday, 11th December			
00	181 47.0	**S22 56.7**	79 21.9
02	211 46.5	**22 57.2**	109 26.9
04	241 45.9	**22 57.6**	139 31.8
06	271 45.3	**22 58.0**	169 36.7
08	301 44.8	**22 58.5**	199 41.7
10	331 44.2	**22 58.9**	229 46.6
12	1 43.6	**22 59.3**	259 51.5
14	31 43.0	**22 59.7**	289 56.4
16	61 42.5	**23 00.1**	320 01.4
18	91 41.9	**23 00.5**	350 06.3
20	121 41.3	**23 00.9**	20 11.2
22	151 40.8	**S23 01.3**	50 16.2
Monday, 12th December			
00	181 40.2	**S23 01.7**	80 21.1
02	211 39.6	**23 02.1**	110 26.0
04	241 39.0	**23 02.5**	140 30.9
06	271 38.4	**23 02.9**	170 35.9
08	301 37.9	**23 03.3**	200 40.8
10	331 37.3	**23 03.7**	230 45.7
12	1 36.7	**23 04.1**	260 50.7
14	31 36.1	**23 04.4**	290 55.6
16	61 35.6	**23 04.8**	321 00.5
18	91 35.0	**23 05.2**	351 05.4
20	121 34.4	**23 05.5**	21 10.4
22	151 33.8	**S23 05.9**	51 15.3
Tuesday, 13th December			
00	181 33.2	**S23 06.3**	81 20.2
02	211 32.6	**23 06.6**	111 25.2
04	241 32.1	**23 07.0**	141 30.1
06	271 31.5	**23 07.3**	171 35.0
08	301 30.9	**23 07.7**	201 39.9
10	331 30.3	**23 08.0**	231 44.9
12	1 29.7	**23 08.4**	261 49.8
14	31 29.1	**23 08.7**	291 54.7
16	61 28.5	**23 09.0**	321 59.6
18	91 28.0	**23 09.4**	352 04.6
20	121 27.4	**23 09.7**	22 09.5
22	151 26.8	**S23 10.0**	52 14.4
Wednesday, 14th December			
00	181 26.2	**S23 10.3**	82 19.4
02	211 25.6	**23 10.7**	112 24.3
04	241 25.0	**23 11.0**	142 29.2
06	271 24.4	**23 11.3**	172 34.1
08	301 23.8	**23 11.6**	202 39.1
10	331 23.2	**23 11.9**	232 44.0
12	1 22.6	**23 12.2**	262 48.9
14	31 22.0	**23 12.5**	292 53.9
16	61 21.4	**23 12.8**	322 58.8
18	91 20.8	**23 13.1**	353 03.7
20	121 20.3	**23 13.4**	23 08.6
22	151 19.7	**S23 13.7**	53 13.6
Thursday, 15th December			
00	181 19.1	**S23 14.0**	83 18.5
02	211 18.5	**23 14.2**	113 23.4
04	241 17.9	**23 14.5**	143 28.4
06	271 17.3	**23 14.8**	173 33.3
08	301 16.7	**23 15.1**	203 38.2
10	331 16.1	**23 15.3**	233 43.1
12	1 15.5	**23 15.6**	263 48.1
14	31 14.9	**23 15.8**	293 53.0
16	61 14.3	**23 16.1**	323 57.9
18	91 13.7	**23 16.4**	354 02.9
20	121 13.1	**23 16.6**	24 07.8
22	151 12.5	**S23 16.9**	54 12.7

GMT	SUN GHA ° ′	SUN Dec ° ′	ARIES GHA ° ′
Friday, 16th December			
00	181 11.9	**S23 17.1**	84 17.6
02	211 11.3	**23 17.3**	114 22.6
04	241 10.7	**23 17.6**	144 27.5
06	271 10.1	**23 17.8**	174 32.4
08	301 09.5	**23 18.1**	204 37.4
10	331 08.8	**23 18.3**	234 42.3
12	1 08.2	**23 18.5**	264 47.2
14	31 07.6	**23 18.7**	294 52.1
16	61 07.0	**23 18.9**	324 57.1
18	91 06.4	**23 19.2**	355 02.0
20	121 05.8	**23 19.4**	25 06.9
22	151 05.2	**S23 19.6**	55 11.9
Saturday, 17th December			
00	181 04.6	**S23 19.8**	85 16.8
02	211 04.0	**23 20.0**	115 21.7
04	241 03.4	**23 20.2**	145 26.6
06	271 02.8	**23 20.4**	175 31.6
08	301 02.2	**23 20.6**	205 36.5
10	331 01.6	**23 20.8**	235 41.4
12	1 01.0	**23 21.0**	265 46.3
14	31 00.3	**23 21.1**	295 51.3
16	60 59.7	**23 21.3**	325 56.2
18	90 59.1	**23 21.5**	356 01.1
20	120 58.5	**23 21.7**	26 06.1
22	150 57.9	**S23 21.8**	56 11.0
Sunday, 18th December			
00	180 57.3	**S23 22.0**	86 15.9
02	210 56.7	**23 22.2**	116 20.8
04	240 56.1	**23 22.3**	146 25.8
06	270 55.4	**23 22.5**	176 30.7
08	300 54.8	**23 22.6**	206 35.6
10	330 54.2	**23 22.8**	236 40.6
12	0 53.6	**23 22.9**	266 45.5
14	30 53.0	**23 23.1**	296 50.4
16	60 52.4	**23 23.2**	326 55.3
18	90 51.8	**23 23.4**	357 00.3
20	120 51.1	**23 23.5**	27 05.2
22	150 50.5	**S23 23.6**	57 10.1
Monday, 19th December			
00	180 49.9	**S23 23.8**	87 15.1
02	210 49.3	**23 23.9**	117 20.0
04	240 48.7	**23 24.0**	147 24.9
06	270 48.1	**23 24.1**	177 29.8
08	300 47.4	**23 24.2**	207 34.8
10	330 46.8	**23 24.4**	237 39.7
12	0 46.2	**23 24.5**	267 44.6
14	30 45.6	**23 24.6**	297 49.6
16	60 45.0	**23 24.7**	327 54.5
18	90 44.4	**23 24.8**	357 59.4
20	120 43.7	**23 24.9**	28 04.3
22	150 43.1	**S23 25.0**	58 09.3
Tuesday, 20th December			
00	180 42.5	**S23 25.0**	88 14.2
02	210 41.9	**23 25.1**	118 19.1
04	240 41.3	**23 25.2**	148 24.1
06	270 40.6	**23 25.3**	178 29.0
08	300 40.0	**23 25.4**	208 33.9
10	330 39.4	**23 25.4**	238 38.8
12	0 38.8	**23 25.5**	268 43.8
14	30 38.2	**23 25.6**	298 48.7
16	60 37.5	**23 25.6**	328 53.6
18	90 36.9	**23 25.7**	358 58.5
20	120 36.3	**23 25.8**	29 03.5
22	150 35.7	**S23 25.8**	59 08.4

GMT	SUN GHA ° ′	SUN Dec ° ′	ARIES GHA ° ′
Wednesday, 21st December			
00	180 35.1	**S23 25.9**	89 13.3
02	210 34.4	**23 25.9**	119 18.3
04	240 33.8	**23 25.9**	149 23.2
06	270 33.2	**23 26.0**	179 28.1
08	300 32.6	**23 26.0**	209 33.0
10	330 31.9	**23 26.1**	239 38.0
12	0 31.3	**23 26.1**	269 42.9
14	30 30.7	**23 26.1**	299 47.8
16	60 30.1	**23 26.1**	329 52.8
18	90 29.5	**23 26.2**	359 57.7
20	120 28.8	**23 26.2**	30 02.6
22	150 28.2	**S23 26.2**	60 07.5
Thursday, 22nd December			
00	180 27.6	**S23 26.2**	90 12.5
02	210 27.0	**23 26.2**	120 17.4
04	240 26.3	**23 26.2**	150 22.3
06	270 25.7	**23 26.2**	180 27.3
08	300 25.1	**23 26.2**	210 32.2
10	330 24.5	**23 26.2**	240 37.1
12	0 23.8	**23 26.2**	270 42.0
14	30 23.2	**23 26.2**	300 47.0
16	60 22.6	**23 26.2**	330 51.9
18	90 22.0	**23 26.1**	0 56.8
20	120 21.4	**23 26.1**	31 01.8
22	150 20.7	**S23 26.1**	61 06.7
Friday, 23rd December			
00	180 20.1	**S23 26.1**	91 11.6
02	210 19.5	**23 26.0**	121 16.5
04	240 18.9	**23 26.0**	151 21.5
06	270 18.2	**23 26.0**	181 26.4
08	300 17.6	**23 25.9**	211 31.3
10	330 17.0	**23 25.9**	241 36.3
12	0 16.4	**23 25.8**	271 41.2
14	30 15.7	**23 25.8**	301 46.1
16	60 15.1	**23 25.7**	331 51.0
18	90 14.5	**23 25.7**	1 56.0
20	120 13.9	**23 25.6**	32 00.9
22	150 13.2	**S23 25.5**	62 05.8
Saturday, 24th December			
00	180 12.6	**S23 25.5**	92 10.8
02	210 12.0	**23 25.4**	122 15.7
04	240 11.4	**23 25.3**	152 20.6
06	270 10.7	**23 25.2**	182 25.5
08	300 10.1	**23 25.2**	212 30.5
10	330 09.5	**23 25.1**	242 35.4
12	0 08.9	**23 25.0**	272 40.3
14	30 08.2	**23 24.9**	302 45.3
16	60 07.6	**23 24.8**	332 50.2
18	90 07.0	**23 24.7**	2 55.1
20	120 06.4	**23 24.6**	33 00.0
22	150 05.8	**S23 24.5**	63 05.0
Sunday, 25th December			
00	180 05.1	**S23 24.4**	93 09.9
02	210 04.5	**23 24.3**	123 14.8
04	240 03.9	**23 24.2**	153 19.8
06	270 03.3	**23 24.0**	183 24.7
08	300 02.6	**23 23.9**	213 29.6
10	330 02.0	**23 23.8**	243 34.5
12	0 01.4	**23 23.7**	273 39.5
14	30 00.8	**23 23.5**	303 44.4
16	60 00.2	**23 23.4**	333 49.3
18	89 59.5	**23 23.3**	3 54.3
20	119 58.9	**23 23.1**	33 59.2
22	149 58.3	**S23 23.0**	64 04.1

GMT	SUN GHA ° ′	SUN Dec ° ′	ARIES GHA ° ′
Monday, 26th December			
00	179 57.7	**S23 22.8**	94 09.0
02	209 57.0	**23 22.7**	124 14.0
04	239 56.4	**23 22.5**	154 18.9
06	269 55.8	**23 22.4**	184 23.8
08	299 55.2	**23 22.2**	214 28.8
10	329 54.6	**23 22.1**	244 33.7
12	359 53.9	**23 21.9**	274 38.6
14	29 53.3	**23 21.7**	304 43.5
16	59 52.7	**23 21.6**	334 48.5
18	89 52.1	**23 21.4**	4 53.4
20	119 51.5	**23 21.2**	34 58.3
22	149 50.8	**S23 21.0**	65 03.3
Tuesday, 27th December			
00	179 50.2	**S23 20.8**	95 08.2
02	209 49.6	**23 20.6**	125 13.1
04	239 49.0	**23 20.5**	155 18.0
06	269 48.4	**23 20.3**	185 23.0
08	299 47.8	**23 20.1**	215 27.9
10	329 47.1	**23 19.9**	245 32.8
12	359 46.5	**23 19.7**	275 37.7
14	29 45.9	**23 19.4**	305 42.7
16	59 45.3	**23 19.2**	335 47.6
18	89 44.7	**23 19.0**	5 52.5
20	119 44.0	**23 18.8**	35 57.5
22	149 43.4	**S23 18.6**	66 02.4
Wednesday, 28th December			
00	179 42.8	**S23 18.4**	96 07.3
02	209 42.2	**23 18.1**	126 12.2
04	239 41.6	**23 17.9**	156 17.2
06	269 41.0	**23 17.7**	186 22.1
08	299 40.4	**23 17.4**	216 27.0
10	329 39.7	**23 17.2**	246 32.0
12	359 39.1	**23 16.9**	276 36.9
14	29 38.5	**23 16.7**	306 41.8
16	59 37.9	**23 16.4**	336 46.7
18	89 37.3	**23 16.2**	6 51.7
20	119 36.7	**23 15.9**	36 56.6
22	149 36.1	**S23 15.7**	67 01.5
Thursday, 29th December			
00	179 35.5	**S23 15.4**	97 06.5
02	209 34.8	**23 15.1**	127 11.4
04	239 34.2	**23 14.9**	157 16.3
06	269 33.6	**23 14.6**	187 21.2
08	299 33.0	**23 14.3**	217 26.2
10	329 32.4	**23 14.0**	247 31.1
12	359 31.8	**23 13.8**	277 36.0
14	29 31.2	**23 13.5**	307 41.0
16	59 30.6	**23 13.2**	337 45.9
18	89 30.0	**23 12.9**	7 50.8
20	119 29.4	**23 12.6**	37 55.7
22	149 28.8	**S23 12.3**	68 00.7
Friday, 30th December			
00	179 28.1	**S23 12.0**	98 05.6
02	209 27.5	**23 11.7**	128 10.5
04	239 26.9	**23 11.4**	158 15.5
06	269 26.3	**23 11.1**	188 20.4
08	299 25.7	**23 10.8**	218 25.3
10	329 25.1	**23 10.4**	248 30.2
12	359 24.5	**23 10.1**	278 35.2
14	29 23.9	**23 09.8**	308 40.1
16	59 23.3	**23 09.5**	338 45.0
18	89 22.7	**23 09.1**	8 49.9
20	119 22.1	**23 08.8**	38 54.9
22	149 21.5	**S23 08.5**	68 59.8

GMT	SUN GHA ° ′	SUN Dec ° ′	ARIES GHA ° ′
Saturday, 31st December			
00	179 20.9	**S23 08.1**	99 04.7
02	209 20.3	**23 07.8**	129 09.7
04	239 19.7	**23 07.4**	159 14.6
06	269 19.1	**S23 07.1**	189 19.5
08	299 18.5	**S23 06.7**	219 24.4
10	329 17.9	**23 06.4**	249 29.4
12	359 17.3	**23 06.0**	279 34.3
14	29 16.7	**S23 05.6**	309 39.2
16	59 16.1	**S23 05.3**	339 44.2
18	89 15.5	**23 04.9**	9 49.1
20	119 14.9	**23 04.5**	39 54.0
22	149 14.3	**S23 04.2**	69 58.9

DECEMBER 2011

MOON

Day	GMT hr	GHA ° ′	Mean Var/hr 14°+	Dec ° ′	Mean Var/hr ′
1 Th	0	106 30.4	32.7	S 9 26.7	11.2
	6	193 46.1	33.0	S 8 19.4	11.3
	12	281 04.1	33.4	S 7 11.3	11.5
	18	8 24.1	33.7	S 6 02.5	11.6
2 Fri	0	95 46.0	33.9	S 4 53.3	11.6
	6	183 09.4	34.2	S 3 43.7	11.6
	12	270 34.3	34.4	S 2 34.0	11.6
	18	358 00.4	34.5	S 1 24.3	11.6
3 Sat	0	85 27.5	34.6	S 0 14.7	11.6
	6	172 55.3	34.7	N 0 54.7	11.5
	12	260 23.6	34.8	N 2 03.6	11.4
	18	347 52.3	34.8	N 3 12.0	11.2
4 Sun	0	75 21.1	34.8	N 4 19.8	11.2
	6	162 49.9	34.7	N 5 26.7	11.0
	12	250 18.4	34.7	N 6 32.8	10.8
	18	337 46.5	34.6	N 7 37.8	10.7
5 Mon	0	65 14.0	34.5	N 8 41.7	10.4
	6	152 40.7	34.3	N 9 44.3	10.2
	12	240 06.4	34.1	N10 45.5	9.9
	18	327 31.1	33.9	N11 45.1	9.7
6 Tu	0	54 54.5	33.7	N12 43.1	9.3
	6	142 16.6	33.4	N13 39.2	9.0
	12	229 37.1	33.2	N14 33.4	8.6
	18	316 56.1	32.8	N15 25.5	8.2
7 Wed	0	44 13.4	32.5	N16 15.3	7.8
	6	131 28.9	32.3	N17 02.7	7.5
	12	218 42.5	31.9	N17 47.5	7.0
	18	305 54.3	31.6	N18 29.6	6.6
8 Th	0	33 04.2	31.3	N19 08.9	6.0
	6	120 12.2	31.0	N19 45.2	5.5
	12	207 18.3	30.7	N20 18.3	4.9
	18	294 22.6	30.4	N20 48.1	4.3
9 Fri	0	21 25.1	30.1	N21 14.4	3.8
	6	108 25.8	29.9	N21 37.1	3.2
	12	195 25.0	29.6	N21 56.2	2.5
	18	282 22.8	29.4	N22 11.4	1.9
10 Sat	0	9 19.2	29.2	N22 22.7	1.2
	6	96 14.4	29.0	N22 30.0	0.4
	12	183 08.7	28.9	N22 33.1	0.3
	18	270 02.2	28.8	N22 32.1	0.9
11 Sun	0	356 55.1	28.8	N22 27.0	1.6
	6	83 47.5	28.7	N22 17.6	2.3
	12	170 39.8	28.7	N22 04.0	3.1
	18	257 32.0	28.7	N21 46.1	3.7
12 Mon	0	344 24.5	28.8	N21 24.1	4.5
	6	71 17.2	28.8	N20 58.0	5.1
	12	158 10.5	29.0	N20 27.8	5.8
	18	245 04.3	29.1	N19 53.6	6.4
13 Tu	0	331 59.0	29.3	N19 15.5	7.0
	6	58 54.5	29.4	N18 33.7	7.6
	12	145 50.9	29.6	N17 48.2	8.2
	18	232 48.4	29.7	N16 59.3	8.8
14 Wed	0	319 46.8	30.0	N16 07.0	9.3
	6	46 46.3	30.1	N15 11.5	9.8
	12	133 46.7	30.3	N14 13.1	10.2
	18	220 48.2	30.4	N13 11.7	10.7
15 Th	0	307 50.5	30.6	N12 07.8	11.1
	6	34 53.6	30.6	N11 01.4	11.5
	12	121 57.4	30.7	N 9 52.7	11.8
	18	209 01.8	30.8	N 8 41.9	12.1
16 Fri	0	296 06.6	30.9	N 7 29.3	12.4
	6	23 11.7	30.9	N 6 15.0	12.6
	12	110 16.9	30.9	N 4 59.3	12.9
	18	197 22.1	30.8	N 3 42.3	13.0
17 Sat	0	284 27.0	30.7	N 2 24.4	13.1
	6	11 31.4	30.7	N 1 05.7	13.2
	12	98 35.3	30.4	S 0 13.6	13.3
	18	185 38.2	30.3	S 1 33.1	13.3
18 Sun	0	272 40.1	30.1	S 2 52.7	13.2
	6	359 40.7	29.8	S 4 12.1	13.1
	12	86 39.9	29.6	S 5 31.0	13.1
	18	173 37.3	29.2	S 6 49.2	12.9
19 Mon	0	260 32.9	28.9	S 8 06.4	12.6
	6	347 26.4	28.5	S 9 22.2	12.3
	12	74 17.7	28.1	S10 36.4	12.0
	18	161 06.5	27.7	S11 48.6	11.6
20 Tu	0	247 52.9	27.3	S12 58.7	11.2
	6	334 36.5	26.8	S14 06.1	10.7
	12	61 17.5	26.3	S15 10.7	10.2
	18	147 55.8	25.8	S16 12.0	9.6
21 Wed	0	234 31.3	25.4	S17 09.8	8.9
	6	321 04.1	25.0	S18 03.7	8.2
	12	47 34.2	24.6	S18 53.5	7.5
	18	134 01.9	24.2	S19 38.8	6.7
22 Th	0	220 27.4	23.9	S20 19.3	5.9
	6	306 50.7	23.6	S20 54.9	5.0
	12	33 12.4	23.3	S21 25.2	4.1
	18	119 32.6	23.2	S21 50.1	3.1
23 Fri	0	205 51.8	23.1	S22 09.4	2.2
	6	292 10.4	23.0	S22 23.1	1.2
	12	18 28.8	23.1	S22 31.0	0.2
	18	104 47.4	23.3	S22 33.2	0.7
24 Sat	0	191 06.7	23.4	S22 29.7	1.6
	6	277 27.1	23.7	S22 20.5	2.5
	12	3 49.1	24.0	S22 05.8	3.4
	18	90 13.0	24.4	S21 45.7	4.3
25 Sun	0	176 39.2	24.8	S21 20.5	5.1
	6	263 08.0	25.3	S20 50.3	5.9
	12	349 39.6	25.8	S20 15.4	6.7
	18	76 14.3	26.3	S19 36.1	7.3
26 Mon	0	162 52.3	26.9	S18 52.6	8.0
	6	249 33.5	27.5	S18 05.4	8.5
	12	336 18.3	28.1	S17 14.6	9.0
	18	63 06.4	28.6	S16 20.7	9.5
27 Tu	0	149 58.0	29.2	S15 23.8	9.9
	6	236 52.9	29.8	S14 24.4	10.3
	12	323 51.1	30.3	S13 22.6	10.6
	18	50 52.5	30.7	S12 18.9	11.0
28 Wed	0	137 56.9	31.2	S11 13.4	11.2
	6	225 04.3	31.8	S10 06.5	11.3
	12	312 14.3	32.1	S 8 58.3	11.5
	18	39 26.8	32.5	S 7 49.2	11.7
29 Th	0	126 41.7	32.8	S 6 39.2	11.7
	6	213 58.7	33.1	S 5 28.7	11.8
	12	301 17.7	33.5	S 4 17.9	11.8
	18	28 38.3	33.7	S 3 06.8	11.9
30 Fri	0	116 00.5	33.9	S 1 55.7	11.9
	6	203 24.0	34.1	S 0 44.8	11.8
	12	290 48.5	34.3	N 0 25.8	11.7
	18	18 13.9	34.4	N 1 35.9	11.6
31 Sat	0	105 40.0	34.4	N 2 45.5	11.5
	6	193 06.6	34.5	N 3 54.3	11.4
	12	280 33.5	34.5	N 5 02.3	11.1
	18	8 00.4	34.5	N 6 09.3	11.0

PLANETS

VENUS / JUPITER

Venus Mer Pass h m	Venus GHA ° ′	Venus Mean Var/hr 14°+	Venus Dec ° ′	Venus Mean Var/hr ′	Day	Jupiter GHA ° ′	Jupiter Mean Var/hr 15°+	Jupiter Dec ° ′	Jupiter Mean Var/hr ′	Jupiter Mer Pass h m
13 47	153 23.8	59.1	S24 44.9	0.1	1 Th	39 44.8	2.7	N10 40.2	0.1	21 17
13 49	153 01.1	59.1	S24 43.2	0.1	2 Fri	40 48.7	2.7	N10 38.8	0.1	21 13
13 50	152 38.4	59.1	S24 40.7	0.1	3 Sat	41 52.4	2.6	N10 37.4	0.1	21 09
13 52	152 15.9	59.1	S24 37.5	0.2	4 SUN	42 55.9	2.6	N10 36.1	0.1	21 05
13 53	151 53.4	59.1	S24 33.5	0.2	5 Mon	43 59.3	2.6	N10 34.9	0.0	21 00
13 55	151 31.0	59.1	S24 28.8	0.2	6 Tu	45 02.5	2.6	N10 33.7	0.0	20 56
13 56	151 08.8	59.1	S24 23.4	0.3	7 Wed	46 05.5	2.6	N10 32.6	0.0	20 52
13 58	150 46.7	59.1	S24 17.2	0.3	8 Th	47 08.3	2.6	N10 31.5	0.0	20 48
13 59	150 24.8	59.1	S24 10.2	0.3	9 Fri	48 11.0	2.6	N10 30.6	0.0	20 44
14 01	150 03.0	59.1	S24 02.6	0.3	10 Sat	49 13.4	2.6	N10 29.7	0.0	20 40
14 02	149 41.5	59.1	S23 54.2	0.4	11 SUN	50 15.7	2.6	N10 28.8	0.0	20 35
14 03	149 20.1	59.1	S23 45.1	0.4	12 Mon	51 17.8	2.6	N10 28.1	0.0	20 31
14 05	148 59.0	59.1	S23 35.3	0.4	13 Tu	52 19.7	2.6	N10 27.4	0.0	20 27
14 06	148 38.1	59.1	S23 24.8	0.5	14 Wed	53 21.4	2.6	N10 26.8	0.0	20 23
14 08	148 17.5	59.2	S23 13.6	0.5	15 Th	54 22.9	2.6	N10 26.2	0.0	20 19
14 09	147 57.1	59.2	S23 01.7	0.5	16 Fri	55 24.2	2.5	N10 25.8	0.0	20 15
14 10	147 37.0	59.2	S22 49.1	0.5	17 Sat	56 25.3	2.5	N10 25.4	0.0	20 11
14 12	147 17.2	59.2	S22 35.9	0.6	18 SUN	57 26.2	2.5	N10 25.0	0.0	20 07
14 13	146 57.6	59.2	S22 22.0	0.6	19 Mon	58 27.0	2.5	N10 24.8	0.0	20 03
14 14	146 38.4	59.2	S22 07.4	0.6	20 Tu	59 27.5	2.5	N10 24.6	0.0	19 59
14 15	146 19.4	59.2	S21 52.3	0.7	21 Wed	60 27.8	2.5	N10 24.5	0.0	19 55
14 17	146 00.8	59.2	S21 36.4	0.7	22 Th	61 28.0	2.5	N10 24.4	0.0	19 51
14 18	145 42.5	59.2	S21 20.0	0.7	23 Fri	62 27.9	2.5	N10 24.5	0.0	19 47
14 19	145 24.5	59.3	S21 02.9	0.7	24 Sat	63 27.7	2.5	N10 24.6	0.0	19 43
14 20	145 06.9	59.3	S20 45.2	0.8	25 SUN	64 27.2	2.5	N10 24.8	0.0	19 39
14 21	144 49.6	59.3	S20 27.0	0.8	26 Mon	65 26.5	2.5	N10 25.0	0.0	19 35
14 22	144 32.6	59.3	S20 08.2	0.8	27 Tu	66 25.7	2.5	N10 25.4	0.0	19 31
14 24	144 16.0	59.3	S19 48.8	0.8	28 Wed	67 24.6	2.4	N10 25.8	0.0	19 27
14 25	143 59.8	59.3	S19 28.8	0.8	29 Th	68 23.3	2.4	N10 26.3	0.0	19 23
14 26	143 43.9	59.4	S19 08.4	0.9	30 Fri	69 21.9	2.4	N10 26.8	0.0	19 19
14 27	143 28.4	59.4	S18 47.4	0.9	31 Sat	70 20.2	2.4	N10 27.5	0.0	19 16

VENUS, Av. Mag. –3.9
SHA December 5 78; 10 72; 15 65; 20 58; 25 52; 30 46

JUPITER, Av. Mag. –2.7
SHA December5 331; 10 331; 15 331; 20 331; 25 331; 30 331

MARS / SATURN

Mars Mer Pass h m	Mars GHA ° ′	Mars Mean Var/hr 15°+	Mars Dec ° ′	Mars Mean Var/hr ′	Day	Saturn GHA ° ′	Saturn Mean Var/hr 15°+	Saturn Dec ° ′	Saturn Mean Var/hr ′	Saturn Mer Pass h m
06 08	267 45.9	1.4	N10 05.6	0.4	1 Th	224 49.2	2.2	S 7 46.7	0.1	08 59
06 06	268 20.3	1.4	N 9 57.1	0.4	2 Fri	225 42.7	2.2	S 7 48.7	0.1	08 56
06 04	268 54.9	1.5	N 9 48.6	0.3	3 Sat	226 36.2	2.2	S 7 50.7	0.1	08 52
06 01	269 29.9	1.5	N 9 40.2	0.3	4 SUN	227 29.7	2.2	S 7 52.6	0.1	08 49
05 59	270 05.1	1.5	N 9 31.9	0.3	5 Mon	228 23.3	2.2	S 7 54.6	0.1	08 45
05 57	270 40.6	1.5	N 9 23.7	0.3	6 Tu	229 17.0	2.2	S 7 56.5	0.1	08 42
05 54	271 16.3	1.5	N 9 15.6	0.3	7 Wed	230 10.7	2.2	S 7 58.4	0.1	08 38
05 52	271 52.4	1.5	N 9 07.6	0.3	8 Th	231 04.5	2.2	S 8 00.3	0.1	08 34
05 49	272 28.7	1.5	N 8 59.7	0.3	9 Fri	231 58.3	2.2	S 8 02.1	0.1	08 31
05 47	273 05.4	1.5	N 8 51.9	0.3	10 Sat	232 52.2	2.2	S 8 03.9	0.1	08 27
05 45	273 42.3	1.6	N 8 44.3	0.3	11 SUN	233 46.1	2.3	S 8 05.7	0.1	08 24
05 42	274 19.5	1.6	N 8 36.7	0.3	12 Mon	234 40.1	2.3	S 8 07.4	0.1	08 20
05 40	274 57.1	1.6	N 8 29.3	0.3	13 Tu	235 34.2	2.3	S 8 09.2	0.1	08 16
05 37	275 35.0	1.6	N 8 22.0	0.3	14 Wed	236 28.3	2.3	S 8 10.8	0.1	08 13
05 35	276 13.2	1.6	N 8 14.9	0.3	15 Th	237 22.5	2.3	S 8 12.5	0.1	08 09
05 32	276 51.7	1.6	N 8 07.8	0.3	16 Fri	238 16.8	2.3	S 8 14.2	0.1	08 06
05 29	277 30.5	1.6	N 8 00.9	0.3	17 Sat	239 11.1	2.3	S 8 15.8	0.1	08 02
05 27	278 09.8	1.6	N 7 54.2	0.3	18 SUN	240 05.5	2.3	S 8 17.3	0.1	07 58
05 24	278 49.3	1.7	N 7 47.6	0.3	19 Mon	241 00.0	2.3	S 8 18.9	0.1	07 55
05 21	279 29.2	1.7	N 7 41.1	0.3	20 Tu	241 54.5	2.3	S 8 20.4	0.1	07 51
05 19	280 09.5	1.7	N 7 34.9	0.3	21 Wed	242 49.1	2.3	S 8 21.9	0.1	07 48
05 16	280 50.2	1.7	N 7 28.7	0.2	22 Th	243 43.8	2.3	S 8 23.4	0.1	07 44
05 13	281 31.2	1.7	N 7 22.8	0.2	23 Fri	244 38.5	2.3	S 8 24.8	0.1	07 40
05 11	282 12.7	1.7	N 7 17.0	0.2	24 Sat	245 33.3	2.3	S 8 26.2	0.1	07 37
05 08	282 54.5	1.8	N 7 11.3	0.2	25 SUN	246 28.2	2.3	S 8 27.5	0.1	07 33
05 05	283 36.7	1.8	N 7 05.9	0.2	26 Mon	247 23.2	2.3	S 8 28.9	0.1	07 29
05 02	284 19.4	1.8	N 7 00.6	0.2	27 Tu	248 18.3	2.3	S 8 30.2	0.0	07 26
04 59	285 02.5	1.8	N 6 55.6	0.2	28 Wed	249 13.4	2.3	S 8 31.4	0.1	07 22
04 56	285 46.0	1.8	N 6 50.7	0.2	29 Th	250 08.6	2.3	S 8 32.7	0.0	07 18
04 53	286 30.0	1.9	N 6 46.0	0.2	30 Fri	251 03.9	2.3	S 8 33.9	0.0	07 15
04 50	287 14.4	1.9	N 6 41.5	0.2	31 Sat	251 59.2	2.3	S 8 35.0	0.1	07 11

MARS, Av. Mag. +0.5
SHA December 5 197; 10 195; 15 193; 20 191; 25 190; 30 188

SATURN, Av. Mag. +0.7
SHA December 5 155; 10 154; 15 154; 20 154; 25 153; 30 153

DECEMBER 2011

STARS

No.	Name	Mag	Transit (h m)	Dec (° ')	SHA (° ')
	0h GMT December 1				
Ψ	ARIES	—	19 19	—	357 44.5
1	Alpheratz	2.1	19 28	N29 09.7	353 16.6
2	Ankaa	2.4	19 46	S42 14.6	349 41.6
3	Schedar	2.2	20 00	N56 36.5	348 56.8
4	Diphda	2.0	20 03	S17 55.3	335 27.2
5	Achernar	0.5	20 57	S57 10.7	335 27.2
6	POLARIS	2.0	22 07	N89 19.2	317 57.0
7	Hamal	2.0	21 26	N23 31.3	328 01.7
8	Acamar	3.2	22 17	S40 15.5	315 18.8
9	Menkar	2.5	22 21	N 4 08.2	314 15.9
10	Mirfak	1.8	22 43	N49 54.3	308 41.4
11	Aldebaran	0.9	23 55	N16 32.0	290 50.2
12	Rigel	0.1	0 37	S 8 11.3	281 12.7
13	Capella	0.1	0 39	N46 00.5	280 35.5
14	Bellatrix	1.6	0 48	N 6 21.6	278 32.8
15	Elnath	1.7	0 49	N28 37.0	278 13.6
16	Alnilam	1.7	0 59	S 1 11.7	275 47.1
17	Betelgeuse	0.1–1.2	1 18	N 7 24.5	271 02.1
18	Canopus	-0.7	1 46	S52 42.1	263 56.1
19	Sirius	-1.5	2 07	S16 44.0	258 34.3
20	Adhara	1.5	2 21	S28 59.3	255 13.0
21	Castor	1.6	2 57	N31 51.5	246 09.0
22	Procyon	0.4	3 01	N 5 11.5	245 00.5
23	Pollux	1.1	3 08	N27 59.6	243 28.7
24	Avior	1.9	3 44	S59 32.8	234 18.0
25	Suhail	2.2	4 30	S43 28.8	222 53.0
26	Miaplacidus	1.7	4 35	S69 45.8	221 39.4
27	Alphard	2.0	4 49	S 8 42.7	217 56.9
28	Regulus	1.4	5 30	N11 54.3	207 44.5
29	Dubhe	1.8	6 25	N61 40.8	193 52.9
30	Denebola	2.1	7 10	N14 30.1	182 34.8
31	Gienah	2.6	7 37	S17 36.5	175 53.5
32	Acrux	1.3	7 48	S63 09.7	173 10.7
33	Gacrux	1.6	7 53	S57 10.6	172 02.3
34	Mimosa	1.3	8 09	S59 45.0	167 53.4
35	Alioth	1.8	8 15	N55 53.4	166 21.8
36	Spica	1.0	8 46	S11 13.4	158 32.5
37	Alkaid	1.9	9 08	N49 15.0	153 00.0
38	Hadar	0.6	9 25	S60 25.6	148 49.9
39	Menkent	2.1	9 28	S36 25.6	148 09.1
40	Arcturus	0.0	9 37	N19 07.1	145 56.9
41	Rigil Kent	-0.3	10 01	S60 52.8	139 53.7
42	Zuben'ubi	2.8	10 12	S16 05.4	137 06.9
43	Kochab	2.1	10 11	N74 06.1	137 20.9
44	Alphecca	2.2	10 55	N26 40.4	126 12.2
45	Antares	1.0	11 50	S26 27.4	112 27.9
46	Atria	1.9	12 10	S69 02.8	107 31.3
47	Sabik	2.4	12 31	S15 44.3	102 14.1
48	Shaula	1.6	12 54	S37 06.6	96 23.8
49	Rasalhague	2.1	12 55	N12 33.2	96 07.8
50	Eltanin	2.2	13 17	N51 29.4	90 47.2
51	Kaus Aust	1.9	13 45	S34 22.6	83 45.6
52	Vega	0.0	13 57	N38 47.9	80 40.1
53	Nunki	2.0	14 16	S26 16.8	76 00.0
54	Altair	0.8	15 11	N 8 54.2	62 09.6
55	Peacock	1.9	15 46	S56 41.8	53 21.3
56	Deneb	1.3	16 01	N45 19.7	49 32.5
57	Enif	2.4	17 04	N 9 56.0	33 48.3
58	Al Na'ir	1.7	17 28	S46 54.3	27 45.2
59	Fomalhaut	1.2	18 17	S29 33.6	15 25.2
60	Markab	2.5	18 24	N15 16.4	13 39.4

SUN AND MOON

SUN

Yr	Day of Mth	Week	Transit (h m)	Semi-Diam	Twilight (h m)	Sunrise (h m)	Lat 52°N Sunset (h m)	Twilight (h m)
335	1	Th	11 49	16.2	07 05	07 45	15 52	16 32
336	2	Fri	11 49	16.2	07 07	07 46	15 52	16 31
337	3	Sat	11 50	16.3	07 08	07 48	15 51	16 31
338	4	Sun	11 50	16.3	07 09	07 49	15 51	16 30
339	5	Mon	11 51	16.3	07 10	07 50	15 50	16 30
340	6	Tu	11 51	16.3	07 12	07 52	15 49	16 29
341	7	Wed	11 51	16.3	07 13	07 53	15 49	16 29
342	8	Th	11 52	16.3	07 15	07 54	15 49	16 29
343	9	Fri	11 52	16.3	07 16	07 55	15 49	16 29
344	10	Sat	11 53	16.3	07 17	07 56	15 48	16 29
345	11	Sun	11 53	16.3	07 18	07 58	15 48	16 29
346	12	Mon	11 54	16.3	07 19	07 59	15 48	16 29
347	13	Tu	11 54	16.3	07 20	08 00	15 48	16 29
348	14	Wed	11 55	16.3	07 21	08 00	15 48	16 29
349	15	Th	11 55	16.3	07 22	08 01	15 49	16 30
350	16	Fri	11 56	16.3	07 23	08 02	15 49	16 30
351	17	Sat	11 56	16.3	07 23	08 03	15 49	16 30
352	18	Sun	11 57	16.3	07 24	08 04	15 49	16 30
353	19	Mon	11 57	16.3	07 25	08 04	15 50	16 31
354	20	Tu	11 58	16.3	07 26	08 05	15 50	16 31
355	21	Wed	11 58	16.3	07 26	08 06	15 51	16 32
356	22	Th	11 59	16.3	07 27	08 06	15 51	16 32
357	23	Fri	11 59	16.3	07 27	08 07	15 52	16 33
358	24	Sat	12 00	16.3	07 27	08 07	15 53	16 33
359	25	Sun	12 00	16.3	07 27	08 08	15 53	16 34
360	26	Mon	12 01	16.3	07 27	08 08	15 54	16 35
361	27	Tu	12 01	16.3	07 27	08 08	15 55	16 36
362	28	Wed	12 02	16.3	07 27	08 08	15 56	16 37
363	29	Th	12 02	16.3	07 27	08 08	15 56	16 37
364	30	Fri	12 03	16.3	07 27	08 08	15 57	16 37
365	31	Sat	12 03	16.3	07 28	08 08	15 58	16 38

MOON

Yr	Day of Mth	Week	Age (days)	Transit (Upper) h m	Diff (m)	Semi-diam	Hor Par	Lat 52°N Moonrise h m	Moonset h m
335	1	Th	06	17 25	43	15.2	55.7	11 54	23 09
336	2	Fri	07	18 08	42	15.0	55.1	12 12	—
337	3	Sat	08	18 50	41	14.9	54.6	12 29	00 17
338	4	Sun	09	19 31	43	14.8	54.3	12 46	01 24
339	5	Mon	10	20 14	44	14.7	54.1	13 05	02 30
340	6	Tu	11	20 58	45	14.8	54.1	13 26	03 36
341	7	Wed	12	21 43	48	14.8	54.2	13 51	04 42
342	8	Th	13	22 31	50	14.9	54.5	14 22	05 47
343	9	Fri	14	23 21	52	14.9	54.8	15 00	06 49
344	10	Sat	15	24 13	—	15.0	55.2	15 48	07 46
345	11	Sun	16	00 13	51	15.1	55.6	16 45	08 35
346	12	Mon	17	01 05	50	15.3	56.0	17 50	09 16
347	13	Tu	18	01 56	50	15.4	56.5	19 00	09 50
348	14	Wed	19	02 46	48	15.5	57.0	20 14	10 18
349	15	Th	20	03 36	48	15.7	57.5	21 30	10 42
350	16	Fri	21	04 24	49	15.8	58.0	22 47	11 03
351	17	Sat	22	05 12	51	16.0	58.6	—	11 23
352	18	Sun	23	06 01	53	16.1	59.1	00 05	11 44
353	19	Mon	24	06 52	57	16.2	59.5	01 25	12 06
354	20	Tu	25	07 45	60	16.3	59.9	02 47	12 33
355	21	Wed	26	08 42	61	16.4	60.1	04 10	13 06
356	22	Th	27	09 42	61	16.4	59.9	05 30	13 49
357	23	Fri	28	10 43	59	16.3	59.5	06 42	14 43
358	24	Sat	29	11 44	55	16.2	58.9	07 42	15 49
359	25	Sun	01	12 43	52	16.2	58.2	08 29	17 03
360	26	Mon	02	13 38	47	16.0	57.4	09 06	18 19
361	27	Tu	03	14 30	45	15.8	56.6	09 34	19 35
362	28	Wed	04	15 17	43	15.6	55.8	09 57	20 49
363	29	Th	05	16 02	42	15.4	55.2	10 16	22 00
364	30	Fri	06	16 45	42	15.2	54.7	10 34	23 09
365	31	Sat	07	17 27		14.9		10 52	—

Lat Corr to Sunrise, Sunset etc.

Lat	Twilight (h m)	Sunrise (h m)	Sunset (h m)	Twilight (h m)
N70	+2 26	SBH	SBH	-2 26
68	+1 52	SBH	SBH	-1 52
66	+1 27	+2 25	-2 26	-1 27
64	+1 08	+1 44	-1 45	-1 08
62	+0 52	+1 16	-1 17	-0 52
N60	+0 38	+0 55	-0 56	-0 38
58	+0 27	+0 38	-0 39	-0 27
56	+0 17	+0 23	-0 24	-0 17
54	+0 08	+0 11	-0 11	-0 08
50	-0 08	-0 10	+0 10	+0 08
N45	-0 23	-0 30	+0 30	+0 23
40	-0 37	-0 47	+0 47	+0 37
35	-0 48	-1 01	+1 01	+0 48
30	-0 59	-1 13	+1 13	+0 59
20	-1 17	-1 34	+1 34	+1 18
N10	-1 34	-1 53	+1 53	+1 35
0	-1 51	-2 10	+2 10	+1 52
S10	-2 09	-2 27	+2 27	+2 10
20	-2 30	-2 46	+2 46	+2 31
30	-2 54	-3 08	+3 08	+2 51
S35	-3 09	-3 21	+3 21	+3 16
40	-3 27	-3 36	+3 36	+3 29
45	-3 49	-3 53	+3 54	+3 51
S50	-4 18	-4 16	+4 16	+4 20

NOTES
The corrections to sunrise etc. are for middle of December. SBH means Sun below Horizon.

Phases of the Moon

	d	h	m
First Quarter	2	09	52
Full Moon	10	14	36
Last Quarter	18	00	48
New Moon	24	18	06

	d	h	
Apogee	6	01	
Perigee	22	03	

☽ ○ ◑ ●

POLARIS (POLE STAR) TABLE – 2011

FOR DETERMINING THE LATITUDE FROM A SEXTANT ALTITUDE

LHA Aries	Q	LHA Aries	Q	LHA Aries	Q	LHA Aries	Q	LHA Aries	Q
359 42	−31	87 24	−28	123 31	−5	156 27	+18	211 42	+41
1 49	−32	89 18	−27	124 55	−4	158 01	+19	231 25	+40
4 02	−33	91 08	−26	126 19	−3	159 35	+20	237 41	+39
6 22	−34	92 56	−25	127 43	−2	161 11	+21	242 10	+38
8 50	−35	94 41	−24	129 07	−1	162 49	+22	245 51	+37
11 29	−36	96 24	−23	130 31	0	164 28	+23	249 05	+36
14 21	−37	98 04	−22	131 55	+1	166 10	+24	251 59	+35
17 33	−38	99 43	−21	133 19	+2	167 53	+25	254 40	+34
21 12	−39	101 20	−20	134 43	+3	169 39	+26	257 09	+33
25 37	−40	102 56	−19	136 07	+4	171 27	+27	259 30	+32
31 49	−41	104 30	−18	137 31	+5	173 18	+28	261 44	+31
51 18	−40	106 02	−17	138 55	+6	175 13	+29	263 52	+30
57 30	−39	107 34	−16	140 20	+7	177 12	+30	265 55	+29
61 55	−38	109 05	−15	141 45	+8	179 15	+31	267 54	+28
65 34	−37	110 34	−14	143 10	+9	181 23	+32	269 49	+27
68 46	−36	112 03	−13	144 36	+10	183 37	+33	271 40	+26
71 38	−35	113 31	−12	146 02	+11	185 58	+34	273 28	+25
74 17	−34	114 58	−11	147 29	+12	188 27	+35	275 14	+24
76 45	−33	116 25	−10	148 57	+13	191 08	+36	276 57	+23
79 05	−32	117 51	−9	150 25	+14	194 02	+37	278 39	+22
81 18	−31	119 17	−8	151 54	+15	197 16	+38	280 18	+21
83 25	−30	120 42	−7	153 24	+16	200 57	+39	281 56	+20
85 27	−29	122 07	−6	154 55	+17	205 26	+40	283 32	+19
87 24	−28	123 31		156 27		211 42		285 06	

LHA Aries	Q	LHA Aries	Q	LHA Aries	Q
285 06	+18	318 12	−5	353 49	−28
286 40	+17	319 36	−6	355 43	−29
288 12	+16	321 00	−7	357 40	−30
289 43	+15	322 25	−8	359 42	−31
291 13	+14	323 50	−9	1 49	−32
292 42	+13	325 16	−10	4 02	−33
294 10	+12	326 42	−11	6 22	−34
295 38	+11	328 09	−12	8 50	−35
297 05	+10	329 36	−13	11 29	−36
298 31	+9	331 04	−14	14 21	−37
299 57	+8	332 33	−15	17 33	−38
301 22	+7	334 02	−16	21 12	−39
302 47	+6	335 33	−17	25 37	−40
304 12	+5	337 05	−18	31 49	−41
305 36	+4	338 37	−19	51 18	−40
307 00	+3	340 11	−20	57 30	−39
308 24	+2	341 47	−21	61 55	−38
309 48	+1	343 24	−22	65 34	−37
311 12	0	345 03	−23	68 46	−36
312 36	−1	346 43	−24	71 38	−35
314 00	−2	348 26	−25	74 17	−34
315 24	−3	350 11	−26	76 45	−33
316 48	−4	351 59	−27	79 05	−32
318 12		353 49		81 18	

In critical cases, ascend

Q, which does not include refraction, is to be applied to the corrected sextant altitude of Polaris.

Polaris: Mag. 2.1, SHA 318 26 Dec N89 18.9

The Pole Star table, above, changes annually and is the last of the ephemeral tables.

ECLIPSES 2011

There are four solar eclipses and two lunar eclipses. All times are UT (GMT).

4th January
Partial eclipse of the Sun, starting at 0640 and ending at 1101. Visible in Europe, northern Africa, Middle East and western Asia.

1st June
Partial eclipse of the Sun, starting at 1925 and ending at 2307. Visible in Europe above 60 degrees North, Arctic regions, northern Asia, Iceland and northern part of North America.

15th June
Total eclipse of the Moon, starting at 1723 and ending at 2302. Visible in South America, Europe, Africa, Indian Ocean, western Asia, Australia and New Zealand.

1st July
Partial eclipse of the Sun, starting at 0754 and ending at 0923. Visible only in the southern Indian Ocean.

25th November
Partial eclipse of the Sun, starting at 0423 and ending at 0817. Visible in Antarctica, southern tip of Africa, oceans around Antarctica, Tasmania, New Zealand.

10th December
Total eclipse of the Moon, starting at 1132 and ending at 1732. Visible in Europe, Africa (except extreme west), Asia, Australia, Pacific Ocean, North America

USABILITY IN 2012 (Leap Year)

The Almanac can be used for Sun and Stars in 2012. It cannot be used for Moon and Planets.

For the Sun take out GHA and DEC for the same date but, in January and February (on 29th February use 1st March) for a time 5h 48m earlier, and in March to December for a time 18h 12m later than the time of observation. In both cases add 87 degrees to the GHA so obtained.

For Stars calculate the GHA for the same date and time but, in January and February (on 29th February use 1st March) subtract 15.1 minutes and, in March to December add 44.0 minutes to the GHA so obtained.

The error, in all cases, is unlikely to exceed 0.4 minutes of arc.

STAR OR PLANET ALTITUDE TOTAL CORRECTION TABLE

ALWAYS SUBTRACTIVE

Height of the eye above the sea. Top line metres, lower line feet

Obs Alt °	1.5 / 5	3.0 / 10	4.6 / 15	6.0 / 20	7.6 / 25	9.0 / 30	10.7 / 35	12 / 40	13.7 / 45	15 / 50	16.8 / 55	18 / 60	21.3 / 70
9	8.0	8.9	9.6	10.3	10.7	11.2	11.6	12.0	12.4	12.8	13.1	13.5	14.1
10	7.4	8.4	9.1	9.7	10.2	10.6	11.1	11.5	11.8	12.2	12.5	12.9	13.5
11	7.0	7.9	8.6	9.2	9.7	10.2	10.6	11.0	11.4	11.8	12.0	12.4	13.0
12	6.6	7.5	8.2	8.8	9.3	9.8	10.2	10.6	11.0	11.4	11.6	12.0	12.6
13	6.2	7.2	7.9	8.4	9.0	9.4	9.9	10.3	10.6	11.0	11.3	11.6	12.3
14	5.9	6.9	7.6	8.1	8.6	9.2	9.6	10.0	10.3	10.7	11.0	11.3	12.0
15	5.7	6.6	7.3	7.9	8.4	8.9	9.3	9.7	10.1	10.4	10.8	11.1	11.7
16	5.5	6.4	7.1	7.7	8.2	8.7	9.1	9.5	9.9	10.2	10.5	10.9	11.5
17	5.3	6.2	6.9	7.5	8.0	8.5	8.9	9.3	9.7	10.0	10.3	10.7	11.3
18	5.1	6.0	6.7	7.3	7.8	8.3	8.7	9.1	9.5	9.8	10.2	10.5	11.1
19	4.9	5.8	6.5	7.1	7.6	8.1	8.5	8.9	9.3	9.7	10.0	10.3	11.0
20	4.8	5.7	6.4	7.0	7.5	8.0	8.4	8.8	9.2	9.6	9.9	10.2	10.8
25	4.2	5.1	5.8	6.4	6.9	7.4	7.8	8.2	8.6	9.0	9.3	9.6	10.2
30	3.8	4.7	5.4	6.0	6.5	7.0	7.4	7.8	8.2	8.6	8.9	9.2	9.8
35	3.5	4.4	5.1	5.7	6.3	6.7	7.2	7.6	7.9	8.3	8.6	8.9	9.5
40	3.3	4.2	4.9	5.5	6.0	6.5	6.9	7.3	7.7	8.1	8.4	8.7	9.3
50	3.0	3.9	4.6	5.2	5.7	6.2	6.6	7.0	7.4	7.7	8.1	8.4	9.0
60	2.7	3.6	4.4	4.9	5.5	5.9	6.4	6.8	7.1	7.5	7.8	8.1	8.8
70	2.5	3.4	4.1	4.7	5.3	5.7	6.2	6.6	6.9	7.3	7.6	7.9	8.6
80	2.3	3.3	4.0	4.6	5.1	5.5	6.0	6.4	6.7	7.1	7.4	7.8	8.4
90	2.2	3.1	3.8	4.4	4.9	5.4	5.8	6.2	6.6	6.9	7.3	7.6	8.2

The above table contains the combined effects of Dip of the Horizon and Refraction and is therefore a total correction table for a Star or Planet. It is always subtractive.

45

SUN ALTITUDE TOTAL CORRECTION TABLE

FOR CORRECTING THE OBSERVED ALTITUDE OF THE SUN'S LOWER LIMB

ALWAYS ADDITIVE

Height of the eye above the sea. Top line metres, lower line feet

Obs Alt °	0.9 / 3	1.8 / 6	2.4 / 8	3.0 / 10	3.7 / 12	4.3 / 14	4.9 / 16	5.5 / 18	6.0 / 20	7.6 / 25	9.0 / 30	12 / 40	15 / 50	18 / 60	21 / 70	24 / 80
9	8.6	8.0	7.6	7.2	6.9	6.6	6.4	6.2	5.9	5.4	4.9	4.1	3.4	2.7	2.1	1.5
10	9.1	8.5	8.1	7.9	7.5	7.2	7.0	6.7	6.6	6.0	5.5	4.7	3.9	3.3	2.7	2.1
11	9.6	9.0	8.6	8.3	8.0	7.7	7.4	7.2	7.0	6.4	6.0	5.2	4.4	3.7	3.1	2.5
12	10.0	9.4	9.0	8.7	8.4	8.1	7.8	7.6	7.4	6.8	6.4	5.6	4.8	4.1	3.5	2.9
13	10.3	9.7	9.3	9.0	8.7	8.4	8.2	7.9	7.7	7.2	6.7	5.9	5.2	4.5	3.9	3.3
14	10.6	10.0	9.6	9.3	9.0	8.7	8.5	8.2	8.0	7.5	7.0	6.2	5.5	4.8	4.2	3.6
15	10.9	10.2	9.9	9.5	9.2	9.0	8.7	8.5	8.2	7.7	7.2	6.4	5.7	5.0	4.4	3.8
16	11.1	10.5	10.1	9.7	9.5	9.2	8.9	8.7	8.5	7.9	7.5	6.7	5.9	5.2	4.6	4.1
17	11.3	10.7	10.3	10.0	9.7	9.4	9.1	8.9	8.7	8.2	7.7	6.9	6.1	5.5	4.9	4.3
18	11.5	10.8	10.5	10.1	9.9	9.6	9.3	9.1	8.9	8.3	7.9	7.0	6.3	5.6	5.0	4.5
19	11.6	11.0	10.6	10.3	10.0	9.7	9.5	9.2	9.0	8.5	8.0	7.2	6.5	5.8	5.2	4.6
20	11.8	11.2	10.8	10.4	10.2	9.9	9.6	9.4	9.2	8.6	8.2	7.4	6.6	5.9	5.3	4.8
21	11.9	11.3	10.9	10.6	10.3	10.0	9.8	9.5	9.3	8.8	8.3	7.5	6.8	6.1	5.5	4.9
22	12.0	11.4	11.0	10.7	10.4	10.1	9.9	9.7	9.4	8.9	8.4	7.6	6.9	6.2	5.6	5.0
23	12.1	11.5	11.1	10.8	10.5	10.2	10.0	9.7	9.5	9.0	8.5	7.7	7.0	6.3	5.7	5.1
24	12.2	11.6	11.2	10.9	10.6	10.3	10.1	9.9	9.6	9.1	8.6	7.8	7.1	6.4	5.8	5.2
25	12.3	11.7	11.3	11.0	10.7	10.4	10.2	9.9	9.7	9.2	8.7	7.9	7.2	6.5	5.9	5.3
26	12.4	11.8	11.4	11.1	10.8	10.5	10.3	10.0	9.8	9.3	8.8	8.0	7.3	6.6	6.0	5.4
27	12.5	11.9	11.5	11.2	10.9	10.6	10.4	10.1	9.9	9.4	8.9	8.1	7.4	6.7	6.1	5.5
28	12.6	12.0	11.6	11.3	11.0	10.7	10.4	10.2	10.0	9.5	9.0	8.2	7.4	6.8	6.2	5.6
30	12.7	12.1	11.7	11.4	11.1	10.8	10.6	10.4	10.1	9.6	9.1	8.3	7.6	6.9	6.3	5.7
32	12.9	12.2	11.9	11.5	11.2	11.0	10.7	10.5	10.2	9.7	9.3	8.4	7.7	7.0	6.4	5.8
34	13.0	12.3	12.0	11.6	11.3	11.1	10.8	10.6	10.3	9.8	9.4	8.5	7.8	7.1	6.5	5.9
36	13.1	12.4	12.1	11.7	11.4	11.2	10.9	10.7	10.4	9.9	9.5	8.6	7.9	7.2	6.6	6.0
38	13.2	12.5	12.1	11.8	11.5	11.2	11.0	10.7	10.5	10.0	9.5	8.7	8.0	7.3	6.7	6.1
40	13.3	12.6	12.2	11.9	11.6	11.3	11.1	10.8	10.6	10.1	9.6	8.8	8.1	7.4	6.8	6.2
42	13.4	12.7	12.3	12.0	11.7	11.4	11.2	10.9	10.7	10.2	9.7	8.9	8.2	7.5	6.9	6.3
44	13.4	12.7	12.4	12.0	11.7	11.5	11.2	11.0	10.7	10.2	9.8	8.9	8.2	7.5	6.9	6.3
46	13.5	12.8	12.4	12.1	11.8	11.5	11.3	11.0	10.8	10.3	9.8	9.0	8.3	7.6	7.0	6.4
48	13.6	12.9	12.5	12.2	11.9	11.6	11.3	11.1	10.9	10.4	9.9	9.1	8.3	7.7	7.1	6.4
50	13.6	12.9	12.5	12.2	11.9	11.6	11.4	11.1	10.9	10.4	9.9	9.1	8.4	7.7	7.1	6.5
52	13.7	13.0	12.6	12.3	12.0	11.7	11.4	11.2	11.0	10.5	10.0	9.2	8.4	7.8	7.2	6.5
54	13.7	13.0	12.7	12.3	12.0	11.7	11.5	11.2	11.0	10.5	10.0	9.2	8.5	7.8	7.2	6.6
56	13.7	13.1	12.7	12.4	12.1	11.8	11.5	11.3	11.1	10.6	10.1	9.3	8.5	7.9	7.3	6.7
58	13.8	13.1	12.7	12.4	12.1	11.8	11.6	11.3	11.1	10.6	10.1	9.3	8.6	7.9	7.3	6.8
60	13.8	13.1	12.8	12.4	12.1	11.9	11.6	11.4	11.1	10.6	10.2	9.3	8.6	7.9	7.3	6.8
62	13.9	13.2	12.8	12.5	12.2	11.9	11.7	11.4	11.2	10.7	10.2	9.4	8.7	8.0	7.4	6.8
64	13.9	13.2	12.8	12.5	12.2	11.9	11.7	11.5	11.2	10.7	10.2	9.4	8.7	8.0	7.4	6.9
66	14.0	13.2	12.9	12.5	12.3	12.0	11.7	11.5	11.3	10.8	10.3	9.5	8.7	8.1	7.5	7.0
70	14.1	13.3	12.9	12.6	12.3	12.0	11.8	11.6	11.3	10.8	10.3	9.5	8.8	8.1	7.5	7.0
80	14.2	13.5	13.1	12.8	12.5	12.2	11.9	11.7	11.5	11.0	10.5	9.7	8.9	8.3	7.7	7.1
90	14.3	13.6	13.2	12.9	12.6	12.3	12.0	11.8	11.5	11.1	10.6	9.8	9.1	8.4	7.8	7.2

MONTHLY CORRECTION

Jan	Feb	Mar	Apr	May	Jun	Jul	Aug	Sep	Oct	Nov	Dec
+0.3'	+0.2'	+0.1'	0.01'	-0.11'	-0.2'	-0.2'	-0.2'	-0.11'	+0.1'	+0.2'	+0.3'

MEAN REFRACTION

ALWAYS SUBTRACTIVE

App. Alt.	Refr.	App. Alt.	Refr.	App. Alt.	Refr.	App. Alt.	Refr.
° ′	′	° ′	′	° ′	′	° ′	′
0 00	34.9	5 00	9.8	10 00	5.3	16 30	3.2
10	32.8	10	9.5	10	5.2	17 00	3.1
20	30.9	20	9.3	20	5.1	30	3.0
30	29.1	30	9.0	30	5.0	18 00	2.9
40	27.4	40	8.8	40	5.0	30	2.9
50	25.8	50	8.6	50	4.9	19 00	2.8
1 00	24.4	6 00	8.4	11 00	4.8	20 00	2.6
10	23.1	10	8.2	10	4.7	21 00	2.5
20	21.9	20	8.0	20	4.7	22 00	2.4
30	20.9	30	7.8	30	4.6	23 00	2.3
40	19.9	40	7.7	40	4.5	24 00	2.2
50	19.0	50	7.5	50	4.5	26 00	2.0
2 00	18.1	7 00	7.3	12 00	4.4	28 00	1.8
10	17.4	10	7.2	10	4.4	30 00	1.7
20	16.7	20	7.0	20	4.3	32 00	1.5
30	16.0	30	6.9	30	4.2	34 00	1.4
40	15.4	40	6.8	40	4.2	36 00	1.3
50	14.8	50	6.6	50	4.1	38 00	1.2
3 00	14.2	8 00	6.5	13 00	4.1	40 00	1.1
10	13.7	10	6.4	10	4.0	43 00	1.0
20	13.3	20	6.3	20	4.0	46 00	0.9
30	12.8	30	6.1	30	3.9	50 00	0.8
40	12.4	40	6.0	40	3.9	55 00	0.7
50	12.0	50	5.9	50	3.8	60 00	0.6
4 00	11.7	9 00	5.8	14 00	3.8	65 00	0.5
10	11.3	10	5.7	10	3.7	70 00	0.4
20	11.0	20	5.6	20	3.6	75 00	0.3
30	10.7	30	5.5	30	3.5	80 00	0.2
40	10.4	40	5.4	40	3.5	85 00	0.1
50	10.1	50	5.4	50	3.4	90 00	0.0
				16 00	3.3		

MOON ALTITUDE TOTAL CORRECTION TABLE

Upper Limb — ADD / SUBTRACT

Obs Alt	Horizontal Parallax 54'	55'	56'	57'	58'	59'	60'	61'
10	23.4	24.0	24.6	25.5	26.0	26.7	27.5	28.3
12	23.8	24.6	25.2	26.0	26.5	27.2	28.0	28.7
14	24.0	24.8	25.4	26.1	26.7	27.5	28.3	29.0
16	24.0	24.8	25.5	26.1	26.7	27.3	28.3	28.8
18	23.8	24.6	25.2	26.0	26.5	27.3	28.0	28.6
20	23.6	24.2	25.0	25.5	26.2	27.0	27.5	28.2
22	23.2	23.8	24.6	25.0	25.7	26.5	27.0	27.8
24	22.7	23.2	24.0	24.5	25.3	25.8	25.0	27.0
26	22.0	22.6	23.4	24.0	24.5	25.0	25.7	26.5
28	21.4	22.0	22.6	23.3	23.8	24.5	25.0	25.5
30	20.6	21.2	21.8	22.3	23.0	23.5	24.3	24.7
32	19.8	20.2	21.0	21.3	22.0	23.2	23.2	23.7
34	19.0	19.4	20.0	20.5	21.0	21.5	22.2	22.7
36	18.0	18.4	19.0	19.5	20.0	20.5	21.0	21.7
38	16.8	17.4	17.8	18.5	19.0	19.5	20.0	20.4
40	15.8	16.2	16.8	17.3	17.7	18.2	18.8	19.2
42	14.7	15.2	15.6	16.0	16.5	17.0	17.5	18.0
44	13.5	13.8	14.2	14.6	15.0	15.5	16.0	16.5
46	12.0	12.6	13.0	13.4	13.8	14.2	14.5	15.0
48	10.5	11.2	11.6	12.0	12.4	12.8	13.2	13.5
50	9.3	10.0	10.2	10.6	11.0	11.3	11.7	12.0
52	8.0	8.4	8.6	9.2	9.5	9.7	10.0	10.5
54	6.7	6.8	7.2	7.5	7.8	8.2	8.5	8.7
56	5.2	5.5	5.6	6.0	6.3	6.5	7.0	7.0
58	3.7	3.7	4.2	4.5	4.5	5.0	5.0	5.5
60	2.0	2.2	2.5	2.7	3.0	3.2	3.5	3.5
62	+0.5	+0.7	+0.8	+1.0	+1.2	+1.5	+1.5	+1.7
64	−1.2	−1.0	−1.0	−0.8	−0.6	−0.5	−0.3	−0.1
66	3.0	2.8	2.6	2.5	2.4	2.3	2.0	2.0
68	4.5	4.5	4.4	4.3	4.2	4.0	4.0	4.0
70	6.3	6.2	6.2	6.1	6.0	6.0	5.8	5.8
72	8.0	8.0	8.0	8.0	8.0	8.0	7.8	7.8
74	9.7	9.7	9.7	9.7	9.7	9.7	9.7	9.7
76	11.5	11.5	11.5	11.5	11.6	11.7	11.7	11.7
78	13.5	13.5	13.5	13.6	13.6	13.7	13.7	13.7
80	15.4	15.4	15.4	15.5	15.6	15.7	15.7	16.0
82	17.0	17.0	17.2	17.3	17.5	17.7	17.8	18.0
84	18.8	19.0	19.2	19.3	19.5	19.7	19.9	20.0
86	20.8	21.0	21.0	21.2	21.5	21.7	22.0	22.0
88	22.6	22.8	23.0	23.2	23.4	23.7	24.0	24.2
90								

Lower Limb — ADD

Obs Alt	Horizontal Parallax 54'	55'	56'	57'	58'	59'	60'	61'
10	52.7	54.0	55.3	56.5	57.7	59.0	60.2	61.5
12	53.2	54.5	55.7	57.0	58.4	59.5	60.7	62.0
14	53.5	54.7	56.0	57.3	58.5	59.8	61.0	62.3
16	53.5	54.6	56.0	57.3	58.5	59.8	61.0	62.2
18	53.4	54.6	55.7	57.0	58.4	59.5	60.6	62.0
20	53.0	54.4	55.5	56.8	58.0	59.0	60.4	61.5
22	52.5	53.7	55.0	56.3	57.5	58.8	60.0	61.0
24	52.0	53.3	54.5	55.7	57.0	58.0	59.4	60.5
26	51.5	52.5	53.7	55.0	56.3	57.5	58.0	59.8
28	50.7	52.0	53.0	54.4	55.5	56.5	57.8	59.0
30	50.0	51.0	52.3	53.5	54.5	55.7	57.0	58.0
32	49.3	50.4	51.3	52.5	53.7	54.8	56.0	57.0
34	48.3	49.5	50.5	51.5	52.7	53.7	55.0	56.0
36	47.3	48.5	49.5	50.5	51.7	52.7	54.0	55.0
38	46.4	47.4	48.5	49.5	50.5	51.5	52.7	53.8
40	45.3	46.3	47.3	48.3	49.5	50.5	51.5	52.5
42	44.0	45.0	46.0	47.0	48.0	49.0	50.0	51.0
44	42.7	43.7	44.7	45.7	46.7	47.7	48.7	49.7
46	41.5	42.5	43.5	44.5	45.5	46.5	47.5	48.5
48	40.2	41.2	42.2	43.5	44.0	45.0	46.0	47.0
50	39.0	40.0	41.0	41.8	43.0	43.6	44.5	45.5
52	37.5	38.5	39.3	40.2	41.0	42.0	42.8	43.7
54	36.0	37.0	38.0	38.8	39.5	40.5	41.3	42.0
56	34.5	35.5	36.2	37.0	38.0	38.7	39.5	40.5
58	33.0	34.0	34.7	35.5	36.3	37.0	38.0	38.8
60	31.5	32.4	33.0	34.0	34.5	35.5	36.0	37.0
62	30.0	30.5	31.5	32.0	33.0	33.5	34.5	35.0
64	28.3	29.0	29.6	30.5	31.0	31.8	32.5	33.3
66	26.5	27.3	28.0	28.5	29.3	30.0	30.7	31.5
68	25.0	25.5	26.3	26.8	27.5	28.0	28.8	29.5
70	23.3	23.8	24.5	25.0	25.5	26.2	27.0	27.5
72	21.5	22.0	22.5	23.3	23.8	24.5	25.0	25.5
74	19.7	20.3	20.7	21.2	22.0	22.5	23.0	23.5
76	18.0	18.5	19.0	19.5	20.0	20.5	21.0	21.5
78	16.0	16.5	17.0	17.5	18.0	18.5	19.0	19.5
80	14.2	14.7	15.3	15.5	16.0	16.5	17.0	17.5
82	12.5	13.0	13.3	13.5	14.0	14.5	15.0	15.5
84	10.5	11.0	11.5	11.7	12.0	12.5	13.0	13.4
86	8.8	9.0	9.5	9.8	10.0	10.5	11.0	11.3
88	7.0	7.2	7.5	8.0	8.3	8.5	8.7	9.0
90								

HEIGHT OF EYE CORRECTION (ADD)

Height of eye (m)	0	1.5	3	4.6	6	7.6	9	10.7
(ft)	0	5	10	15	20	25	30	35
Correction +	9.8'	7.6'	6.7'	6.0'	5.5'	5.0'	4.5'	4.0'

Height of eye (m)	12	14	15	17	18	20	21	23	24	26	30
(ft)	40	45	50	55	60	65	70	75	80	85	100
Correction +	3.5'	3.2'	3.0'	2.5'	2.3'	2.0'	1.7'	1.3'	1.0'	0.8'	0.0'

SUN GHA CORRECTION TABLE

Min. or Sec.	Add for Minutes	Add for 1 Hour + Minutes	Add for Secs.	Min. or Sec.	Add for Minutes	Add for 1 Hour + Minutes	Add for Secs.
0	0 0.0	15 0.0	0.0	30	7 30.0	22 30.0	7.5
1	0 15.0	15 15.0	0.3	31	7 45.0	22 45.0	7.8
2	0 30.0	15 30.0	0.5	32	8 0.0	23 0.0	8.0
3	0 45.0	15 45.0	0.8	33	8 15.0	23 15.0	8.3
4	1 0.0	16 0.0	1.0	34	8 30.0	23 30.0	8.5
5	1 15.0	16 15.0	1.3	35	8 45.0	23 45.0	8.8
6	1 30.0	16 30.0	1.5	36	9 0.0	24 0.0	9.0
7	1 45.0	16 45.0	1.8	37	9 15.0	24 15.0	9.3
8	2 0.0	17 0.0	2.0	38	9 30.0	24 30.0	9.5
9	2 15.0	17 15.0	2.3	39	9 45.0	24 45.0	9.8
10	2 30.0	17 30.0	2.5	40	10 0.0	25 0.0	10.0
11	2 45.0	17 45.0	2.8	41	10 15.0	25 15.0	10.3
12	3 0.0	18 0.0	3.0	42	10 30.0	25 30.0	10.5
13	3 15.0	18 15.0	3.3	43	10 45.0	25 45.0	10.8
14	3 30.0	18 30.0	3.5	44	11 0.0	26 0.0	11.0
15	3 45.0	18 45.0	3.8	45	11 15.0	26 15.0	11.3
16	4 0.0	19 0.0	4.0	46	11 30.0	26 30.0	11.5
17	4 15.0	19 15.0	4.3	47	11 45.0	26 45.0	11.8
18	4 30.0	19 30.0	4.5	48	12 0.0	27 0.0	12.0
19	4 45.0	19 45.0	4.8	49	12 15.0	27 15.0	12.3
20	5 0.0	20 0.0	5.0	50	12 30.0	27 30.0	12.5
21	5 15.0	20 15.0	5.3	51	12 45.0	27 45.0	12.8
22	5 30.0	20 30.0	5.5	52	13 0.0	28 0.0	13.0
23	5 45.0	20 45.0	5.8	53	13 15.0	28 15.0	13.3
24	6 0.0	21 0.0	6.0	54	13 30.0	28 30.0	13.5
25	6 15.0	21 15.0	6.3	55	13 45.0	28 45.0	13.8
26	6 30.0	21 30.0	6.5	56	14 0.0	29 0.0	14.0
27	6 45.0	21 45.0	6.8	57	14 15.0	29 15.0	14.3
28	7 0.0	22 0.0	7.0	58	14 30.0	29 30.0	14.5
29	7 15.0	22 15.0	7.3	59	14 45.0	29 45.0	14.8
				60	15 0.0	30 0.0	15.0

ARIES GHA CORRECTION TABLE

Correction for 1 HOUR + MINS			Correction for MINS			Correction for SECONDS		
1 Hour + Mins		Correction	Mins		Correction	Seconds		Correction
0	+	15 2.5	0	+	0 0.0	0	+	0.0
1	+	15 17.5	1	+	0 15.0	1	+	0.3
2	+	15 32.6	2	+	0 30.1	2	+	0.5
3	+	15 47.6	3	+	0 45.1	3	+	0.8
4	+	16 2.7	4	+	1 0.2	4	+	1.0
5	+	16 17.7	5	+	1 15.2	5	+	1.3
6	+	16 32.7	6	+	1 30.2	6	+	1.5
7	+	16 47.8	7	+	1 45.3	7	+	1.8
8	+	17 2.8	8	+	2 0.3	8	+	2.0
9	+	17 17.9	9	+	2 15.4	9	+	2.3
10	+	17 32.9	10	+	2 30.4	10	+	2.5
11	+	17 48.0	11	+	2 45.5	11	+	2.8
12	+	18 3.0	12	+	3 0.5	12	+	3.0
13	+	18 18.0	13	+	3 15.5	13	+	3.3
14	+	18 33.1	14	+	3 30.6	14	+	3.5
15	+	18 48.1	15	+	3 45.6	15	+	3.8
16	+	19 3.2	16	+	4 0.7	16	+	4.0
17	+	19 18.2	17	+	4 15.7	17	+	4.3
18	+	19 33.2	18	+	4 30.7	18	+	4.5
19	+	19 48.3	19	+	4 45.8	19	+	4.8
20	+	20 3.3	20	+	5 0.8	20	+	5.0
21	+	20 18.4	21	+	5 15.9	21	+	5.3
22	+	20 33.4	22	+	5 30.9	22	+	5.5
23	+	20 48.4	23	+	5 45.9	23	+	5.8
24	+	21 3.5	24	+	6 1.0	24	+	6.0
25	+	21 18.5	25	+	6 16.0	25	+	6.3
26	+	21 33.6	26	+	6 31.1	26	+	6.5
27	+	21 48.6	27	+	6 46.1	27	+	6.8
28	+	22 3.6	28	+	7 1.1	28	+	7.0
29	+	22 18.7	29	+	7 16.2	29	+	7.3
30	+	22 33.7	30	+	7 31.2	30	+	7.5
31	+	22 48.8	31	+	7 46.3	31	+	7.8
32	+	23 3.8	32	+	8 1.3	32	+	8.0
33	+	23 18.9	33	+	8 16.4	33	+	8.3
34	+	23 33.9	34	+	8 31.4	34	+	8.5
35	+	23 48.9	35	+	8 46.4	35	+	8.8
36	+	24 4.0	36	+	9 1.5	36	+	9.0
37	+	24 19.0	37	+	9 16.5	37	+	9.3
38	+	24 34.1	38	+	9 31.6	38	+	9.5
39	+	24 49.1	39	+	9 46.6	39	+	9.8
40	+	25 4.1	40	+	10 1.6	40	+	10.0
41	+	25 19.2	41	+	10 16.7	41	+	10.3
42	+	25 34.2	42	+	10 31.7	42	+	10.5
43	+	25 49.3	43	+	10 46.8	43	+	10.8
44	+	26 4.3	44	+	11 1.8	44	+	11.0
45	+	26 19.3	45	+	11 16.8	45	+	11.3
46	+	26 34.4	46	+	11 31.9	46	+	11.5
47	+	26 49.4	47	+	11 46.9	47	+	11.8
48	+	27 4.5	48	+	12 2.0	48	+	12.0
49	+	27 19.5	49	+	12 17.0	49	+	12.3
50	+	27 34.6	50	+	12 32.1	50	+	12.5
51	+	27 49.6	51	+	12 47.1	51	+	12.8
52	+	28 4.6	52	+	13 2.1	52	+	13.0
53	+	28 19.7	53	+	13 17.2	53	+	13.3
54	+	28 34.7	54	+	13 32.2	54	+	13.5
55	+	28 49.8	55	+	13 47.3	55	+	13.8
56	+	29 4.8	56	+	14 2.3	56	+	14.0
57	+	29 19.8	57	+	14 17.3	57	+	14.3
58	+	29 34.9	58	+	14 32.4	58	+	14.5
59	+	29 49.9	59	+	14 47.4	59	+	14.8
60	+	30 4.9	60	+	15 2.5	60	+	15.0

47

PLANETS GHA CORRECTION TABLE (HOURS)

ALWAYS ADD

HOURS

Var/hr	15°01.5'	15°01.7'	15°01.8'	15°02.0'	15°02.1'	15°02.3'	15°02.4'	15°02.6'	15°02.7'
0	0 00.0	0 00.0	0 00.0	0 00.0	0 00.0	0 00.0	0 00.0	0 00.0	0 00.0
1	15 01.5	15 01.7	15 01.8	15 02.0	15 02.1	15 02.3	15 02.4	15 02.6	15 02.7
2	30 03.0	30 03.4	30 03.6	30 04.0	30 04.2	30 04.6	30 04.8	30 05.2	30 05.4
3	45 04.5	45 05.1	45 05.4	45 06.0	45 06.3	45 06.9	45 07.2	45 07.8	45 08.1
4	60 06.0	60 06.8	60 07.2	60 08.0	60 08.4	60 09.2	60 09.6	60 10.4	60 10.8
5	75 07.5	75 08.5	75 09.0	75 10.0	75 10.5	75 11.5	75 12.0	75 13.0	75 13.5
6	90 09.0	90 10.2	90 10.8	90 12.0	90 12.6	90 13.8	90 14.4	90 15.6	90 16.2
7	105 10.5	105 11.9	105 12.6	105 14.0	105 14.7	105 16.1	105 16.8	105 18.2	105 18.9
8	120 12.0	120 13.6	120 14.4	120 16.0	120 16.8	120 18.4	120 19.2	120 20.8	120 21.6
9	135 13.5	135 15.3	135 16.2	135 18.0	135 18.9	135 20.7	135 21.6	135 23.4	135 24.3
10	150 15.0	150 17.0	150 18.0	150 20.0	150 21.0	150 23.0	150 24.0	150 26.0	150 27.0
11	165 16.5	165 18.7	165 19.8	165 22.0	165 23.1	165 25.3	165 26.4	165 28.6	165 29.7
12	180 18.0	180 20.4	180 21.6	180 24.0	180 25.2	180 27.6	180 28.8	180 31.2	180 32.4
13	195 19.5	195 22.1	195 23.4	195 26.0	195 27.3	195 29.9	195 31.2	195 33.8	195 35.1
14	210 21.0	210 23.8	210 25.2	210 28.0	210 29.4	210 32.2	210 33.6	210 36.4	210 37.8
15	225 22.5	225 25.5	225 27.0	225 30.0	225 31.5	225 34.5	225 36.0	225 39.0	225 40.5
16	240 24.0	240 27.2	240 28.8	240 32.0	240 33.6	240 36.8	240 38.4	240 41.6	240 43.2
17	255 25.5	255 28.9	255 30.6	255 34.0	255 35.7	255 39.1	255 40.8	255 44.8	255 45.9
18	270 27.0	270 30.6	270 32.4	270 36.0	270 37.8	270 41.4	270 43.2	270 46.8	270 48.6
19	285 28.5	285 32.3	285 34.2	285 38.0	285 39.9	285 43.7	285 45.6	285 49.4	285 51.3
20	300 30.0	300 34.0	300 36.0	300 40.0	300 42.0	300 46.0	300 48.0	300 52.0	300 54.0
21	315 31.5	315 35.7	315 37.8	315 42.0	315 44.1	315 48.3	315 50.4	315 54.6	315 56.7
22	330 33.0	330 37.4	330 39.6	330 44.0	330 46.2	330 50.6	330 52.8	330 57.2	330 59.4
23	345 34.5	345 39.1	345 41.4	345 46.0	345 48.3	345 52.9	345 55.2	345 59.8	346 02.1
24	0 36.0	0 40.8	0 43.2	0 48.0	0 50.4	0 55.2	0 57.6	1 02.4	1 04.8

Var/hr	15°02.9'	15°03.0'	15°03.2'	15°03.3'	15°03.5'	15°03.6'	15°03.8'	15°03.9'	15°04.1'
0	0 00.0	0 00.0	0 00.0	0 00.0	0 00.0	0 00.0	0 00.0	0 00.0	0 00.0
1	15 02.9	15 03.0	15 03.2	15 03.3	15 03.5	15 03.6	15 03.8	15 03.9	15 04.1
2	30 05.8	30 06.0	30 06.4	30 06.6	30 07.0	30 07.2	30 07.6	30 07.8	30 08.2
3	45 08.7	45 09.0	45 09.6	45 09.9	45 10.5	45 10.8	45 11.4	45 11.7	45 12.3
4	60 11.6	60 12.0	60 12.8	60 13.2	60 14.0	60 14.4	60 15.2	60 15.6	60 16.4
5	75 14.5	75 15.0	75 16.0	75 16.5	75 17.5	75 18.0	75 19.0	75 19.5	75 20.5
6	90 17.4	90 18.0	90 19.2	90 19.8	90 21.0	90 21.6	90 22.8	90 23.4	90 24.6
7	105 20.3	105 21.0	105 22.4	105 23.1	105 24.5	105 25.2	105 26.6	105 27.3	105 28.7
8	120 23.2	120 24.0	120 25.6	120 26.4	120 28.0	120 28.8	120 30.4	120 31.2	120 32.8
9	135 26.1	135 27.0	135 28.8	135 29.7	135 31.5	135 32.4	135 34.2	135 35.1	135 36.9
10	150 29.0	150 30.0	150 32.0	150 33.0	150 35.0	150 36.0	150 38.0	150 39.0	150 41.0
11	165 31.9	165 33.0	165 35.2	165 36.3	165 38.5	165 39.6	165 41.8	165 42.9	165 45.1
12	180 34.8	180 36.0	180 38.4	180 39.6	180 42.0	180 43.2	180 45.6	180 46.8	180 49.2
13	195 37.7	195 39.0	195 41.6	195 42.9	195 45.5	195 46.8	195 49.4	195 50.7	195 53.3
14	210 40.6	210 42.0	210 44.8	210 46.2	210 49.0	210 50.4	210 53.2	210 54.6	210 57.4
15	225 43.5	225 45.0	225 48.0	225 49.5	225 52.5	225 54.0	225 57.0	225 58.5	226 01.5
16	240 46.4	240 48.0	240 51.2	240 52.8	240 56.0	240 57.6	241 00.8	241 02.4	241 05.6
17	255 49.3	255 51.0	255 54.4	255 56.1	255 59.5	256 01.2	256 04.6	256 06.3	256 09.7
18	270 52.2	270 54.0	270 57.6	270 59.4	271 03.0	271 04.8	271 08.4	271 10.2	271 13.8
19	285 55.1	285 57.0	286 00.8	286 02.7	286 06.5	286 08.4	286 12.2	286 14.1	286 17.9
20	300 58.0	301 00.0	301 04.0	301 06.0	301 10.0	301 12.0	301 16.0	301 18.0	301 22.0
21	316 00.9	316 03.0	316 07.2	316 09.3	316 13.5	316 15.6	316 19.8	316 21.9	316 26.1
22	331 03.8	331 06.0	331 10.4	331 12.6	331 17.0	331 19.2	331 23.6	331 25.8	331 30.2
23	346 06.7	346 09.0	346 13.6	346 15.9	346 20.5	346 22.8	346 27.4	346 29.7	346 34.3
24	001 09.6	001 12.0	001 16.8	001 19.2	001 24.0	001 26.4	001 31.2	001 33.6	001 38.4

PLANETS GHA CORRECTION TABLE (HOURS)

ALWAYS ADD

HOURS

Var/hr	14°58.8'	14°59.0'	14°59.1'	14°59.3'	14°59.4'	14°59.6'	14°59.7'	14°59.9'	15°00.0'
0	0 00.0	0 00.0	0 00.0	0 00.0	0 00.0	0 00.0	0 00.0	0 00.0	0 00.0
1	14 58.8	14 59.0	14 59.1	14 59.3	14 59.4	14 59.6	14 59.7	14 59.9	15 00.0
2	29 57.6	29 58.0	29 58.2	29 58.6	29 58.8	29 59.2	29 59.4	29 59.8	30 00.0
3	44 56.4	44 57.0	44 57.3	44 57.9	44 58.2	44 58.8	44 59.1	44 59.7	45 00.0
4	59 55.2	59 56.0	59 56.4	59 57.2	59 57.6	59 58.4	59 58.8	59 59.6	60 00.0
5	74 54.0	74 55.0	74 55.5	74 56.5	74 57.0	74 58.0	74 58.5	74 59.5	75 00.0
6	89 52.8	89 54.0	89 54.6	89 55.8	89 56.4	89 57.6	89 58.2	89 59.4	90 00.0
7	104 51.6	104 53.0	104 53.7	104 55.1	104 55.8	104 57.2	104 57.9	104 59.3	105 00.0
8	119 50.4	119 52.0	119 52.8	119 54.4	119 55.2	119 56.8	119 57.6	119 59.2	120 00.0
9	134 49.2	134 51.0	134 51.9	134 53.7	134 54.6	134 56.4	134 57.3	134 59.1	135 00.0
10	149 48.0	149 50.0	149 51.0	149 53.0	149 54.0	149 56.0	149 57.0	149 59.0	150 00.0
11	164 46.8	164 49.0	164 50.1	164 52.3	164 53.4	164 55.6	164 56.7	164 58.9	165 00.0
12	179 45.6	179 48.0	179 49.2	179 51.6	179 52.8	179 55.2	179 56.4	179 58.7	180 00.0
13	194 44.4	194 47.0	194 48.3	194 50.9	194 52.2	194 54.8	194 56.1	194 58.7	195 00.0
14	209 43.2	209 46.0	209 47.4	209 50.2	209 51.6	209 54.4	209 55.8	209 58.6	210 00.0
15	224 42.0	224 45.0	224 46.5	224 49.5	224 51.0	224 54.0	224 55.5	224 58.5	225 00.0
16	239 40.8	239 44.0	239 45.6	239 48.8	239 50.4	239 53.6	239 55.2	239 58.4	240 00.0
17	254 39.6	254 43.0	254 44.7	254 48.1	254 49.8	254 53.2	254 54.9	254 58.3	255 00.0
18	269 38.4	269 42.0	269 43.8	269 47.4	269 49.2	269 52.8	269 54.6	269 58.2	270 00.0
19	284 37.2	284 41.0	284 42.9	284 46.7	284 48.6	284 52.4	284 54.3	284 58.1	285 00.0
20	299 36.0	299 40.0	299 42.0	299 46.0	299 48.0	299 52.0	299 54.0	299 58.0	300 00.0
21	314 34.8	314 39.0	314 41.1	314 45.3	314 47.4	314 51.6	314 53.7	314 57.9	315 00.0
22	329 33.6	329 38.0	329 40.2	329 44.6	329 46.8	329 51.2	329 53.4	329 57.8	330 00.0
23	344 32.4	344 37.0	344 39.3	344 43.7	344 46.2	344 50.8	344 53.1	344 57.7	345 00.0
24	359 31.2	359 36.0	359 38.4	359 43.2	359 45.6	359 50.4	359 52.8	359 57.6	0 00.0

Var/hr	15°00.2'	15°00.3'	15°00.5'	15°00.6'	15°00.8'	15°00.9'	15°01.1'	15°01.2'	15°01.4'
0	0 00.0	0 00.0	0 00.0	0 00.0	0 00.0	0 00.0	0 00.0	0 00.0	0 00.0
1	15 00.2	15 00.3	15 00.5	15 00.6	15 00.8	15 00.9	15 01.1	15 01.2	15 01.4
2	30 00.4	30 00.6	30 01.0	30 01.2	30 01.6	30 01.8	30 02.2	30 02.4	30 02.8
3	45 00.6	45 00.9	45 01.5	45 01.8	45 02.4	45 02.7	45 03.3	45 03.6	45 04.2
4	60 00.8	60 01.2	60 02.0	60 02.4	60 03.2	60 03.6	60 04.4	60 04.8	60 05.6
5	75 01.0	75 01.5	75 02.5	75 03.0	75 04.0	75 04.5	75 05.5	75 06.0	75 07.0
6	90 01.2	90 01.8	90 03.0	90 03.6	90 04.8	90 05.4	90 06.6	90 07.2	90 08.4
7	105 01.4	105 02.1	105 03.5	105 04.2	105 05.6	105 06.3	105 07.7	105 08.4	105 09.8
8	120 01.6	120 02.4	120 04.0	120 04.8	120 06.4	120 07.2	120 08.8	120 09.6	120 11.2
9	135 01.8	135 02.7	135 04.5	135 05.4	135 07.2	135 08.1	135 09.9	135 10.8	135 12.6
10	150 02.0	150 03.0	150 05.0	150 06.0	150 08.0	150 09.0	150 11.0	150 12.0	150 14.0
11	165 02.2	165 03.3	165 05.5	165 06.6	165 08.8	165 09.9	165 12.1	165 13.2	165 15.4
12	180 02.4	180 03.6	180 06.0	180 07.2	180 09.6	180 10.8	180 13.2	180 14.4	180 16.8
13	195 02.6	195 03.9	195 06.5	195 07.8	195 10.4	195 11.7	195 14.3	195 15.6	195 18.2
14	210 02.8	210 04.2	210 07.0	210 08.4	210 11.2	210 12.6	210 15.4	210 16.8	210 19.6
15	225 03.0	225 04.5	225 07.5	225 09.0	225 12.0	225 13.5	225 16.5	225 18.0	225 21.0
16	240 03.2	240 04.8	240 08.0	240 09.6	240 12.8	240 14.4	240 17.6	240 19.2	240 22.4
17	255 03.4	255 05.1	255 08.5	255 10.2	255 13.6	255 15.3	255 18.7	255 20.4	255 23.8
18	270 03.6	270 05.4	270 09.0	270 10.8	270 14.4	270 16.2	270 19.8	270 21.6	270 25.2
19	285 03.8	285 05.7	285 09.5	285 11.4	285 15.2	285 17.1	285 20.9	285 22.8	285 26.6
20	300 04.0	300 06.0	300 10.0	300 12.0	300 16.0	300 18.0	300 22.0	300 24.0	300 28.0
21	315 04.2	315 06.3	315 10.5	315 12.6	315 16.8	315 18.9	315 23.1	315 25.2	315 29.4
22	330 04.4	330 06.6	330 11.0	330 13.2	330 17.6	330 19.8	330 24.2	330 26.4	330 30.8
23	345 04.6	345 06.9	345 11.5	345 13.8	345 18.4	345 20.7	345 25.3	345 27.6	345 32.2
24	0 04.8	0 07.2	0 12.0	0 14.4	0 19.2	0 21.6	0 26.4	0 28.8	0 33.6

48

PLANETS NOTES 2011

VENUS shines brightly in the morning sky until mid July when it becomes too close to the Sun for observation. In mid September it reappears in the evening sky where it remains until year end. Venus is in conjunction with Jupiter on 11th May and with Mars on 22nd May.

JUPITER is in the evening sky until late March when it becomes too close to the Sun for observation. In mid April it reappears in the morning sky; its westward elongation increases, and at end October it is in opposition and can be seen throughout the night. Jupiter is in conjunction with Mercury on 16th March and 10th May, with Mars on 1st May and with Venus on 11th May.

MARS is too close to the Sun for observation until mid April, when it appears in the morning sky. Its westward elongation increases and it remains in the morning sky until year end.

SATURN rises shortly after midnight at the beginning of the year. It is in opposition on 4th April when it can be seen throughout the night. In late September it becomes too close to the Sun for observation. It reappears in the morning sky at the end of October where it remains until year end.

VISIBILITY OF PLANETS IN MORNING AND EVENING TWILIGHT

	Morning	Evening
VENUS	1st January to 11th July	23rd September to 31st December
JUPITER	21st April to 29th October	1st January to 24th March 29th October to 31st December
MARS	17th April to 31st December	
SATURN	1st January to 4th April 31st October to 31st December	4th April to 26th September

PLANETS GHA CORRECTION TABLE (MINUTES)

var/hr	14°58.8'	14°59.4'	15°00.0'	15°00.6'	15°01.2'	15°01.8'	15°02.4'	15°03.0'	15°03.6'	Sec
0	0 00.0	0 00.0	0 00.0	0 00.0	0 00.0	0 00.0	0 00.0	0 00.0	0 00.0	0 0.0
1	0 15.0	0 15.0	0 15.0	0 15.0	0 15.0	0 15.0	0 15.0	0 15.0	0 15.1	1 0.3
2	0 30.0	0 30.0	0 30.0	0 30.0	0 30.0	0 30.1	0 30.1	0 30.1	0 30.1	2 0.5
3	0 44.9	0 45.0	0 45.0	0 45.0	0 45.1	0 45.1	0 45.1	0 45.1	0 45.2	3 0.8
4	0 59.9	1 00.0	1 00.0	1 00.0	1 00.1	1 00.1	1 00.2	1 00.2	1 00.2	4 1.0
5	1 14.9	1 14.9	1 15.0	1 15.0	1 15.1	1 15.2	1 15.2	1 15.2	1 15.3	5 1.3
6	1 29.9	1 29.9	1 30.0	1 30.1	1 30.1	1 30.2	1 30.2	1 30.3	1 30.4	6 1.5
7	1 44.9	1 44.9	1 45.0	1 45.1	1 45.1	1 45.2	1 45.3	1 45.3	1 45.4	7 1.8
8	1 59.8	1 59.9	2 00.0	2 00.1	2 00.2	2 00.2	2 00.3	2 00.4	2 00.5	8 2.0
9	2 14.8	2 14.9	2 15.0	2 15.1	2 15.2	2 15.3	2 15.4	2 15.4	2 15.5	9 2.3
10	2 29.8	2 29.9	2 30.0	2 30.1	2 30.2	2 30.3	2 30.4	2 30.5	2 30.6	10 2.5
11	2 44.8	2 44.9	2 45.0	2 45.1	2 45.2	2 45.3	2 45.4	2 45.5	2 45.7	11 2.8
12	2 59.8	2 59.9	3 00.0	3 00.1	3 00.2	3 00.4	3 00.5	3 00.6	3 00.7	12 3.0
13	3 14.7	3 14.9	3 15.0	3 15.1	3 15.3	3 15.4	3 15.5	3 15.6	3 15.8	13 3.3
14	3 29.7	3 29.9	3 30.0	3 30.1	3 30.3	3 30.4	3 30.6	3 30.7	3 30.8	14 3.5
15	3 44.7	3 44.8	3 45.0	3 45.2	3 45.3	3 45.4	3 45.6	3 45.7	3 45.9	15 3.8
16	3 59.7	3 59.8	4 00.0	4 00.2	4 00.3	4 00.5	4 00.6	4 00.8	4 01.0	16 4.0
17	4 14.7	4 14.8	4 15.0	4 15.2	4 15.3	4 15.5	4 15.7	4 15.8	4 16.0	17 4.3
18	4 29.6	4 29.8	4 30.0	4 30.2	4 30.4	4 30.5	4 30.7	4 30.9	4 31.1	18 4.5
19	4 44.6	4 44.8	4 45.0	4 45.2	4 45.4	4 45.6	4 45.8	4 45.9	4 46.1	19 4.8
20	4 59.6	4 59.8	5 00.0	5 00.2	5 00.4	5 00.6	5 00.8	5 01.0	5 01.2	20 5.0
21	5 14.6	5 14.8	5 15.0	5 15.2	5 15.4	5 15.6	5 15.8	5 16.0	5 16.3	21 5.3
22	5 29.6	5 29.8	5 30.0	5 30.2	5 30.4	5 30.7	5 30.9	5 31.1	5 31.3	22 5.5
23	5 44.5	5 44.8	5 45.0	5 45.2	5 45.5	5 45.7	5 45.9	5 46.1	5 46.4	23 5.8
24	5 59.5	5 59.8	6 00.0	6 00.2	6 00.5	6 00.7	6 01.0	6 01.2	6 01.4	24 6.0
25	6 14.5	6 14.8	6 15.0	6 15.2	6 15.5	6 15.7	6 16.0	6 16.2	6 16.5	25 6.3
26	6 29.5	6 29.7	6 30.0	6 30.3	6 30.5	6 30.8	6 31.0	6 31.3	6 31.6	26 6.5
27	6 04.5	6 44.7	6 45.0	6 45.3	6 45.5	6 45.8	6 46.1	6 46.3	6 46.6	27 6.8
28	6 59.4	6 59.7	7 00.0	7 00.3	7 00.6	7 00.8	7 01.1	7 01.4	7 01.7	28 7.0
29	7 14.4	7 14.7	7 15.0	7 15.3	7 15.6	7 15.9	7 16.2	7 16.4	7 16.7	29 7.3
30	7 29.4	7 29.7	7 30.0	7 30.3	7 30.6	7 30.9	7 31.2	7 31.5	7 31.8	30 7.5
31	7 44.4	7 44.7	7 45.0	7 45.3	7 45.6	7 45.9	7 46.2	7 46.5	7 46.9	31 7.8
32	7 59.4	7 59.7	8 00.0	8 00.3	8 00.6	8 01.0	8 01.3	8 01.6	8 01.9	32 8.0
33	8 14.3	8 14.7	8 15.0	8 15.3	8 15.7	8 16.0	8 16.3	8 16.6	8 17.0	33 8.3
34	8 29.3	8 29.7	8 30.0	8 30.3	8 30.7	8 31.0	8 31.4	8 31.7	8 32.0	34 8.5
35	8 44.3	8 44.6	8 45.0	8 45.5	8 45.7	8 46.0	8 46.4	8 46.7	8 47.1	35 8.8
36	8 59.3	8 59.6	9 00.0	9 00.4	9 00.7	9 01.1	9 01.4	9 01.8	9 02.2	36 9.0
37	9 14.3	9 14.6	9 15.0	9 15.4	9 15.7	9 16.1	9 16.5	9 16.8	9 17.2	37 9.3
38	9 29.2	9 29.6	9 30.0	9 30.4	9 30.8	9 31.1	9 31.5	9 31.9	9 32.3	38 9.5
39	9 44.2	9 44.6	9 45.0	9 45.4	9 45.8	9 46.2	9 46.6	9 46.9	9 47.3	39 9.8
40	9 59.2	9 59.6	10 00.0	10 00.4	10 00.8	10 01.2	10 01.6	10 02.0	10 02.4	40 10.0
41	10 14.2	10 14.6	10 15.0	10 15.4	10 15.8	10 16.2	10 16.6	10 17.0	10 17.5	41 10.3
42	10 29.2	10 29.6	10 30.0	10 30.4	10 30.8	10 31.3	10 31.7	10 32.1	10 32.5	42 10.5
43	10 44.1	10 44.6	10 45.0	10 45.4	10 45.9	10 46.3	10 46.7	10 47.1	10 47.6	43 10.8
44	10 59.1	10 59.6	11 00.0	11 00.4	11 00.9	11 01.3	11 01.8	11 02.2	11 02.6	44 11.0
45	11 14.1	11 14.6	11 15.0	11 15.4	11 15.9	11 16.3	11 16.8	11 17.2	11 17.7	45 11.3
46	11 29.1	11 29.5	11 30.0	11 30.5	11 30.9	11 31.4	11 31.8	11 32.3	11 32.8	46 11.5
47	11 44.1	11 44.5	11 45.0	11 45.5	11 45.9	11 46.4	11 46.9	11 47.3	11 47.8	47 11.8
48	11 59.0	11 59.5	12 00.0	12 00.5	12 01.0	12 01.4	12 01.9	12 02.4	12 02.9	48 12.0
49	12 14.0	12 14.5	12 15.0	12 15.5	12 16.0	12 16.5	12 17.0	12 17.4	12 17.9	49 12.3
50	12 29.0	12 29.5	12 30.0	12 30.5	12 31.0	12 31.5	12 32.0	12 32.5	12 33.0	50 12.5
51	12 44.0	12 44.5	12 45.0	12 45.5	12 46.0	12 46.5	12 47.0	12 47.5	12 48.1	51 12.8
52	12 59.0	12 59.5	13 00.0	13 00.5	13 01.0	13 01.6	13 02.1	13 02.6	13 03.1	52 13.1
53	13 13.9	13 14.5	13 15.0	13 15.5	13 16.1	13 16.6	13 17.1	13 17.6	13 18.2	53 13.3
54	13 28.9	13 29.5	13 30.0	13 30.5	13 31.1	13 31.6	13 32.2	13 32.7	13 33.2	54 13.5
55	13 43.9	13 44.4	13 45.0	13 45.6	13 46.1	13 46.6	13 47.2	13 47.7	13 48.3	55 13.8
56	13 58.9	13 59.4	14 00.0	14 00.6	14 01.1	14 01.7	14 02.2	14 02.8	14 03.4	56 14.0
57	14 13.9	14 14.4	14 15.0	14 15.6	14 16.1	14 16.7	14 17.3	14 17.8	14 18.4	57 14.3
58	14 28.8	14 29.4	14 30.0	14 30.6	14 31.2	14 31.7	14 32.3	14 32.9	14 33.5	58 14.5
59	14 43.8	14 44.4	14 45.0	14 45.6	14 46.2	14 46.8	14 47.4	14 47.9	14 48.5	59 14.8
60	14 58.8	14 59.4	15 00.0	15 00.6	15 01.2	15 01.8	15 02.4	15 03.0	15 03.6	60 15.0

PLANETS DECLINATION CORRECTION TABLE

Var/hr h m	0.0'	0.1'	0.2'	0.3'	0.4'	0.5'	0.6'	0.7'	0.8'	0.9'	1.0'	1.1'	1.2'	1.3'	1.4'	1.5'
0 00	0.0	0.0	0.0	0.0	0.0	0.0	0.0	0.0	0.0	0.0	0.0	0.0	0.0	0.0	0.0	0.0
12	0.0	0.0	0.0	0.1	0.1	0.1	0.1	0.1	0.2	0.2	02	0.2	0.2	0.3	0.3	0.3
24	0.0	0.0	0.1	0.1	0.2	0.2	0.2	0.3	0.3	0.4	0.4	0.4	0.5	0.6	0.6	0.6
36	0.0	0.1	0.1	0.2	0.2	0.3	0.4	0.4	0.5	0.5	0.6	0.7	0.7	0.8	0.8	0.9
48	0.0	0.1	0.2	0.2	0.3	0.4	0.5	0.6	0.6	0.7	0.8	0.9	1.0	1.0	1.1	1.2
1 00	0.0	0.1	0.2	0.3	0.4	0.5	0.6	0.7	0.8	0.9	1.0	1.1	1.2	1.3	1.4	1.5
12	0.0	0.1	0.2	0.4	0.5	0.6	0.7	0.8	1.0	1.1	1.2	1.3	1.4	1.6	1.7	1.8
24	0.0	0.1	0.3	0.4	0.6	0.7	0.8	1.0	1.1	1.3	1.4	1.5	1.7	1.8	2.0	2.1
36	0.0	0.2	0.3	0.5	0.6	0.8	1.0	1.1	1.3	1.4	1.6	1.8	1.9	2.1	2.2	2.4
48	0.0	0.2	0.4	0.5	0.7	0.9	1.1	1.3	1.4	1.6	1.8	2.0	2.2	2.3	2.5	2.7
2 00	0.0	0.2	0.4	0.6	0.8	1.0	1.2	1.4	1.6	1.8	2.0	2.2	2.4	2.6	2.8	3.0
12	0.0	0.2	0.4	0.7	0.9	1.1	1.3	1.5	1.8	2.0	2.2	2.4	2.6	2.9	3.1	3.3
24	0.0	02	0.5	0.7	1.0	1.2	1.4	1.7	1.9	2.2	2.4	2.6	2.9	3.1	3.4	3.6
36	0.0	0.3	0.5	0.8	1.0	1.3	1.6	1.8	2.1	2.3	2.6	2.9	3.1	3.4	3.6	3.9
48	0.0	0.3	0.6	0.8	1.1	1.4	1.7	2.0	2.2	2.5	2.8	3.1	3.4	3.6	3.9	4.2
3 00	0.0	0.3	0.6	0.9	1.2	1.5	1.8	2.1	2.4	2.7	3.0	3.3	3.6	3.9	4.2	4.5
12	0.0	0.3	0.6	1.0	1.3	1.6	1.9	2.2	2.6	2.9	3.2	3.5	3.8	4.2	4.5	4.8
24	0.0	0.3	0.7	1.0	1.4	1.7	2.0	2.4	2.7	3.1	3.4	3.7	4.1	4.4	4.8	5.1
36	0.0	0.4	0.7	1.1	1.4	1.8	2.2	2.5	2.9	3.2	3.6	4.0	4.3	4.7	5.0	5.4
48	0.0	0.4	0.8	1.1	1.5	1.9	2.3	2.7	3.0	3.4	3.8	4.2	4.6	4.9	5.3	5.7
4 00	0.0	0.4	0.8	1.2	1.6	2.0	2.4	2.8	3.2	3.6	4.0	4.4	4.8	5.2	5.6	6.0
12	0.0	0.4	0.8	1.3	1.7	2.1	2.5	2.9	3.4	3.8	4.2	4.6	5.0	5.5	5.9	6.3
24	0.0	0.4	0.9	1.3	1.8	2.2	2.6	3.1	3.5	4.0	4.4	4.8	5.3	5.7	6.2	6.6
36	0.0	0.5	0.9	1.4	1.8	2.3	2.8	3.2	3.7	4.1	4.6	5.1	5.5	6.0	6.4	6.9
48	0.0	0.5	1.0	1.4	1.9	2.4	2.9	3.4	3.8	4.3	4.8	5.3	5.8	6.2	6.7	7.2
5 00	0.0	0.5	1.0	1.5	2.0	2.5	3.0	3.5	4.0	4.5	5.0	5.5	6.0	6.5	7.0	7.5
12	0.0	0.5	1.0	1.6	2.1	2.6	3.1	3.6	4.2	4.7	5.2	5.7	6.2	6.8	7.3	7.8
24	0.0	0.5	1.1	1.6	2.2	2.7	3.2	3.8	4.3	4.9	5.4	5.9	6.5	7.0	7.6	8.1
36	0.0	0.6	1.1	1.7	2.2	2.8	3.4	3.9	4.5	5.0	5.6	6.2	6.7	7.3	7.8	8.4
48	0.0	0.6	1.2	1.7	2.3	2.9	3.5	4.1	4.6	5.2	5.8	6.4	7.0	7.5	8.1	8.7
6 00	0.0	0.6	1.2	1.8	2.4	3.0	3.6	4.2	4.8	5.4	6.0	6.6	7.2	7.8	8.4	9.0
12	0.0	0.6	1.2	1.9	2.5	3.1	3.7	4.3	5.0	5.6	6.2	6.8	7.4	8.1	8.7	9.3
24	0.0	0.6	1.3	1.9	2.6	3.2	3.8	4.5	5.1	5.8	6.4	7.0	7.7	8.3	9.0	9.6
36	0.0	0.7	1.3	2.0	2.6	3.3	4.0	4.6	5.3	5.9	6.6	7.3	7.9	8.6	9.2	9.9
48	0.0	0.7	1.4	2.0	2.7	3.4	4.1	4.8	5.4	6.1	6.8	7.5	8.2	B.8	9.5	10.2
7 00	0.0	0.7	1.4	2.1	2.8	3.5	4.2	4.9	5.6	6.3	7.0	7.7	8.4	9.1	9.8	10.5
12	0.0	0.7	1.4	2.2	2.9	3.6	4.3	5.0	5.8	6.5	7.2	7.9	8.6	9.4	10.1	10.8
24	0.0	0.7	1.5	2.2	3.0	3.7	4.4	5.2	5.9	6.7	7.4	8.1	8.9	9.6	10.4	11.1
36	0.0	0.8	1.5	2.3	3.0	3.8	4.6	5.3	6.1	6.8	7.6	8.4	9.1	9.9	10.6	11.4
48	0.0	0.8	1.6	2.3	3.1	3.9	4.7	5.5	6.2	7.0	7.8	8.6	9.4	10.1	10.9	11.7
8 00	0.0	0.8	1.6	2.4	3.2	4.0	4.8	5.6	6.4	7.2	8.0	8.8	9.6	10.4	11.2	12.0
12	0.0	0.8	1.6	2.5	3.3	4.1	4.9	5.7	6.6	7.4	8.2	9.0	9.8	10.7	11.5	12.3
24	0.0	0.8	1.7	2.5	3.4	4.2	5.0	5.9	6.7	7.6	8.4	9.2	10.1	10.9	11.8	12.6
36	0.0	0.9	1.7	2.6	3.4	4.3	5.2	6.0	6.9	7.7	8.6	9.5	10.3	11.2	12.0	12.9
48	0.0	0.9	1.8	2.6	3.5	4.4	5.3	6.2	7.0	7.9	8.8	9.7	10.6	11.4	12.3	13.2
9 00	0.0	0.9	1.8	2.7	3.6	4.5	5.4	6.3	7.2	8.1	9.0	9.9	10.8	11.7	12.6	13.5
12	0.0	0.9	1.8	2.8	3.7	4.6	5.5	6.4	7.4	8.3	9.2	10.1	11.0	12.0	12.9	13.8
24	0.0	0.9	1.9	2.8	3.8	4.7	5.6	6.6	7.5	8.5	9.4	10.3	11.3	12.2	13.2	14.1
36	0.0	1.0	1.9	2.9	3.8	4.8	5.8	6.7	7.7	8.6	9.6	10.6	11.5	12.5	13.4	14.4
48	0.0	1.0	2.0	2.9	3.9	4.9	5.9	6.9	7.8	8.8	9.8	10.8	11.8	12.7	13.7	14.7
10 00	0.0	1.0	2.0	3.0	4.0	5.0	6.0	7.0	8.0	9.0	10.0	11.0	12.0	13.0	14.0	15.0
12	0.0	1.0	2.0	3.1	4.1	5.1	6.1	7.1	8.2	9.2	10.2	11.2	12.2	13.2	14.3	15.3
24	0.0	1.0	2.1	3.1	4.2	5.2	6.2	7.3	8.3	9.4	10.4	11.4	12.5	13.5	14.6	15.6
36	0.0	1.1	2.2	3.2	4.2	5.3	6.4	7.4	8.5	9.5	10.6	11.7	12.7	13.8	14.8	15.9
48	0.0	1.1	2.2	3.2	4.3	5.4	6.5	7.6	8.6	9.7	10.8	11.9	13.0	14.0	15.1	16.2
11 00	0.0	1.1	2.2	3.3	4.4	5.5	6.6	7.7	8.8	9.9	11.0	12.1	13.2	14.3	15.4	16.5
12	0.0	1.1	2.2	3.4	4.5	5.6	6.7	7.8	9.0	10.1	11.2	12.3	13.4	14.6	15.7	16.8
24	0.0	1.1	2.3	3.4	4.6	5.7	6.8	8.0	9.1	10.3	11.4	12.5	13.7	14.8	16.0	17.1
36	0.0	1.2	2.3	3.5	4.6	5.8	7.0	8.1	9.3	10.4	11.6	12.8	13.9	15.1	16.2	17.4
48	0.0	1.2	2.4	3.5	4.7	5.9	7.1	8.3	9.4	10.6	11.8	13.0	14.2	15.3	16.5	17.7
12 00	0.0	1.2	2.4	3.6	4.8	6.0	7.2	8.4	9.6	10.8	12.0	13.2	14.4	15.6	16.8	18.0

PLANETS DECLINATION CORRECTION TABLE

Var/hr h m	0.0'	0.1'	0.2'	0.3'	0.4'	0.5'	0.6'	0.7'	0.8'	0.9'	1.0'	1.1'	1.2'	1.3'	1.4'	1.5'
12 00	0.0	1.2	2.4	3.6	4.8	6.0	7.2	8.4	9.6	10.8	12.0	13.2	14.4	15.6	16.8	18.0
12	0.0	1.2	2.4	3.7	4.9	6.1	7.3	8.5	9.8	11.0	12.2	13.4	14.6	15.9	17.1	18 3
24	0.0	1.2	2.5	3.7	5.0	6.2	7.4	8.7	9.9	11.2	12.4	13.6	14.9	16.1	17.4	18 6
36	0.0	1.3	2.5	3.8	5.0	6.3	7.6	8.8	10.1	11.3	12.6	13.9	15.1	16.4	17.6	18.9
48	0.0	1.3	2.6	3.8	5.1	6.4	7.7	9.0	10.2	11.5	12.8	14.1	15.4	16.6	17.9	19.2
13 00	0.0	1.3	2.6	3.9	5.2	6.5	7.8	9.1	10.4	11.7	13.0	14.3	15.6	16.9	18.2	19.5
12	0.0	1.3	2.6	4.0	5.3	6.6	7.9	9.2	10.6	11.9	13.2	14.5	15.8	17.2	18.5	19.8
24	0.0	1.3	2.7	4.0	5.4	6.7	8.0	9.4	10.7	12.1	13.4	14.7	16.1	17.4	18.8	20.1
36	0.0	1.4	2.7	4.1	5.4	6.8	8.2	9.5	10.9	12.2	13.6	15.0	16.3	17.7	19.0	20.4
48	0.0	1.4	2.8	4.1	5.5	6.9	8.3	9.7	11.0	12.4	13.8	15.2	16.6	17.9	19.3	20.7
14 00	0.0	1.4	2.8	4.2	5.6	7.0	8.4	9.8	11.2	12.6	14.0	15.4	16.8	18.2	19.6	21.0
12	0.0	1.4	2.8	4.3	5.7	7.1	8.5	9.9	11.4	12.8	14.2	15.6	17.0	18.5	19.9	21.3
24	0.0	1.4	2.9	4.3	5.8	7.2	8.6	10.1	11.5	13.0	14.4	15.8	17.3	18.7	20.2	21.6
36	0.0	1.5	2.9	4.4	5.8	7.3	8.8	10.2	11.7	13.1	14.6	16.1	17.5	19.0	20.4	21.9
48	0.0	1.5	3.0	4.4	5.9	7.4	8.9	10.4	11.8	13.3	14.8	16.3	17.8	19.2	20.7	22.2
15 00	0.0	1.5	3.0	4.5	6.0	7.5	9.0	10.5	12.0	13.5	15.0	16.5	18.0	19.5	21.0	22.5
12	0.0	1.5	3.0	4.6	6.1	7.6	9.1	10.6	12.2	13.7	15.2	16.7	18.2	19.8	21.3	22.8
24	0.0	1.5	3.1	4.6	6.2	7.7	9.2	10.8	12.3	13.9	15.4	16.9	18.5	20.0	21.6	23.1
36	0.0	1.6	3.1	4.7	6.2	7.8	9.4	10.9	12.5	14.0	15.6	17.2	18.7	20.3	21.8	23.4
48	0.0	1.6	3.2	4.7	6.3	7.9	9.5	11.1	12.6	14.2	15.8	17.4	19.0	20.5	22.1	23.7
16 00	0.0	1.6	3.2	4.8	6.4	8.0	9.6	11.2	12.8	14.4	16.0	17.6	19.2	20.8	22.4	24.0
12	0.0	1.6	3.2	4.9	6.5	8.1	9.7	11.3	13.0	14.6	16.2	17.8	19.4	21.1	22.7	24.3
24	0.0	1.6	3.3	4.9	6.6	8.2	9.8	11.5	13.1	14.8	16.4	18.0	19.7	21.3	23.0	24.6
36	0.0	1.7	3.3	5.0	6.6	8.3	10.0	11.6	13.3	14.9	16.6	18.3	19.9	21.6	23.2	24.9
48	0.0	1.7	3.4	5.0	6.7	8.4	10.1	11.8	13.4	15.1	16.8	18.5	20.2	21.8	23.5	25.2
17 00	0.0	1.7	3.4	5.1	6.8	8.5	10.2	11.9	13.6	15.3	17.0	18.7	20.4	22.1	23.8	25.5
12	0.0	1.7	3.4	5.2	6.9	8.6	10.3	12.0	13.8	15.5	17.2	18.9	20.6	22.4	24.1	25.8
24	0.0	1.7	3.5	5.2	7.0	8.7	10.4	12.2	13.9	15.7	17.4	19.1	20.9	22.6	24.4	26.1
36	0.0	1.8	3.5	5.3	7.0	8.8	10.6	12.3	14.1	15.8	17.6	19.4	21.1	22.9	24.6	26.4
48	0.0	1.8	3.6	5.3	7.1	8.9	10.7	12.5	14.2	16.0	17.8	19.6	21.4	23.1	24.9	26.7
18 00	0.0	1.8	3.6	5.4	7.2	9.0	10.8	12.6	14.4	16.2	18.0	19.8	21.6	23.4	25.2	27.0
12	0.0	1.8	3.6	5.5	7.3	9.1	10.9	12.7	14.6	16.4	18.2	20.0	21.8	23.7	25.5	27.3
24	0.0	1.8	3.7	5.5	7.4	9.2	11.0	12.9	14.7	16.6	18.4	20.2	22.1	23.9	25.8	27.6
36	0.0	1.9	3.7	5.6	7.4	9.3	11.2	13.0	14.9	16.7	18.6	20.5	22.3	24.2	26.0	27.9
48	0.0	1.9	3.8	5.6	7.5	9.4	11.3	13.2	15.0	16.9	18.8	20.7	22.6	24.4	26.3	28.2
19 00	0.0	1.9	3.8	5.7	7.6	9.5	11.4	13.3	15.2	17.1	19.0	20.9	22.8	24.7	26.6	28.5
12	0.0	1.9	3.8	5.8	7.7	9.6	11.5	13.4	15.4	17.3	19.2	21.1	23.0	25.0	26.9	28.8
24	0.0	1.9	3.9	5.8	7.8	9.7	11.6	13.6	15.5	17.5	19.4	21.3	23.3	25.2	27.2	29.1
36	0.0	2.0	3.9	5.9	7.8	9.8	11.8	13.7	15.7	17.6	19.6	21.6	23.5	25.5	27.4	29.4
48	0.0	2.0	4.0	5.9	7.9	9.9	11.9	13.9	15.8	17.8	19.8	21.8	23.8	25.7	27.7	29.7
20 00	0.0	2.0	4.0	6.0	8.0	10.0	12.0	14.0	16.0	18.0	20.0	22.0	24.0	26.0	28.0	30.0
12	0.0	2.0	4.0	6.1	8.1	10.1	12.1	14.1	16.2	18.2	20.2	22.2	24.2	26.3	28.3	30.3
24	0.0	2.0	4.1	6.1	8.2	10.2	12.2	14.3	16.3	18.4	20.4	22.4	24.5	26.5	28.6	30.6
36	0.0	2.1	4.1	6.2	8.2	10.3	12.4	14.4	16.5	18.5	20.6	22.7	24.7	26.8	28.8	30.9
48	0.0	2.1	4.2	6.2	8.3	10.4	12.5	14.6	16.6	18.7	20.8	22.9	25.0	27.0	29.1	31.2
21 00	0.0	2.1	4.2	6.3	8.4	10.5	12.6	14.7	16.8	18.9	21.0	23.1	25.2	27.3	29.4	31.5
12	0.0	2.1	4.2	6.4	8.5	10.6	12.7	14.8	17.0	19.1	21.2	23.3	25.4	27.6	29.7	31.8
24	0.0	2.1	4.3	6.4	8.6	10.7	12.8	15.0	17.1	19.3	21.4	23.5	25.7	27.8	30.0	32.1
36	0.0	2.2	4.3	6.5	8.6	10.8	13.0	15.1	17.3	19.4	21.6	23.8	25.9	28.1	30.2	32.4
48	0.0	2.2	4.4	6.5	8.7	10.9	13.1	15.3	17.4	19.6	21.8	24.0	26.2	28.3	30.5	32.7
22 00	0.0	2.2	4.4	6.6	8.8	11.0	13.2	15.4	17.6	19.8	22.0	24.2	26.4	28.6	30.8	33.0
12	0.0	2.2	4.4	6.7	8.9	11.1	13.3	15.5	17.8	20.0	22.2	24.4	26.6	28.9	31.1	33.3
24	0.0	2.2	4.5	6.7	9.0	11.2	13.4	15.7	17.9	20.2	22.4	24.6	26.9	29.1	31.4	33.6
36	0.0	2.3	4.5	6.8	9.0	11.3	13.6	15.8	18.1	20.3	22.6	24.9	27.1	29.4	31.6	33.9
48	0.0	2.3	4.6	6.8	9.1	11.4	13.7	16.0	18.2	20.5	22.8	25.1	27.4	29.6	31.9	34.2
23 00	0.0	2.3	4.6	6.9	9.2	11.5	13.8	16.1	18.4	20.7	23.0	25.3	27.6	29.9	32.2	34.5
12	0.0	2.3	4.6	7.0	9.3	11.6	13.9	16.2	18.6	20.9	23.2	25.5	27.8	30.2	32.5	34.8
24	0.0	2.3	4.7	7.0	9.4	11.7	14.0	16.4	18.7	21.1	23.4	25.7	28.1	30.4	32.8	35.1
36	0.0	2.4	4.7	7.1	9.4	11.8	14.2	16.5	18.9	21.2	23.6	26.0	28.3	30.7	33.0	35.4
48	0.0	2.4	4.8	7.1	9.5	11.9	14.3	16.7	19.0	21.4	23.8	26.2	28.6	30.9	33.3	35.7
24 00	0.0	2.4	4.8	7.2	9.6	12.0	14.4	16.8	19.2	21.6	24.0	26.4	28.8	31.2	33.6	36.0

50

NOTES ON PHASES OF THE MOON

What follows below is approximate. The season of the year, the latitude of the observer and the declination of the Moon all have an influence.

The Moon's phases are shown on the third monthly pages. It is useful to be aware of them, not least because they will indicate if the Moon is likely to be available for sights. Certainly the Moon should normally be used, if visible at the right altitude, for morning and evening Stars; and don't overlook its value during the day.

1 Technically the 'New Moon' cannot be seen; it rises and sets with the Sun. As it begins its monthly journey around the Earth it will be seen as a conventional 'New Moon', setting at dusk and not yet useful to us.

2 By the time the Moon reaches its first quarter (Half Moon) it will be rising at midday and setting at midnight – very useful for afternoon sights and evening Stars.

3 Nearing Full Moon it will be rising and setting ever later until, at Full Moon, it rises at dusk and sets at dawn; it is then often too low in the sky for morning or evening Stars.

4 As the Moon approaches its last quarter (Half Moon) it will start setting after dawn and therefore be available for morning Stars and for daylight sights in the forenoon. At its last quarter it will rise at midnight and set at midday. Soon, however, it will be rising too late to be high enough for morning Stars and in two or three days will start its cycle again as the New Moon.

MOON GHA CORRECTION TABLE (HOURS)
ALWAYS ADD

Var/hr	14°20'	14°20.5'	14°21'	14°21.5'	14°22'	14°22.5'	14°23'	14°23.5'	14°24'	14°24.5'	14°25'	14°25.5'
	° '	° '	° '	° '	° '	° '	° '	° '	° '	° '	° '	° '
1	14 20.0	14 20.5	14 21.0	14 21.5	14 22.0	14 22.5	14 23.0	14 23.5	14 24.0	14 24.5	14 25.0	14 25.5
2	28 40.0	28 41.0	28 42.0	28 43.0	28 44.0	28 45.0	28 46.0	28 47.0	28 48.0	28 49.0	28 50.0	28 51.0
3	43 00.0	43 01.5	43 03.0	43 04.5	43 06.0	43 07.5	43 09.0	43 10.5	43 12.0	43 13.5	43 15.0	43 16.5
4	57 20.0	57 22.0	57 24.0	57 26.0	57 28.0	57 30.0	57 32.0	57 34.0	57 36.0	57 38.0	57 40.0	57 42.0
5	71 40.0	71 42.5	71 45.0	71 47.5	71 50.0	71 52.5	71 55.0	71 57.5	72 00.0	72 02.5	72 05.0	72 07.5

(HOURS)

Var/hr	14°26'	14°26.5'	14°27'	14°27.5'	14°28'	14°28.5'	14°29'	14°29.5'	14°30'	14°30.5'	14°31'	14°31.5'
	° '	° '	° '	° '	° '	° '	° '	° '	° '	° '	° '	° '
1	14 26.0	14 26.5	14 27.0	14 27.5	14 28.0	14 28.5	14 29.0	14 29.5	14 30.0	14 30.5	14 31.0	14 31.5
2	28 52.0	28 53.0	28 54.0	28 55.0	28 56.0	28 57.0	28 58.0	28 59.0	29 00.0	29 01.0	29 02.0	29 03.0
3	43 18.0	43 19.5	43 21.0	43 22.5	43 24.0	43 25.5	43 27.0	43 28.5	43 30.0	43 31.5	43 33.0	43 34.5
4	57 44.0	57 46.0	57 48.0	57 50.0	57 52.0	57 54.0	57 56.0	57 58.0	58 00.0	58 02.0	58 04.0	58 06.0
5	72 10.0	72 12.5	72 15.0	72 17.5	72 20.0	72 22.5	72 25.0	72 27.5	72 30.0	72 32.5	72 35.0	72 37.5

(HOURS)

Var/hr	14°32'	14°32.5'	14°33'	14°33.5'	14°34'	14°34.5'	14°35'	14°35.5'	14°36'	14°36.5'	14°37'	14°37.5'
	° '	° '	° '	° '	° '	° '	° '	° '	° '	° '	° '	° '
1	14 32.0	14 32.5	14 33.0	14 33.5	14 34.0	14 34.5	14 35.0	14 35.5	14 36.0	14 36.5	14 37.0	14 37.5
2	29 04.0	29 05.0	29 06.0	29 07.0	29 08.0	29 09.0	29 10.0	29 11.0	29 12.0	29 13.0	29 14.0	29 15.0
3	43 36.0	43 37.5	43 39.0	43 40.5	43 42.0	43 43.5	43 45.0	43 46.5	43 48.0	43 49.5	43 51.0	43 52.5
4	58 08.0	58 10.0	58 12.0	58 14.0	58 16.0	58 18.0	58 20.0	58 22.0	58 24.0	58 26.0	58 28.0	58 30.0
5	72 40.0	72 42.5	72 45.0	72 47.5	72 50.0	72 52.5	72 55.0	72 57.5	73 00.0	73 02.5	73 05.0	73 07.5

(HOURS)

NOTE: The correction for Moon GHA is taken in three parts. Using the figure for variation per hour given in the monthly tables, you get the correction for hours from the table above and the correction for minutes from the tables on the next two pages. Finally, take the correction for seconds from the right-hand side of whichever minutes table you used.

MOON GHA CORRECTION TABLE (MINUTES)

MINUTES

Var/hr	14°20'	14°21'	14°22'	14°23'	14°24'	14°25'	14°26'	14°27'	14°28'	Diff 1'	Sec	Corr
	° '	° '	° '	° '	° '	° '	° '	° '	° '	'	'	'
0	0 00.0	0 00.0	0 00.0	0 00.0	0 00.0	0 00.0	0 00.0	0 00.0	0 00.0	0.0	0	0.0
1	0 14.3	0 14.4	0 14.4	0 14.4	0 14.4	0 14.4	0 14.4	0 14.4	0 14.5	0.0	1	0.2
2	0 28.7	0 28.7	0 28.7	0 28.8	0 28.8	0 28.8	0 28.9	0 28.9	0 28.9	0.0	2	0.5
3	0 43.0	0 43.0	0 43.1	0 43.2	0 43.2	0 43.2	0 43.3	0 43.4	0 43.4	0.0	3	0.7
4	0 57.3	0 57.4	0 57.5	0 57.5	0 57.6	0 57.7	0 57.7	0 57.8	0 57.9	0.1	4	1.0
5	1 11.7	1 11.8	1 11.8	1 11.9	1 12.0	1 12.1	1 12.2	1 12.2	1 12.3	0.1	5	1.2
6	1 26.0	1 26.1	1 26.2	1 26.3	1 26.4	1 26.5	1 26.6	1 26.7	1 26.8	0.1	6	1.4
7	1 40.3	1 40.4	1 40.6	1 40.7	1 40.8	1 40.9	1 41.2	1 41.2	1 41.3	0.1	7	1.7
8	1 54.7	1 54.8	1 54.9	1 55.1	1 55.2	1 55.3	1 55.5	1 55.6	1 55.7	0.1	8	1.9
9	2 09.0	2 09.2	2 09.3	2 09.4	2 09.6	2 09.8	2 09.9	2 10.0	2 10.2	0.2	9	2.2
10	2 23.3	2 23.5	2 23.7	2 23.8	2 24.0	2 24.2	2 24.3	2 24.5	2 24.7	0.2	10	2.4
11	2 37.7	2 37.8	2 38.0	2 38.2	2 38.4	2 38.6	2 38.8	2 39.0	2 39.1	0.2	11	2.6
12	2 52.0	2 52.2	2 52.4	2 52.6	2 52.8	2 53.0	2 53.2	2 53.4	2 53.6	0.2	12	2.9
13	3 06.3	3 06.6	3 06.8	3 07.0	3 07.2	3 07.4	3 07.6	3 07.8	3 08.1	0.2	13	3.1
14	3 20.7	3 20.9	3 21.1	3 21.4	3 21.6	3 21.8	3 22.1	3 22.3	3 22.5	0.2	14	3.4
15	3 35.0	3 35.2	3 35.5	3 35.8	3 36.0	3 36.2	3 36.5	3 36.8	3 37.0	0.2	15	3.6
16	3 49.3	3 49.6	3 49.9	3 50.1	3 50.4	3 50.7	3 50.9	3 51.2	3 51.5	0.3	16	3.8
17	4 03.7	4 04.0	4 04.2	4 04.5	4 04.8	4 05.1	4 05.4	4 05.6	4 05.9	0.3	17	4.1
18	4 18.0	4 18.3	4 18.6	4 18.9	4 19.2	4 19.5	4 19.8	4 20.1	4 20.4	0.3	18	4.3
19	4 32.3	4 32.6	4 33.0	4 33.3	4 33.6	4 33.9	4 34.2	4 34.6	4 34.9	0.3	19	4.6
20	4 46.7	4 47.0	4 47.3	4 47.7	4 48.0	4 48.3	4 48.7	4 49.0	4 49.3	0.3	20	4.8
21	5 01.0	5 01.4	5 01.7	5 02.0	5 02.4	5 02.8	5 03.1	5 03.4	5 03.8	0.4	21	5.0
22	5 15.3	5 15.7	5 16.1	5 16.4	5 16.8	5 17.2	5 17.5	5 17.9	5 18.3	0.4	22	5.3
23	5 29.7	5 30.0	5 30.4	5 30.8	5 31.2	5 31.6	5 32.0	5 32.4	5 32.7	0.4	23	5.5
24	5 44.0	5 44.4	5 44.8	5 45.2	5 45.6	5 46.0	5 46.4	5 46.8	5 47.2	0.4	24	5.8
25	5 58.3	5 58.8	5 59.2	5 59.6	6 00.0	6 00.4	6 00.8	6 01.2	6 01.7	0.4	25	6.0
26	6 12.7	6 13.1	6 13.5	6 14.0	6 14.4	6 14.8	6 15.3	6 15.7	5 16.1	0.4	26	6.2
27	6 27.0	6 27.4	6 27.9	6 28.4	6 28.8	6 29.2	6 29.7	6 30.2	6 30.6	0.4	27	6.5
28	6 41.3	6 41.8	6 42.3	6 42.7	6 43.2	6 43.7	6 44.1	6 44.6	6 45.1	0.5	28	6.7
29	6 55.7	6 56.2	6 56.6	6 57.1	6 57.6	6 58.1	6 58.6	6 59.0	6 59.5	0.5	29	7.0
30	7 10.0	7 10.5	7 11.0	7 11.5	7 12.0	7 12.5	7 13.0	7 13.5	7 14.0	0.5	30	7.2
31	7 24.3	7 24.8	7 25.4	7 25.9	7 26.4	7 26.9	7 27.4	7 28.0	7 28.5	0.5	31	7.4
32	7 38.7	7 39.2	7 39.7	7 40.3	7 40.8	7 41.3	7 41.9	7 42.4	7 42.9	0.5	32	7.7
33	7 53.0	7 53.6	7 54.1	7 54.6	7 55.2	7 55.8	7 56.3	7 56.8	7 57.4	0.6	33	7.9
34	8 07.3	8 07.9	8 08.5	8 09.0	8 09.6	8 10.2	8 10.7	8 11.3	8 11.9	0.6	34	8.2
35	8 21.7	8 22.2	8 22.8	8 23.4	8 24.0	8 24.6	8 25.2	8 25.8	8 26.3	0.6	35	8.4
36	8 36.0	8 36.6	8 37.2	8 37.8	8 38.4	8 39.0	8 39.6	8 40.2	8 40.8	0.6	36	8.6
37	8 50.3	8 51.0	8 51.6	8 52.2	8 52.8	8 53.4	8 54.0	8 54.6	8 55.3	0.6	37	8.9
38	9 04.7	9 05.3	9 05.9	9 06.6	9 07.2	9 07.8	9 08.5	9 09.1	9 09.7	0.6	38	9.1
39	9 19.0	9 19.6	9 20.3	9 21.0	9 21.6	9 22.2	9 22.9	9 23.6	9 24.2	0.6	39	9.4
40	9 33.3	9 34.0	9 34.7	9 35.3	9 36.0	9 36.7	9 37.3	9 38.0	9 38.7	0.7	40	9.6
41	9 47.7	9 48.4	9 49.0	9 49.7	9 50.4	9 51.1	9 51.8	9 52.4	9 53.1	0.7	41	9.8
42	10 02.0	10 02.7	10 03.4	10 04.1	10 04.8	10 05.5	10 06.2	10 06.9	10 07.6	0.7	42	10.1
43	10 16.3	10 17.0	10 17.8	10 18.5	10 19.2	10 19.9	10 20.6	10 21.4	10 22.1	0.7	43	10.3
44	10 30.7	10 31.4	10 32.1	10 32.9	10 33.6	10 34.3	10 35.1	10 35.8	10 36.5	0.7	44	10.6
45	10 45.0	10 45.8	10 46.5	10 47.2	10 48.0	10 48.8	10 49.5	10 50.2	10 51.0	0.8	45	10.8
46	10 59.3	11 00.1	11 00.9	11 01.6	11 02.4	11 03.2	11 03.9	11 04.7	11 05.5	0.8	46	11.0
47	11 13.7	11 14.4	11 15.2	11 16.0	11 16.8	11 17.6	11 18.4	11 19.2	11 19.9	0.8	47	11.3
48	11 28.0	11 28.8	11 29.6	11 30.4	11 31.2	11 32.0	11 32.8	11 33.6	11 34.4	0.8	48	11.5
49	11 42.3	11 43.2	11 44.0	11 44.8	11 45.6	11 46.4	11 47.2	11 48.0	11 48.9	0.8	49	11.8
50	11 56.7	11 57.5	11 58.3	11 59.2	12 00.0	12 00.8	12 01.7	12 02.5	12 03.3	0.8	50	12.0
51	12 11.0	12 11.8	12 12.7	12 13.6	12 14.4	12 15.2	12 16.1	12 17.0	12 17.8	0.8	51	12.2
52	12 25.3	12 26.2	12 27.1	12 27.9	12 28.8	12 29.7	12 30.5	12 31.4	12 32.3	0.9	52	12.5
53	12 39.7	12 40.6	12 41.4	12 42.3	12 43.2	12 44.1	12 45.0	12 45.8	12 46.7	0.9	53	12.7
54	12 54.0	12 54.9	12 55.8	12 56.7	12 57.6	12 58.5	12 59.4	13 00.3	13 01.2	0.9	54	13.0
55	13 08.3	13 09.2	13 10.2	13 11.1	13 12.0	13 12.9	13 13.8	13 14.8	13 15.7	0.9	55	13.2
56	13 22.7	13 23.6	13 24.5	13 25.5	13 26.4	13 27.3	13 28.3	13 29.2	13 30.1	0.9	56	13.4
57	13 37.0	13 38.0	13 38.9	13 39.8	13 40.8	13 41.8	13 42.7	13 43.6	13 44.6	1.0	57	13.7
58	13 51.3	13 52.3	13 53.3	13 54.2	13 55.2	13 56.2	13 57.1	13 58.1	13 59.1	1.0	58	13.9
59	14 05.7	14 06.6	14 07.6	14 08.6	14 09.6	14 10.6	14 11.6	14 12.6	14 13.5	1.0	59	14.1
60	14 20.0	14 21.0	14 22.0	14 23.0	14 24.0	14 25.0	14 26.0	14 27.0	14 28.0	1.0	60	14.4

MOON GHA CORRECTION TABLE (MINUTES)

MINUTES

Var/hr	14°290'	14°30'	14°31'	14°32'	14°33'	14°34'	14°35'	14°36'	14°37'	Diff 1'	Sec	Corr
	° '	° '	° '	° '	° '	° '	° '	° '	° '	'	'	'
0	0 00.0	0 00.0	0 00.0	0 00.0	0 00.0	0 00.0	0 00.0	0 00.0	0 00.0	0.0	0	0.0
1	0 14.5	0 14.5	0 14.5	0 14.6	0 14.6	0 14.6	0 14.6	0 14.6	0 14.6	0.0	1	0.2
2	0 29.0	0 29.0	0 29.0	0 29.1	0 29.1	0 29.1	0 29.2	0 29.2	0 29.2	0.0	2	0.5
3	0 43.4	0 43.5	0 43.6	0 43.6	0 43.6	0 43.7	0 43.8	0 43.8	0 43.8	0.0	3	0.7
4	0 57.9	0 58.0	0 58.1	0 58.1	0 58.2	0 58.3	0 58.3	0 58.4	0 58.5	0.1	4	1.0
5	1 12.4	1 12.5	1 12.6	1 12.7	1 12.8	1 12.8	1 12.9	1 13.0	1 13.1	0.1	5	1.2
6	1 26.9	1 27.0	1 27.1	1 27.2	1 27.3	1 27.4	1 27.5	1 27.6	1 27.7	0.1	6	1.5
7	1 41.4	1 41.5	1 41.6	1 41.7	1 41.8	1 42.0	1 42.1	1 42.2	1 42.3	0.1	7	1.7
8	1 55.9	1 56.0	1 56.1	1 56.3	1 56.4	1 56.5	1 56.7	1 56.8	1 56.9	0.1	8	1.9
9	2 10.4	2 10.5	2 10.6	2 10.8	2 11.0	2 11.1	2 11.2	2 11.4	2 11.6	0.2	9	2.2
10	2 24.8	2 25.0	2 25.2	2 25.2	2 25.5	2 25.7	2 25.8	2 26.0	2 26.2	0.2	10	2.4
11	2 39.3	2 39.5	2 39.7	2 39.9	2 40.0	2 40.2	2 40.4	2 40.6	2 40.8	0.2	11	2.7
12	2 53.8	2 54.0	2 54.2	2 54.4	2 54.6	2 54.8	2 55.0	2 55.2	2 55.4	0.2	12	2.9
13	3 08.3	3 08.5	3 08.7	3 08.9	3 09.2	3 09.4	3 09.6	3 09.8	3 10.0	0.2	13	3.2
14	3 22.8	3 23.0	3 23.2	3 23.5	3 23.7	3 23.9	3 24.2	3 24.4	3 24.6	0.2	14	3.4
15	3 37.2	3 37.5	3 37.8	3 38.0	3 38.2	3 38.5	3 38.8	3 39.0	3 39.2	0.2	15	3.6
16	3 51.7	3 52.0	3 52.3	3 52.5	3 52.8	3 53.1	3 53.3	3 53.6	3 53.9	0.3	16	3.9
17	4 06.2	4 06.5	4 06.8	4 07.1	4 07.4	4 07.6	4 07.9	4 08.2	4 08.5	0.3	17	4.1
18	4 20.7	4 21.0	4 21.3	4 21.6	4 21.9	4 22.2	4 22.5	4 22.8	4 23.1	0.3	18	4.4
19	4 35.2	4 35.5	4 35.8	4 36.1	4 36.4	4 36.8	4 37.1	4 37.4	4 37.7	0.3	19	4.6
20	4 49.7	4 50.0	4 50.3	4 50.7	4 51.0	4 51.3	4 51.7	4 52.0	4 52.3	0.3	20	4.9
21	5 04.2	5 04.5	5 04.8	5 05.2	5 05.6	5 05.9	5 06.2	5 06.6	5 07.0	0.4	21	5.1
22	5 18.6	5 19.0	5 19.4	5 19.7	5 20.1	5 20.5	5 20.8	5 21.2	5 21.6	0.4	22	5.3
23	5 33.1	5 33.5	5 33.9	5 34.3	5 34.6	5 35.0	5 35.4	5 35.8	5 36.2	0.4	23	5.6
24	5 47.6	5 48.0	5 48.4	5 48.8	5 49.2	5 49.6	5 50.0	5 50.4	5 50.8	0.4	24	5.8
25	6 02.1	6 02.5	6 02.9	6 03.3	6 03.8	6 04.2	6 04.6	6 05.0	6 05.4	0.4	25	6.1
26	6 16.6	6 17.0	6 17.4	6 17.9	6 18.3	6 18.7	6 19.2	6 19.6	6 20.0	0.4	26	6.3
27	6 31.0	6 31.5	6 32.0	6 32.4	6 32.8	6 33.3	6 33.8	6 34.2	6 34.6	0.4	27	6.5
28	6 45.5	6 46.0	6 46.5	6 46.9	6 47.4	6 47.9	6 4&3	6 48.8	6 49.3	0.5	28	6.8
29	7 00.0	7 00.5	7 01.0	7 01.5	7 02.0	7 02.4	7 02.9	7 03.4	7 03.9	0.5	29	7.0
30	7 14.5	7 15.0	7 15.5	7 16.0	7 16.5	7 17.0	7 17.5	7 18.0	7 18.5	0.5	30	7.3
31	7 29.0	7 29.5	7 30.0	7 30.5	7 31.0	7 31.6	7 32.1	7 32.6	7 33.1	0.5	31	7.5
32	7 43.5	7 44.0	7 44.5	7 45.1	7 45.6	7 46.1	7 46.7	7 47.2	7 47.7	0.5	32	7.8
33	7 58.0	7 58.5	7 59.0	7 59.6	8 00.2	8 00.7	8 01.2	8 01.8	8 02.4	0.6	33	8.0
34	8 12.4	8 13.0	8 13.6	8 14.1	8 14.7	8 15.3	8 15.8	8 16.4	8 17.0	0.6	34	8.2
35	8 26.9	8 27.5	8 28.1	8 28.7	8 29.2	8 29.8	8 30.4	8 31.0	8 31.6	0.6	35	8.5
36	8 41.4	8 42.0	8 42.6	8 43.2	8 43.8	8 44.4	8 45.0	8 45.6	8 46.2	0.6	36	8.7
37	8 55.9	8 56.5	8 57.1	8 57.7	8 58.4	8 59.0	8 59.6	9 00.2	9 00.8	0.6	37	9.0
38	9 10.4	9 11.0	9 11.6	9 12.3	9 12.9	9 13.5	9 14.2	9 14.8	9 15.4	0.6	38	9.2
39	9 24.8	9 25.5	9 26.2	9 26.8	9 27.4	9 28.1	9 28.8	9 29.4	9 30.0	0.6	39	9.5
40	9 39.3	9 40.0	9 40.7	9 41.3	9 42.0	9 42.7	9 43.3	9 44.0	9 44.7	0.7	40	9.7
41	9 53.8	9 54.5	9 55.2	9 55.9	9 56.6	9 57.2	9 57.9	9 58.6	9 59.3	0.7	41	9.9
42	10 08.3	10 09.0	10 09.7	10 10.4	10 11.1	10 11.8	10 12.5	10 13.2	10 13.9	0.7	42	10.2
43	10 22.8	10 23.5	10 24.2	10 24.9	10 25.6	10 26.4	10 27.1	10 27.8	10 28.5	0.7	43	10.4
44	10 37.3	10 38.0	10 38.7	10 39.5	10 40.2	10 40.9	10 41.7	10 42.2	10 43.1	0.7	44	10.7
45	10 51.8	10 52.5	10 53.2	10 54.0	10 54.8	10 55.5	10 56.2	10 57.0	10 57.8	0.8	45	10.9
46	11 06.2	11 07.0	11 07.8	11 08.5	11 09.3	11 10.1	11 10.8	11 11.6	11 12.4	0.8	46	11.2
47	11 20.7	11 21.5	11 22.3	11 23.1	11 23.8	11 24.6	11 25.4	11 26.2	11 27.0	0.8	47	11.4
48	11 35.2	11 36.0	11 36.8	11 37.6	11 38.4	11 39.2	11 40.0	11 40.8	11 41.6	0.8	48	11.6
49	11 49.7	11 50.5	11 51.3	11 52.1	11 53.0	11 53.8	11 54.6	11 55.4	11 56.2	0.8	49	11.9
50	12 04.2	12 05.0	12 05.8	12 06.7	12 07.5	12 08.3	12 09.2	12 10.0	12 10.8	0.8	50	12.1
51	12 18.6	12 19.5	12 20.4	12 21.2	12 22.0	12 22.9	12 23.8	12 24.6	12 25.4	0.8	51	12.4
52	12 33.1	12 34.0	12 34.9	12 35.7	12 36.6	12 37.5	12 38.3	12 39.2	12 40.1	0.9	52	12.6
53	12 47.6	12 48.5	12 49.4	12 50.3	12 51.2	12 52.0	12 52.9	12 53.8	12 54.7	0.9	53	12.9
54	13 02.1	13 03.0	13 03.9	13 04.8	13 05.7	13 06.6	13 07.5	13 08.4	13 09.3	0.9	54	13.1
55	13 16.6	13 17.5	13 18.4	13 19.3	13 20.2	13 21.2	13 22.1	13 23.0	13 23.9	0.9	55	13.3
56	13 31.1	13 32.0	13 32.9	13 33.9	13 34.8	13 35.7	13 36.7	13 37.6	13 38.5	0.9	56	13.6
57	13 45.6	13 46.5	13 47.4	13 48.4	13 49.4	13 50.3	13 51.2	13 52.2	13 53.2	1.0	57	13.9
58	14 00.0	14 01.0	14 02.0	14 02.9	14 03.9	14 04.9	14 05.8	14 06.8	14 07.8	1.0	58	14.1
59	14 14.5	14 15.5	14 16.5	14 17.5	14 18.4	14 19.4	14 20.4	14 21.4	14 22.4	1.0	59	14.3
60	14 29.0	14 30.0	14 31.0	14 32.0	14 33.0	14 34.0	14 35.0	14 36.0	14 37.0	1.0	60	14.6

MOON DECLINATION CORRECTION TABLE

Var/hr	0.0'	1.0'	2.0'	3.0'	4.0'	5.0'	6.0'	7.0'	8.0'	9.0'	10.0'	11.0'	12.0'	13.0'	14.0'	15.0'	16.0'	17.0'	18.0'
MINUTES	'	'	'	'	'	'	'	'	'	'	'	'	'	'	'	'	'	'	'
0	0.0	0.0	0.0	0.0	0.0	0.0	0.0	0.0	0.0	0.0	0.0	0.0	0.0	0.0	0.0	0.0	0.0	0.0	0.0
1	0.0	0.0	0.0	0.0	0.1	0.1	0.1	0.1	0.1	0.2	0.2	0.2	0.2	0.2	0.2	0.2	0.3	0.3	0.3
2	0.0	0.0	0.1	0.1	0.1	0.2	0.2	0.2	0.3	0.3	0.3	0.4	0.4	0.4	0.5	0.5	0.5	0.6	0.6
3	0.0	0.0	0.1	0.2	0.2	0.2	0.3	0.4	0.4	0.4	0.5	0.6	0.6	0.6	0.7	0.8	0.8	0.8	0.9
4	0.0	0.1	0.1	0.2	0.3	0.3	0.4	0.5	0.5	0.6	0.7	0.7	0.8	0.9	0.9	1.0	1.1	1.1	1.2
5	0.0	0.1	0.2	0.2	0.3	0.4	0.5	0.6	0.7	0.8	0.8	0.9	1.0	1.1	1.2	1.2	1.3	1.4	1.5
6	0.0	0.1	0.2	0.3	0.4	0.5	0.6	0.7	0.8	0.9	1.0	1.1	1.2	1.3	1.4	1.5	1.6	1.7	1.8
7	0.0	0.1	0.2	0.4	0.5	0.6	0.7	0.8	0.9	1.0	1.2	1.3	1.4	1.5	1.6	1.8	1.9	2.0	2.1
8	0.0	0.1	0.3	0.4	0.5	0.7	0.8	0.9	1.1	1.2	1.3	1.5	1.6	1.7	1.9	2.0	2.1	2.3	2.4
9	0.0	0.2	0.3	0.4	0.6	0.8	0.9	1.0	1.2	1.4	1.5	1.6	1.8	2.0	2.1	2.2	2.4	2.6	2.7
10	0.0	0.2	0.3	0.5	0.7	0.8	1.0	1.2	1.3	1.5	1.7	1.8	2.0	2.2	2.3	2.5	2.7	2.8	3.0
11	0.0	0.2	0.4	0.6	0.7	0.9	1.1	1.3	1.5	1.6	1.8	2.0	2.2	2.4	2.6	2.8	2.9	3.1	3.3
12	0.0	0.2	0.4	0.6	0.8	1.0	1.2	1.4	1.6	1.8	2.0	2.2	2.4	2.6	2.8	3.0	3.2	3.4	3.6
13	0.0	0.2	0.4	0.6	0.9	1.1	1.3	1.5	1.7	2.0	2.2	2.4	2.6	2.8	3.0	3.2	3.5	3.7	3.9
14	0.0	0.2	0.5	0.7	0.9	1.2	1.4	1.6	1.9	2.1	2.3	2.6	2.8	3.0	3.3	3.5	3.7	4.0	4.2
15	0.0	0.2	0.5	0.8	1.0	1.2	1.5	1.8	2.0	2.2	2.5	2.8	3.0	3.2	3.5	3.8	4.0	4.2	4.5
16	0.0	0.3	0.5	0.8	1.1	1.3	1.6	1.9	2.1	2.4	2.7	2.9	3.2	3.5	3.7	4.0	4.3	4.5	4.8
17	0.0	0.3	0.6	0.8	1.1	1.4	1.7	2.0	2.3	2.6	2.8	3.1	3.4	3.7	4.0	4.2	4.5	4.8	5.1
18	0.0	0.3	0.6	0.9	1.2	1.5	1.8	2.1	2.4	2.7	3.0	3.3	3.6	3.9	4.2	4.5	4.8	5.1	5.4
19	0.0	0.3	0.6	1.0	1.3	1.6	1.9	2.2	2.5	2.8	3.2	3.5	3.8	4.1	4.4	4.8	5.1	5.4	5.7
20	0.0	0.3	0.7	1.0	1.3	1.7	2.0	2.3	2.7	3.0	3.3	3.7	4.0	4.3	4.7	5.0	5.3	5.7	6.0
21	0.0	0.4	0.7	1.0	1.4	1.8	2.1	2.4	2.8	3.2	3.5	3.8	4.2	4.6	4.9	5.2	5.6	6.0	6.3
22	0.0	0.4	0.7	1.1	1.5	1.8	2.2	2.6	2.9	3.3	3.7	4.0	4.4	4.8	5.1	5.5	5.9	6.2	6.6
23	0.0	0.4	0.8	1.2	1.5	1.9	2.3	2.7	3.1	3.4	3.8	4.2	4.6	5.0	5.4	5.8	6.1	6.5	6.9
24	0.0	0.4	0.8	1.2	1.6	2.0	2.4	2.8	3.2	3.6	4.0	4.4	4.8	5.2	5.6	6.0	6.4	6.8	7.2
25	0.0	0.4	0.8	1.2	1.7	2.1	2.5	2.9	3.3	3.8	4.2	4.6	5.0	5.4	5.8	6.2	6.7	7.1	7.5
26	0.0	0.4	0.9	1.3	1.7	2.2	2.6	3.0	3.5	3.9	4.3	4.8	5.2	5.6	6.1	6.5	6.9	7.4	7.8
27	0.0	0.4	0.9	1.3	1.8	2.2	2.7	3.2	3.6	4.0	4.5	5.0	5.4	5.8	6.3	6.8	7.2	7.6	8.1
28	0.0	0.5	0.9	1.4	1.9	2.3	2.8	3.3	3.7	4.2	4.7	5.1	5.6	6.1	6.5	7.0	7.5	7.9	8.4
29	0.0	0.5	1.0	1.4	1.9	2.4	2.9	3.4	3.9	4.4	4.8	5.3	5.8	6.3	6.8	7.2	7.7	8.2	8.7
30	0.0	0.5	1.0	1.5	2.0	2.5	3.0	3.5	4.0	4.5	5.0	5.5	6.0	6.5	7.0	7.5	8.0	8.5	9.0
31	0.0	0.5	1.0	1.6	2.1	2.6	3.1	3.6	4.1	4.6	5.2	5.7	6.2	6.7	7.2	7.8	8.3	8.8	9.3
32	0.0	0.5	1.1	1.6	2.1	2.7	3.2	3.7	4.3	4.8	5.3	5.9	6.4	6.9	7.5	8.0	8.5	9.1	9.6
33	0.0	0.6	1.1	1.6	2.2	2.8	3.3	3.8	4.4	5.0	5.5	6.0	6.6	7.2	7.7	8.2	8.8	9.4	9.9
34	0.0	0.6	1.1	1.7	2.3	2.8	3.4	4.0	4.5	5.1	5.7	6.2	6.8	7.4	7.9	8.5	9.1	9.6	10.2
35	0.0	0.6	1.2	1.8	2.3	2.9	3.5	4.1	4.7	5.2	5.8	6.4	7.0	7.6	8.2	8.8	9.3	9.9	10.5
36	0.0	0.6	1.2	1.8	2.4	3.0	3.6	4.2	4.8	5.4	6.0	6.6	7.2	7.8	8.4	9.0	9.6	10.2	10.8
37	0.0	0.6	1.2	1.8	2.5	3.1	3.7	4.3	4.9	5.6	6.2	6.8	7.4	8.0	8.6	9.2	9.9	10.5	11.1
38	0.0	0.6	1.3	1.9	2.5	3.2	3.8	4.4	5.1	5.7	6.3	7.0	7.6	8.2	8.9	9.5	10.1	10.8	11.4
39	0.0	0.6	1.3	2.0	2.6	3.2	3.9	4.6	5.2	5.8	6.5	7.2	7.8	8.4	9.1	9.8	10.4	11.0	11.7
40	0.0	0.7	1.3	2.0	2.7	3.3	4.0	4.7	5.3	6.0	6.7	7.3	8.0	8.7	9.3	10.0	10.7	11.3	12.0
41	0.0	0.7	1.4	2.0	2.7	3.4	4.1	4.8	5.5	6.2	6.8	7.5	8.2	8.9	9.6	10.2	10.9	11.6	12.3
42	0.0	0.7	1.4	2.1	2.8	3.5	4.2	4.9	5.6	6.3	7.0	7.7	8.4	9.1	9.8	10.5	11.2	11.9	12.6
43	0.0	0.7	1.4	2.2	2.9	3.6	4.3	5.0	5.7	6.4	7.2	7.9	8.6	9.3	10.0	10.8	11.5	12.2	12.9
44	0.0	0.7	1.5	2.2	2.9	3.7	4.4	5.1	5.9	6.6	7.3	8.1	8.8	9.5	10.3	11.0	11.7	12.5	13.2
45	0.0	0.7	1.5	2.2	3.0	3.8	4.5	5.2	6.0	6.8	7.5	8.2	9.0	9.8	10.5	11.2	12.0	12.8	13.5
46	0.0	0.8	1.5	2.3	3.1	3.8	4.6	5.4	6.1	6.9	7.7	8.4	9.2	10.0	10.7	11.5	12.3	13.0	13.8
47	0.0	0.8	1.6	2.4	3.1	3.9	4.7	5.5	6.3	7.0	7.8	8.6	9.4	10.2	11.0	11.8	12.5	13.3	14.1
48	0.0	0.8	1.6	2.4	3.2	4.0	4.8	5.6	6.4	7.2	8.0	8.8	9.6	10.4	11.2	12.0	12.8	13.6	14.4
49	0.0	0.8	1.6	2.4	3.3	4.1	4.9	5.7	6.5	7.4	8.2	9.0	9.8	10.6	11.4	12.2	13.1	13.9	14.7
50	0.0	0.8	1.7	2.5	3.3	4.2	5.0	5.8	6.7	7.5	8.3	9.2	10.0	10.8	11.7	12.5	13.3	14.2	15.0
51	0.0	0.8	1.7	2.6	3.4	4.2	5.1	6.0	6.8	7.6	8.5	9.4	10.2	11.0	11.9	12.8	13.6	14.4	15.3
52	0.0	0.9	1.7	2.6	3.5	4.3	5.2	6.1	6.9	7.8	8.7	9.5	10.4	11.3	12.1	13.0	13.9	14.7	15.6
53	0.0	0.9	1.8	2.6	3.5	4.4	5.3	6.2	7.1	8.0	8.8	9.7	10.6	11.5	12.4	13.2	14.1	15.0	15.9
54	0.0	0.9	1.8	2.7	3.6	4.5	5.4	6.3	7.2	8.1	9.0	9.9	10.8	11.7	12.6	13.5	14.4	15.3	16.2
55	0.0	0.9	1.8	2.8	3.7	4.6	5.5	6.4	7.3	8.2	9.2	10.1	11.0	11.9	12.8	13.8	14.7	15.6	16.5
56	0.0	0.9	1.9	2.8	3.7	4.7	5.6	6.5	7.5	8.4	9.3	10.3	11.2	12.1	13.1	14.0	14.9	15.9	16.8
57	0.0	1.0	1.9	2.8	3.8	4.8	5.7	6.6	7.6	8.6	9.5	10.4	11.4	12.4	13.3	14.2	15.2	16.2	17.1
58	0.0	1.0	1.9	2.9	3.9	4.8	5.8	6.8	7.7	8.7	9.7	10.6	11.6	12.6	13.5	14.5	15.5	16.4	17.4
59	0.0	1.0	2.0	3.0	3.9	4.9	5.9	6.9	7.9	8.8	9.8	10.8	11.8	12.8	13.8	14.8	15.7	16.7	17.7
60	0.0	1.0	2.0	3.0	4.0	5.0	6.0	7.0	8.0	9.0	10.0	11.0	12.0	13.0	14.0	15.0	16.0	17.0	18.0

ARC TO TIME CONVERSION TABLE

Arc °	Time h m	Arc °	Time h m	Arc °	Time h m	Arc °	Time h m	Arc °	Time h m	Arc °	Time h m	Arc '	Time m s	Arc " '	Time s
0	0 00	**60**	4 00	**120**	8 00	**180**	12 00	**240**	16 00	**300**	20 00	**0**	0 00	**0 = 0.0**	0.00
1	0 04	**61**	4 04	**121**	8 04	**181**	12 04	**241**	16 04	**301**	20 04	**1**	0 04	**1**	0.07
2	0 08	**62**	4 08	**122**	8 08	**182**	12 08	**242**	16 08	**302**	20 08	**2**	0 08	**2**	0.13
3	0 12	**63**	4 12	**123**	8 12	**183**	12 12	**243**	16 12	**303**	20 12	**3**	0 12	**3**	0.20
4	0 16	**64**	4 16	**124**	8 16	**184**	12 16	**244**	16 16	**304**	20 16	**4**	0 16	**4**	0.27
5	0 20	**65**	4 20	**125**	8 20	**185**	12 20	**245**	16 20	**305**	20 20	**5**	0 20	**5**	0.33
6	0 24	**66**	4 24	**126**	8 24	**186**	12 24	**246**	16 24	**306**	20 24	**6**	0 24	**6 = 0.1**	0.40
7	0 28	**67**	4 28	**127**	8 28	**187**	12 28	**247**	16 28	**307**	20 28	**7**	0 28	**7**	0.47
8	0 32	**68**	4 32	**128**	8 32	**188**	12 32	**248**	16 32	**308**	20 32	**8**	0 32	**8**	0.53
9	0 36	**69**	4 36	**129**	8 36	**189**	12 36	**249**	16 36	**309**	20 36	**9**	0 36	**9**	0.60
10	0 40	**70**	4 40	**130**	8 40	**190**	12 40	**250**	16 40	**310**	20 40	**10**	0 40	**10**	0.67
11	0 44	**71**	4 44	**131**	8 44	**191**	12 44	**251**	16 44	**311**	20 44	**11**	0 44	**11**	0.73
12	0 48	**72**	4 48	**132**	8 48	**192**	12 48	**252**	16 48	**312**	20 48	**12**	0 48	**12 = 0.2**	0.80
13	0 52	**73**	4 52	**133**	8 52	**193**	12 52	**253**	16 52	**313**	20 52	**13**	0 52	**13**	0.87
14	0 56	**74**	4 56	**134**	8 56	**194**	12 56	**254**	16 56	**314**	20 56	**14**	0 56	**14**	0.93
15	1 00	**75**	5 00	**135**	9 00	**195**	13 00	**255**	17 00	**315**	21 00	**15**	1 00	**15**	1.00
16	1 04	**76**	5 04	**136**	9 04	**196**	13 04	**256**	17 04	**316**	21 04	**16**	1 04	**16**	1.07
17	1 08	**77**	5 08	**137**	9 08	**197**	13 08	**257**	17 08	**317**	21 08	**17**	1 08	**17**	1.13
18	1 12	**78**	5 12	**138**	9 12	**198**	13 12	**258**	17 12	**318**	21 12	**18**	1 12	**18 = 0.3**	1.20
19	1 16	**79**	5 16	**139**	9 16	**199**	13 16	**259**	17 16	**319**	21 16	**19**	1 16	**19**	1.27
20	1 20	**80**	5 20	**140**	9 20	**200**	13 20	**260**	17 20	**320**	21 20	**20**	1 20	**20**	1.33
21	1 24	**81**	5 24	**141**	9 24	**201**	13 24	**261**	17 24	**321**	21 24	**21**	1 24	**21**	1.40
22	1 28	**82**	5 28	**142**	9 28	**202**	13 28	**262**	17 28	**322**	21 28	**22**	1 28	**22**	1.47
23	1 32	**83**	5 32	**143**	9 32	**203**	13 32	**263**	17 32	**323**	21 32	**23**	1 32	**23**	1.53
24	1 36	**84**	5 36	**144**	9 36	**204**	13 36	**264**	17 36	**324**	21 36	**24**	1 36	**24 = 0.4**	1.60
25	1 40	**85**	5 40	**145**	9 40	**205**	13 40	**265**	17 40	**325**	21 40	**25**	1 40	**25**	1.67
26	1 44	**86**	5 44	**146**	9 44	**206**	13 44	**266**	17 44	**326**	21 44	**26**	1 44	**26**	1.73
27	1 48	**87**	5 48	**147**	9 48	**207**	13 48	**267**	17 48	**327**	21 48	**27**	1 48	**27**	1.80
28	1 52	**88**	5 52	**148**	9 52	**208**	13 52	**268**	17 52	**328**	21 52	**28**	1 52	**28**	1.87
29	1 56	**89**	5 56	**149**	9 56	**209**	13 56	**269**	17 56	**329**	21 56	**29**	1 56	**29**	1.93
30	2 00	**90**	6 00	**150**	10 00	**210**	14 00	**270**	18 00	**330**	22 00	**30**	2 00	**30 = 0.5**	2.00
31	2 04	**91**	6 04	**151**	10 04	**211**	14 04	**271**	18 04	**331**	22 04	**31**	2 04	**31**	2.07
32	2 08	**92**	6 08	**152**	10 08	**212**	14 08	**272**	18 08	**332**	22 08	**32**	2 08	**32**	2.13
33	2 12	**93**	6 12	**153**	10 12	**213**	14 12	**273**	18 12	**333**	22 12	**33**	2 12	**33**	2.20
34	2 16	**94**	6 16	**154**	10 16	**214**	14 16	**274**	18 16	**334**	22 16	**34**	2 16	**34**	2.27
35	2 20	**95**	6 20	**155**	10 20	**215**	14 20	**275**	18 20	**335**	22 20	**35**	2 20	**35**	2.33
36	2 24	**96**	6 24	**156**	10 24	**216**	14 24	**276**	18 24	**336**	22 24	**36**	2 24	**36 = 0.6**	2.40
37	2 28	**97**	6 28	**157**	10 28	**217**	14 28	**277**	18 28	**337**	22 28	**37**	2 28	**37**	2.47
38	2 32	**98**	6 32	**158**	10 32	**218**	14 32	**278**	18 32	**338**	22 32	**38**	2 32	**38**	2.53
39	2 36	**99**	6 36	**159**	10 36	**219**	14 36	**279**	18 36	**339**	22 36	**39**	2 36	**39**	2.60
40	2 40	**100**	6 40	**160**	10 40	**220**	14 40	**280**	18 40	**340**	22 40	**40**	2 40	**40**	2.67
41	2 44	**101**	6 44	**161**	10 44	**221**	14 44	**281**	18 44	**341**	22 44	**41**	2 44	**41**	2.73
42	2 48	**102**	6 48	**162**	10 48	**222**	14 48	**282**	18 48	**342**	22 48	**42**	2 48	**42 = 0.7**	2.80
43	2 52	**103**	6 52	**163**	10 52	**223**	14 52	**283**	18 52	**343**	22 52	**43**	2 52	**43**	2.87
44	2 56	**104**	6 56	**164**	10 56	**224**	14 56	**284**	18 56	**344**	22 56	**44**	2 56	**44**	2.93
45	3 00	**105**	7 00	**165**	11 00	**225**	15 00	**285**	19 00	**345**	23 00	**45**	3 00	**45**	3.00
46	3 04	**106**	7 04	**166**	11 04	**226**	15 04	**286**	19 04	**346**	23 04	**46**	3 04	**46**	3.07
47	3 08	**107**	7 08	**167**	11 08	**227**	15 08	**287**	19 08	**347**	23 08	**47**	3 08	**47**	3.13
48	3 12	**108**	7 12	**168**	11 12	**228**	15.12	**288**	19 12	**348**	23 12	**48**	3 12	**48 = 0.8**	3.20
49	3 16	**109**	7 16	**169**	11 16	**229**	15 16	**289**	19 16	**349**	23 16	**49**	3 16	**49**	3.27
50	3 20	**110**	7 20	**170**	11 20	**230**	15 20	**290**	19 20	**350**	23 20	**50**	3 20	**50**	3.33
51	3 24	**111**	7 24	**171**	11 24	**231**	15 24	**291**	19 24	**351**	23 24	**51**	3 24	**51**	3.40
52	3 28	**112**	7 28	**172**	11 28	**232**	15 28	**292**	19 28	**353**	23 28	**52**	3 28	**52**	3.47
53	3 32	**113**	7 32	**173**	11 32	**233**	15 32	**293**	19 32	**353**	23 32	**53**	3 32	**53**	3.53
54	3 36	**114**	7 36	**174**	11 36	**234**	15 36	**294**	19 36	**354**	23 36	**54**	3 36	**54 = 0.9**	3.60
55	3 40	**115**	7 40	**175**	11 40	**235**	15 40	**295**	19 40	**355**	23 40	**55**	3 40	**55**	3.67
56	3 44	**116**	7 44	**176**	11 44	**236**	15 44	**296**	19 44	**356**	23 44	**56**	3 44	**56**	3.73
57	3 48	**117**	7 48	**177**	11 48	**237**	15 48	**297**	19 48	**357**	23 48	**57**	3 48	**57**	3.80
58	3 52	**118**	7 52	**178**	11 52	**238**	15 52	**298**	19 52	**358**	23 52	**58**	3 52	**58**	3.87
59	3 56	**119**	7 56	**179**	11 56	**239**	15 56	**299**	19 56	**359**	23 56	**59**	3 56	**59**	3.93
60	4 00	**120**	8 00	**180**	12 00	**240**	16 00	**300**	20 00	**360**	24 00	**60**	4 00	**60 = 1.0**	4.00

53

MOON MERIDIAN PASSAGE CORRECTION TABLE

Long.	DAILY DIFFERENCE OF MERIDIAN PASSAGE											Long.
	39	42	45	48	51	54	57	60	63	66	69	
	min	min	min	min	min	min	min	min	min	min	min	
0°	0	0	0	0	0	0	0	0	0	0	0	0°
10°	1	1	1	1	1	1	2	2	2	2	2	10°
20°	2	2	2	3	3	3	3	3	3	4	4	20°
30°	3	3	4	4	4	4	5	5	5	5	6	30°
40°	4	5	5	5	6	6	6	7	7	7	8	40°
50°	5	6	6	7	7	7	8	8	9	9	10	50°
60°	6	7	7	8	8	9	9	10	10	11	11	60°
70°	8	8	9	9	10	10	11	12	12	13	13	70°
80°	9	9	10	11	11	12	13	13	14	15	15	80°
90°	10	10	11	12	13	13	14	15	16	16	17	90°
100°	11	12	12	13	14	15	16	17	17	18	19	100°
110°	12	13	14	15	16	16	17	18	19	20	21	110°
120°	13	14	15	16	17	18	19	20	21	22	23	120°
130°	14	15	16	17	18	19	21	22	23	24	25	130°
140°	15	16	17	19	20	21	22	23	24	26	27	140°
150°	16	17	19	20	21	22	24	25	26	27	29	150°
160°	17	19	20	21	23	24	25	27	28	29	31	160°
170°	18	20	21	23	24	25	27	28	30	31	33	170°
180°	19	21	22	24	25	27	28	30	31	33	34	180°

NOTE: Apply correction in **minutes** to the time of meridian passage given in the monthly tables. **Add** the correction if longitude is West; **subtract** if longitude is East.

54

ALPHABETICAL INDEX OF PRINCIPAL STARS
WITH THEIR APPROXIMATE PLACES – 2011

Proper Name		Constellation Name	Mag	RA	Dec	SHA	No
Acamar	θ	Eridani	3.2	2 59	S 40	315	8
Achernar	α	Eridan	0.5	1 38	S 57	335	5
Acrux	α	Crucis	1.3	12 27	S 63	173	32
Adhara	ε	Canis Majoris	1.5	6 59	S 29	255	20
Aldebaran	α	Tauri	0.9	4 37	N 17	291	11
Alioth	ε	Ursae Majoris	1.8	12 55	N 56	166	35
Alkaid	η	Ursae Majoris	1.9	13 48	N 49	153	37
Al Na'ir	α	Gruis	1.7	22 09	S 47	28	58
Alnilam	ε	Orionis	1.7	5 37	S 1	276	16
Alphard	α	Hydrae	2.0	9 28	S 9	218	27
Alphecca	α	Coronae Bor	2.2	15 35	N 27	126	44
Alpheratz	α	Andromedae	2.1	0 09	N 29	358	1
Altair	α	Aquilae	0.8	19 51	N 9	62	54
Ankaa	α	Phoenicis	2.4	0 27	S 42	353	2
Antares	α	Scorpii	1.0	16 30	S 26	112	45
Arcturus	α	Bootis	0.0	14 16	N 19	146	40
Atria	α	Triang Aust	1.9	16 50	S 69	108	46
Avior	ε	Carinae	1.9	8 23	S 60	234	24
Bellatrix	γ	Orionis	1.6	5 26	N 6	279	14
Betelgeuse	α	Orionis	0.1–1.2	5 56	N 7	271	17
Canopus	α	Carinae	-0.7	6 24	S 53	264	18
Capella	α	Aurigae	0.1	5 18	N 46	281	13
Castor	α	Geminorum	1.6	7 35	N 32	246	21
Deneb	α	Cygni	1.3	20 42	N 45	50	56
Denebola	β	Leonis	2.1	11 50	N 15	183	30
Diphda	β	Ceti	2.0	0 44	S 18	349	4
Dubhe	α	Ursae Majoris	1.8	11 04	N 62	194	29
Elnath	β	Tauri	1.7	5 27	N 29	278	15
Eltanin	γ	Draconis	2.4	17 57	N 51	91	50
Enif	ε	Pegasi	2.4	21 45	N 10	34	57
Fomalhaut	α	Piscis Aust	1.2	22 58	S 30	15	59
Gacrux	γ	Crucis	1.6	12 32	S 57	172	33
Gienah	γ	Corvi	2.6	12 16	S 18	176	31
Hadar	β	Centauri	0.6	14 05	S 60	149	38
Hamal	α	Arietis	2.0	2 08	N 24	328	7
Kaus Aust.	ε	Sagittarii	1.9	18 25	S 34	84	51
Kochab	β	Ursae Minoris	2.1	14 51	N 74	137	43
Markab	α	Pegasi	2.5	23 05	N 15	14	60
Menkar	α	Ceti	2.5	3 03	N 4	314	9
Menkent	θ	Centauri	2.1	14 07	S 36	148	39
Miaplacidus	β	Carinae	1.7	9 13	S 70	222	26
Mimosa	β	Crucis	1.3	12 48	S 60	168	34
Mirfak	α	Persei	1.8	3 25	N 50	309	10
Nunki	σ	Sagittarii	2.0	18 56	S 26	76	53
Peacock	α	Pavonis	1.9	20 27	S 57	53	55
POLARIS	α	Ursae Minoris	2.0	2 46	N 89	318	6
Pollux	β	Geminorum	1.1	7 46	N 28	243	23
Procyon	α	Canis Minoris	0.4	7 40	N 5	245	22
Rasalhague	α	Ophiuchi	2.1	17 35	N 13	96	49
Regulus	α	Leonis	1.4	10 09	N 12	208	28
Rigel	β	Orionis	0.1	5 15	S 8	281	12
Rigil Kent.	α	Centauri	-0.3	14 40	S 61	140	41
Sabik	η	Ophiuchi	2.4	17 11	S 16	102	47
Schedar	α	Cassiopeiae	2.2	0 41	N 57	350	3
Shaula	γ	Scorpii	1.6	17 34	S 37	96	48
Sirius	α	Canis Majoris	-1.5	6 46	S 17	259	19
Spica	α	Virginis	1.0	13 26	S 11	159	36
Suhail	γ	Velorum	2.2	9 08	S 43	223	25
Vega	α	Lyrae	0.0	18 37	N 39	81	52
Zuben'ubi	α	Librae	2.8	14 52	S 16	137	42

The last column refers to the number given to the star in this almanac. The star's exact position may be found according to this number on the monthly pages.

′	0° Log	0° Nat	1° Log	1° Nat	2° Log	2° Nat	3° Log	3° Nat	4° Log	4° Nat	5° Log	5° Nat	6° Log	6° Nat	′
0	∞	0.0000	1827	0002	6.7847	0006	1369	0014	7.3867	0024	5804	0038	7.7386	0.0055	60
1	2.6264	0.0000	1971	0002	6.7919	0006	1417	0014	7.3903	0025	5833	0038	7.7410	0.0055	59
2	3.2285	0.0000	2112	0002	6.7991	0006	1465	0014	7.3939	0025	5862	0039	7.7434	0.0055	58
3	3.5807	0.0000	2251	0002	6.8062	0006	1512	0014	7.3975	0025	5890	0039	7.7458	0.0056	57
4	3.8305	0.0000	2388	0002	6.8132	0007	1560	0014	7.4010	0025	5919	0039	7.7482	0.0056	56
5	4.0244	0.0000	2522	0002	6.8202	0007	1607	0014	7.4046	0025	5947	0039	7.7506	0.0056	55
6	4.1827	0.0000	2655	0002	6.8271	0007	1653	0015	7.4081	0026	5976	0040	7.7530	0.0057	54
7	4.3166	0.0000	2786	0002	6.8340	0007	1700	0015	7.4116	0026	6004	0040	7.7553	0.0057	53
8	4.4326	0.0000	2914	0002	6.8408	0007	1746	0015	7.4151	0026	6032	0040	7.7577	0.0057	52
9	4.5349	0.0000	3041	0002	6.8476	0007	1792	0015	7.4186	0026	6060	0040	7.7601	0.0058	51
10	4.6264	0.0000	3166	0002	6.8543	0007	1838	0015	7.4221	0026	6089	0041	7.7624	0.0058	50
11	4.7092	0.0000	3289	0002	6.8609	0007	1884	0015	7.4256	0027	6116	0041	7.7647	0.0058	49
12	4.7848	0.0000	3411	0002	6.8675	0007	1929	0016	7.4290	0027	6144	0041	7.7671	0.0058	48
13	4.8543	0.0000	3531	0002	6.8741	0007	1974	0016	7.4325	0027	6172	0041	7.7694	0.0059	47
14	4.9187	0.0000	3649	0002	6.8806	0008	2019	0016	7.4359	0027	6200	0042	7.7717	0.0059	46
15	4.9786	0.0000	3765	0002	6.8870	0008	2064	0016	7.4393	0027	6227	0042	7.7741	0.0059	45
16	5.0347	0.0000	3880	0002	6.8934	0008	2108	0016	7.4427	0028	6255	0042	7.7764	0.0060	44
17	5.0873	0.0000	3994	0003	6.8998	0008	2152	0016	7.4461	0028	6282	0042	7.7787	0.0060	43
18	5.1370	0.0000	4106	0003	6.9061	0008	2196	0017	7.4495	0028	6310	0043	7.7810	0.0060	42
19	5.1839	0.0000	4217	0003	6.9124	0008	2240	0017	7.4528	0028	6337	0043	7.7833	0.0061	41
20	5.2285	0.0000	4326	0003	6.9186	0008	2284	0017	7.4562	0029	6364	0043	7.7855	0.0061	40
21	5.2709	0.0000	4434	0003	6.9248	0008	2327	0017	7.4595	0029	6391	0044	7.7878	0.0061	39
22	5.3113	0.0000	4540	0003	6.9309	0009	2370	0017	7.4628	0029	6418	0044	7.7901	0.0062	38
23	5.3499	0.0000	4646	0003	6.9370	0009	2413	0017	7.4661	0029	6445	0044	7.7924	0.0062	37
24	5.3868	0.0000	4750	0003	6.9431	0009	2456	0018	7.4694	0029	6472	0044	7.7946	0.0062	36
25	5.4223	0.0000	4852	0003	6.9491	0009	2498	0018	7.4727	0030	6499	0045	7.7969	0.0063	35
26	5.4564	0.0000	4954	0003	6.9551	0009	2540	0018	7.4760	0030	6525	0045	7.7991	0.0063	34
27	5.4891	0.0000	5054	0003	6.9610	0009	2582	0018	7.4792	0030	6552	0045	7.8014	0.0063	33
28	5.5207	0.0000	5154	0003	6.9669	0009	2624	0018	7.4825	0030	6578	0045	7.8036	0.0064	32
29	5.5512	0.0000	5252	0003	6.9727	0009	2666	0018	7.4857	0031	6605	0046	7.8059	0.0064	31
30	5.5807	0.0000	5349	0003	6.9785	0010	2707	0019	7.4889	0031	6631	0046	7.8081	0.0064	30
31	5.6091	0.0000	5445	0004	6.9843	0010	2749	0019	7.4921	0031	6657	0046	7.8103	0.0065	29
32	5.6367	0.0000	5540	0004	6.9900	0010	2790	0019	7.4953	0031	6684	0047	7.8125	0.0065	28
33	5.6634	0.0000	5634	0004	6.9957	0010	2830	0019	7.4985	0032	6710	0047	7.8147	0.0065	27
34	5.6894	0.0000	5727	0004	7.0014	0010	2871	0019	7.5017	0032	6736	0047	7.8169	0.0066	26
35	5.7146	0.0001	5818	0004	7.0070	0010	2912	0020	7.5049	0032	6762	0047	7.8191	0.0066	25
36	5.7390	0.0001	5909	0004	7.0126	0010	2952	0020	7.5080	0032	6788	0048	7.8213	0.0066	24
37	5.7628	0.0001	5999	0004	7.0181	0010	2992	0020	7.5111	0032	6813	0048	7.8235	0.0067	23
38	5.7860	0.0001	6088	0004	7.0237	0011	3032	0020	7.5143	0033	6839	0048	7.8257	0.0067	22
39	5.8085	0.0001	6177	0004	7.0291	0011	3072	0020	7.5174	0033	6865	0049	7.8279	0.0067	21
40	5.8305	0.0001	6264	0004	7.0346	0011	3111	0020	7.5205	0033	6890	0049	7.8301	0.0068	20
41	5.8520	0.0001	6350	0004	7.0400	0011	3151	0021	7.5236	0033	6916	0049	7.8322	0.0068	19
42	5.8729	0.0001	6436	0004	7.0454	0011	3190	0021	7.5267	0034	6941	0049	7.8344	0.0068	18
43	5.8934	0.0001	6521	0004	7.0507	0011	3229	0021	7.5297	0034	6967	0050	7.8365	0.0069	17
44	5.9133	0.0001	6605	0005	7.0560	0011	3268	0021	7.5328	0034	6992	0050	7.8387	0.0069	16
45	5.9328	0.0001	6688	0005	7.0613	0012	3306	0021	7.5359	0034	7017	0050	7.8408	0.0069	15
46	5.9519	0.0001	6770	0005	7.0666	0012	3345	0022	7.5389	0035	7042	0051	7.8430	0.0070	14
47	5.9706	0.0001	6852	0005	7.0718	0012	3383	0022	7.5419	0035	7067	0051	7.8451	0.0070	13
48	5.9889	0.0001	6932	0005	7.0770	0012	3421	0022	7.5450	0035	7092	0051	7.8472	0.0070	12
49	6.0068	0.0001	7012	0005	7.0821	0012	3459	0022	7.5480	0035	7117	0051	7.8494	0.0071	11
50	6.0244	0.0001	7092	0005	7.0872	0012	3497	0022	7.5510	0036	7142	0052	7.8515	0.0071	10
51	6.0416	0.0001	7170	0005	7.0923	0012	3535	0023	7.5539	0036	7167	0052	7.8536	0.0071	9
52	6.0584	0.0001	7248	0005	7.0974	0013	3572	0023	7.5569	0036	7191	0052	7.8557	0.0072	8
53	6.0750	0.0001	7325	0005	7.1024	0013	3610	0023	7.5599	0036	7216	0053	7.8578	0.0072	7
54	6.0912	0.0001	7402	0005	7.1074	0013	3647	0023	7.5629	0037	7240	0053	7.8599	0.0072	6
55	6.1071	0.0001	7478	0006	7.1124	0013	3684	0023	7.5658	0037	7265	0053	7.8620	0.0073	5
56	6.1228	0.0001	7553	0006	7.1174	0013	3721	0024	7.5687	0037	7289	0054	7.8641	0.0073	4
57	6.1382	0.0001	7628	0006	7.1223	0013	3757	0024	7.5717	0037	7314	0054	7.8662	0.0073	3
58	6.1533	0.0001	7701	0006	7.1272	0013	3794	0024	7.5746	0038	7338	0054	7.8682	0.0074	2
59	6.1681	0.0001	7775	0006	7.1320	0014	3830	0024	7.5775	0038	7362	0054	7.8703	0.0074	1
60	6.1827	0.0002	7847	0006	7.1369	0014	3867	0024	7.5804	0038	7386	0055	7.8724	0.0075	0
′	Log 359°	Nat	Log 358°	Nat	Log 357°	Nat	Log 356°	Nat	Log 355°	Nat	Log 354°	Nat	Log 353°	Nat	′

NOTE: For compactness. the leading digit is not displayed in every column. Scan other columns for necessary digit.

VERSINES

′	7° Log	7° Nat	8° Log	8° Nat	9° Log	9° Nat	10° Log	10° Nat	11° Log	11° Nat	12° Log	12° Nat	13° Log	13° Nat	′
0	7.8724	0.0075	9882	0097	8.0903	0123	1816	0152	8.2642	0184	3395	0219	8.4087	0.0256	60
1	7.8744	0.0075	9900	0098	8.0919	0124	1831	0152	8.2655	0184	3407	0219	8.4099	0.0257	59
2	7.8765	0.0075	9918	0098	8.0935	0124	1845	0153	8.2668	0185	3419	0220	8.4110	0.0258	58
3	7.8786	0.0076	9936	0099	8.0951	0124	1859	0153	8.2681	0185	3431	0220	8.4121	0.0258	57
4	7.8806	0.0076	9954	0099	8.0967	0125	1874	0154	8.2694	0186	3443	0221	8.4132	0.0259	56
5	7.8826	0.0076	9972	0099	8.0983	0125	1888	0154	8.2707	0187	3455	0222	8.4143	0.0260	55
6	7.8847	0.0077	9990	0100	8.0999	0126	1902	0155	8.2720	0187	3467	0222	8.4154	0.0260	54
7	7.8867	0.0077	0008	0100	8.1015	0126	1917	0155	8.2733	0188	3479	0223	8.4165	0.0261	53
8	7.8887	0.0077	0025	0101	8.1031	0127	1931	0156	8.2746	0188	3491	0223	8.4176	0.0262	52
9	7.8908	0.0078	0043	0101	8.1046	0127	1945	0157	8.2759	0189	3502	0224	8.4187	0.0262	51
10	7.8928	0.0078	0061	0101	8.1062	0128	1959	0157	8.2772	0189	3514	0225	8.4198	0.0263	50
11	7.8948	0.0078	0078	0102	8.1078	0128	1974	0158	8.2785	0190	3526	0225	8.4209	0.0264	49
12	7.8968	0.0079	0096	0102	8.1094	0129	1988	0158	8.2798	0190	3538	0226	8.4220	0.0264	48
13	7.8988	0.0079	0114	0103	8.1109	0129	2002	0159	8.2811	0191	3550	0226	8.4230	0.0265	47
14	7.9008	0.0080	0131	0103	8.1125	0130	2016	0159	8.2824	0192	3562	0227	8.4241	0.0266	46
15	7.9028	0.0080	0149	0103	8.1141	0130	2030	0160	8.2836	0192	3573	0228	8.4252	0.0266	45
16	7.9048	0.0080	0166	0104	8.1156	0131	2044	0160	8.2849	0193	3585	0228	8.4263	0.0267	44
17	7.9068	0.0081	0184	0104	8.1172	0131	2058	0161	8.2862	0193	3597	0229	8.4274	0.0268	43
18	7.9088	0.0081	0201	0105	8.1187	0131	2072	0161	8.2875	0194	3609	0230	8.4285	0.0268	42
19	7.9108	0.0081	0219	0105	8.1203	0132	2086	0162	8.2887	0194	3620	0230	8.4296	0.0269	41
20	7.9127	0.0082	0236	0106	8.1218	0132	2100	0162	8.2900	0195	3632	0231	8.4306	0.0270	40
21	7.9147	0.0082	0253	0106	8.1234	0133	2114	0163	8.2913	0196	3644	0231	8.4317	0.0270	39
22	7.9167	0.0083	0271	0106	8.1249	0133	2128	0163	8.2926	0196	3655	0232	8.4328	0.0271	38
23	7.9186	0.0083	0288	0107	8.1265	0134	2142	0164	8.2938	0197	3667	0233	8.4339	0.0272	37
24	7.9206	0.0083	0305	0107	8.1280	0134	2156	0164	8.2951	0197	3679	0233	8.4350	0.0272	36
25	7.9225	0.0084	0322	0108	8.1295	0135	2170	0165	8.2964	0198	3690	0234	8.4360	0.0273	35
26	7.9245	0.0084	0339	0108	8.1311	0135	2184	0165	8.2976	0198	3702	0235	8.4371	0.0274	34
27	7.9264	0.0084	0357	0109	8.1326	0136	2198	0166	8.2989	0199	3714	0235	8.4382	0.0274	33
28	7.9284	0.0085	0374	0109	8.1341	0136	2211	0166	8.3001	0200	3725	0236	8.4392	0.0275	32
29	7.9303	0.0085	0391	0109	8.1357	0137	2225	0167	8.3014	0200	3737	0236	8.4403	0.0276	31
30	7.9322	0.0086	0408	0110	8.1372	0137	2239	0167	8.3027	0201	3748	0237	8.4414	0.0276	30
31	7.9342	0.0086	0425	0110	8.1387	0138	2253	0168	8.3039	0201	3760	0238	8.4424	0.0277	29
32	7.9361	0.0086	0442	0111	8.1402	0138	2266	0169	8.3052	0202	3771	0238	8.4435	0.0278	28
33	7.9380	0.0087	0459	0111	8.1417	0139	2280	0169	8.3064	0202	3783	0239	8.4446	0.0278	27
34	7.9399	0.0087	0475	0112	8.1432	0139	2294	0170	8.3077	0203	3794	0240	8.4456	0.0279	26
35	7.9418	0.0087	0492	0112	8.1447	0140	2307	0170	8.3089	0204	3806	0240	8.4467	0.0280	25
36	7.9437	0.0088	0509	0112	8.1463	0140	2321	0171	8.3102	0204	3817	0241	8.4478	0.0280	24
37	7.9456	0.0088	0526	0113	8.1478	0141	2335	0171	8.3114	0205	3829	0241	8.4488	0.0281	23
38	7.9475	0.0089	0543	0113	8.1493	0141	2348	0172	8.3126	0205	3840	0242	8.4499	0.0282	22
39	7.9494	0.0089	0559	0114	8.1508	0141	2362	0172	8.3139	0206	3851	0243	8.4509	0.0282	21
40	7.9513	0.0089	0576	0114	8.1522	0142	2375	0173	8.3151	0207	3863	0243	8.4520	0.0283	20
41	7.9532	0.0090	0593	0115	8.1537	0142	2389	0173	8.3164	0207	3874	0244	8.4530	0.0284	19
42	7.9551	0.0090	0609	0115	8.1552	0143	2402	0174	8.3176	0208	3886	0245	8.4541	0.0285	18
43	7.9569	0.0091	0626	0116	8.1567	0143	2416	0174	8.3188	0208	3897	0245	8.4551	0.0285	17
44	7.9588	0.0091	0642	0116	8.1582	0144	2429	0175	8.3200	0209	3908	0246	8.4562	0.0286	16
45	7.9607	0.0091	0659	0116	8.1597	0144	2443	0175	8.3213	0210	3920	0247	8.4572	0.0287	15
46	7.9625	0.0092	0675	0117	8.1612	0145	2456	0176	8.3225	0210	3931	0247	8.4583	0.0287	14
47	7.9644	0.0092	0692	0117	8.1626	0145	2469	0177	8.3237	0211	3942	0248	8.4593	0.0288	13
48	7.9662	0.0093	0708	0118	8.1641	0146	2483	0177	8.3250	0211	3953	0249	8.4604	0.0289	12
49	7.9681	0.0093	0725	0118	8.1656	0146	2496	0178	8.3262	0212	3965	0249	8.4614	0.0289	11
50	7.9699	0.0093	0741	0119	8.1671	0147	2510	0178	8.3274	0213	3976	0250	8.4625	0.0290	10
51	7.9718	0.0094	0757	0119	8.1685	0147	2523	0179	8.3286	0213	3987	0250	8.4635	0.0291	9
52	7.9736	0.0094	0774	0120	8.1700	0148	2536	0179	8.3298	0214	3998	0251	8.4645	0.0291	8
53	7.9755	0.0095	0790	0120	8.1715	0148	2549	0180	8.3310	0214	4010	0252	8.4656	0.0292	7
54	7.9773	0.0095	0806	0120	8.1729	0149	2563	0180	8.3323	0215	4021	0252	8.4666	0.0293	6
55	7.9791	0.0095	0823	0121	8.1744	0149	2576	0181	8.3335	0216	4032	0253	8.4677	0.0294	5
56	7.9809	0.0096	0839	0121	8.1758	0150	2589	0182	8.3347	0216	4043	0254	8.4687	0.0294	4
57	7.9828	0.0096	0855	0122	8.1773	0150	2602	0182	8.3359	0217	4054	0254	8.4697	0.0295	3
58	7.9846	0.0097	0871	0122	8.1787	0151	2615	0183	8.3371	0217	4065	0255	8.4708	0.0296	2
59	7.9864	0.0097	0887	0123	8.1802	0151	2629	0183	8.3383	0218	4076	0256	8.4718	0.0296	1
60	7.9882	0.0097	0903	0123	8.1816	0152	2642	0184	8.3395	0219	4087	0256	8.4728	0.0297	0
′	Log 352°	Nat	Log 351°	Nat	Log 350°	Nat	Log 349°	Nat	Log 348°	Nat	Log 347°	Nat	Log 346°	Nat	′

NOTE: For compactness. the leading digit is not displayed in every column. Scan other columns for necessary digit.

55

VERSINES

′	14° Log	14° Nat	15° Log	15° Nat	16° Log	16° Nat	17° Log	17° Nat	18° Log	18° Nat	19° Log	19° Nat	20° Log	20° Nat	′
0	8.4728	0.0297	5324	0341	8.5881	0387	6404	0437	8.6897	0489	7362	0545	8.7804	0.0603	60
1	8.4738	0.0298	5334	0341	8.5890	0388	6413	0438	8.6905	0490	7370	0546	8.7811	0.0604	59
2	8.4749	0.0298	5343	0342	8.5899	0389	6421	0439	8.6913	0491	7378	0547	8.7818	0.0605	58
3	8.4759	0.0299	5353	0343	8.5908	0390	6430	0440	8.6921	0492	7385	0548	8.7825	0.0606	57
4	8.4769	0.0300	5363	0344	8.5917	0391	6438	0440	8.6929	0493	7393	0549	8.7832	0.0607	56
5	8.4779	0.0301	5372	0345	8.5926	0391	6447	0441	8.6937	0494	7400	0550	8.7839	0.0608	55
6	8.4790	0.0301	5382	0345	8.5935	0392	6455	0442	8.6945	0495	7408	0551	8.7847	0.0609	54
7	8.4800	0.0302	5391	0346	8.5944	0393	6463	0443	8.6953	0496	7415	0551	8.7854	0.0610	53
8	8.4810	0.0303	5401	0347	8.5953	0394	6472	0444	8.6961	0497	7423	0552	8.7861	0.0611	52
9	8.4820	0.0303	5410	0348	8.5962	0395	6480	0445	8.6968	0498	7430	0553	8.7868	0.0612	51
10	8.4830	0.0304	5420	0348	8.5971	0395	6488	0445	8.6976	0496	7438	0554	8.7875	0.0613	50
11	8.4841	0.0305	5429	0349	8.5980	0396	6497	0446	8.6984	0499	7445	0555	8.7882	0.0614	49
12	8.4851	0.0306	5439	0350	8.5989	0397	6505	0447	8.6992	0500	7453	0556	8.7889	0.0615	48
13	8.4861	0.0306	5448	0351	8.5997	0398	6514	0448	8.7000	0501	7460	0557	8.7896	0.0616	47
14	8.4871	0.0307	5458	0351	8.6006	0399	6522	0449	8.7008	0502	7468	0558	8.7903	0.0617	46
15	8.4881	0.0308	5467	0352	8.6015	0400	6530	0450	8.7016	0503	7475	0559	8.7910	0.0618	45
16	8.4891	0.0308	5476	0353	8.6024	0400	6539	0451	8.7024	0504	7482	0560	8.7918	0.0619	44
17	8.4901	0.0309	5486	0354	8.6033	0401	6547	0452	8.7031	0505	7490	0561	8.7925	0.0620	43
18	8.4911	0.0310	5495	0354	8.6042	0402	6555	0452	8.7039	0506	7497	0562	8.7932	0.0621	42
19	8.4921	0.0311	5505	0355	8.6051	0403	6563	0453	8.7047	0507	7505	0563	8.7939	0.0622	41
20	8.4932	0.0311	5514	0356	8.6059	0404	6572	0454	8.7055	0508	7512	0564	8.7946	0.0623	40
21	8.4942	0.0312	5523	0357	8.6068	0404	6580	0455	8.7063	0508	7520	0565	8.7963	0.0624	39
22	8.4952	0.0313	5533	0358	8.6077	0405	6588	0456	8.7071	0509	7527	0566	8.7960	0.0625	38
23	8.4962	0.0313	5542	0358	8.6086	0406	6597	0457	8.7078	0510	7534	0567	8.7967	0.0626	37
24	8.4972	0.0314	5552	0359	8.6094	0407	6605	0458	8.7086	0511	7542	0568	8.7974	0.0627	36
25	8.4982	0.0315	5561	0360	8.6103	0408	6613	0458	8.7094	0512	7549	0569	8.7981	0.0628	35
26	8.4992	0.0316	5570	0361	8.6112	0409	6621	0459	8.7102	0513	7557	0570	8.7988	0.0629	34
27	8.5002	0.0316	5579	0361	8.6121	0409	6630	0460	8.7110	0514	7564	0571	8.7995	0.0630	33
28	8.5012	0.0317	5589	0362	8.6129	0410	6638	0461	8.7117	0515	7571	0572	8.8002	0.0631	32
29	8.5021	0.0318	5598	0363	8.6138	0411	6646	0462	8.7125	0516	7579	0573	8.8009	0.0632	31
30	8.5031	0.0319	5607	0364	8.6147	0412	6654	0463	8.7133	0517	7586	0574	8.8016	0.0633	30
31	8.5041	0.0319	5617	0364	8.6156	0413	6662	0464	8.7141	0518	7593	0575	8.8023	0.0634	29
32	8.5051	0.0320	5626	0365	8.6164	0413	6671	0465	8.7148	0519	7601	0576	8.8030	0.0635	28
33	8.5061	0.0321	5635	0366	8.6173	0414	6679	0465	8.7156	0520	7608	0577	8.8037	0.0636	27
34	8.5071	0.0321	5644	0367	8.6182	0415	6687	0466	8.7164	0520	7615	0577	8.8044	0.0637	26
35	8.5081	0.0322	5654	0368	8.6190	0416	6695	0467	8.7172	0521	7623	0578	8.8051	0.0638	25
36	8.5091	0.0323	5663	0368	8.6199	0417	6703	0468	8.7179	0522	7630	0579	8.8058	0.0639	24
37	8.5101	0.0324	5672	0369	8.6208	0418	6711	0469	8.7187	0523	7637	0580	8.8065	0.0640	23
38	8.5110	0.0324	5681	0370	8.6216	0418	6720	0470	8.7195	0524	7645	0581	8.8072	0.0641	22
39	8.5120	0.0325	5691	0371	8.6225	0419	6728	0471	8.7202	0525	7652	0582	8.8079	0.0642	21
40	8.5130	0.0326	5700	0372	8.6234	0420	6736	0472	8.7210	0526	7659	0583	8.8086	0.0644	20
41	8.5140	0.0327	5709	0372	8.6242	0421	6744	0473	8.7218	0527	7666	0584	8.8092	0.0645	19
42	8.5150	0.0327	5718	0373	8.6251	0422	6752	0473	8.7225	0528	7674	0585	8.8099	0.0646	18
43	8.5160	0.0328	5727	0374	8.6259	0423	6760	0474	8.7233	0529	7681	0586	8.8106	0.0647	17
44	8.5169	0.0329	5736	0375	8.6268	0423	6768	0475	8.7241	0530	7688	0587	8.8113	0.0648	16
45	8.5179	0.0330	5745	0375	8.6277	0424	6776	0476	8.7248	0531	7696	0588	8.8120	0.0649	15
46	8.5189	0.0330	5755	0376	8.6285	0425	6785	0477	8.7256	0532	7703	0589	8.8127	0.0650	14
47	8.5199	0.0331	5764	0377	8.6294	0426	6793	0478	8.7264	0533	7710	0590	8.8134	0.0651	13
48	8.5208	0.0332	5773	0378	8.6302	0427	6801	0479	8.7271	0534	7717	0591	8.8141	0.0652	12
49	8.5218	0.0333	5782	0379	8.6311	0428	6809	0480	8.7279	0534	7725	0592	8.8148	0.0653	11
50	8.5228	0.0333	5791	0379	8.6319	0428	6817	0480	8.7287	0535	7732	0593	8.8155	0.0654	10
51	8.5237	0.0334	5800	0380	8.6328	0429	6825	0481	8.7294	0536	7739	0594	8.8161	0.0655	9
52	8.5247	0.0335	5809	0381	8.6336	0430	6833	0482	8.7302	0537	7746	0595	8.8168	0.0656	8
53	8.5257	0.0335	5818	0382	8.6345	0431	6841	0483	8.7309	0538	7753	0596	8.8175	0.0657	7
54	8.5266	0.0336	5827	0383	8.6353	0432	6849	0484	8.7317	0539	7761	0597	8.8182	0.0658	6
55	8.5276	0.0337	5836	0383	8.6362	0433	6857	0485	8.7325	0540	7768	0598	8.8189	0.0659	5
56	8.5286	0.0338	5845	0384	8.6370	0434	6865	0486	8.7332	0541	7775	0599	8.8196	0.0660	4
57	8.5295	0.0338	5854	0385	8.6379	0434	6873	0487	8.7340	0542	7782	0600	8.8202	0.0661	3
58	8.5305	0.0339	5863	0386	8.6387	0435	6881	0488	8.7347	0543	7789	0601	8.8209	0.0662	2
59	8.5315	0.0340	5872	0387	8.6396	0436	6889	0489	8.7355	0544	7797	0602	8.8216	0.0663	1
60	8.5324	0.0341	5881	0387	8.6404	0437	6897	0489	8.7362	0545	7804	0603	8.8223	0.0664	0
′	Log 345°	Nat 345°	Log 344°	Nat 344°	Log 343°	Nat 343°	Log 342°	Nat 342°	Log 341°	Nat 341°	Log 340°	Nat 340°	Log 339°	Nat 339°	′

NOTE: For compactness. the leading digit is not displayed in every column. Scan other columns for necessary digit.

VERSINES

′	21° Log	21° Nat	22° Log	22° Nat	23° Log	23° Nat	24° Log	24° Nat	25° Log	25° Nat	26° Log	26° Nat	27° Log	27° Nat	′
0	8.8223	0.0664	8622	0728	8.9003	0795	9368	0865	8.9717	0937	0052	1012	9.0374	0.1090	60
1	8.8230	0.0665	8629	0729	8.9010	0796	9374	0866	8.9723	0938	0058	1013	9.0379	0.1091	59
2	8.8237	0.0666	8635	0730	8.9016	0797	9380	0867	8.9728	0939	0063	1015	9.0385	0.1093	58
3	8.8243	0.0667	8642	0731	8.9022	0798	9386	0868	8.9734	0941	0068	1016	9.0390	0.1094	57
4	8.8250	0.0668	8648	0733	8.9028	0800	9392	0869	8.9740	0942	0074	1017	9.0395	0.1095	56
5	8.8257	0.0669	8655	0734	8.9034	0801	9398	0870	8.9745	0943	0079	1018	9.0400	0.1097	55
6	8.8264	0.0670	8661	0735	8.9041	0802	9403	0872	8.9751	0944	0085	1020	9.0406	0.1098	54
7	8.8271	0.0672	8668	0736	8.9047	0803	9409	0873	8.9757	0946	0090	1021	9.0411	0.1099	53
8	8.8277	0.0673	8674	0737	8.9053	0804	9415	0874	8.9762	0947	0096	1022	9.0416	0.1101	52
9	8.8284	0.0674	8681	0738	8.9059	0805	9421	0875	8.9768	0948	0101	1024	9.0421	0.1102	51
10	8.8291	0.0675	8687	0739	8.9065	0806	9427	0876	8.9774	0949	0107	1025	9.0426	0.1103	50
11	8.8298	0.0676	8693	0740	8.9071	0808	9433	0878	8.9779	0950	0112	1026	9.0432	0.1105	49
12	8.8304	0.0677	8700	0741	8.9078	0809	9439	0879	8.9785	0952	0117	1027	9.0437	0.1106	48
13	8.8311	0.0678	8706	0742	8.9084	0810	9445	0880	8.9791	0953	0123	1029	9.0442	0.1107	47
14	8.8318	0.0679	8713	0743	8.9090	0811	9451	0881	8.9796	0954	0128	1030	9.0447	0.1108	46
15	8.8325	0.0680	8719	0745	8.9096	0812	9457	0882	8.9802	0955	0134	1031	9.0453	0.1110	45
16	8.8331	0.0681	8726	0746	8.9102	0813	9462	0884	8.9808	0957	0139	1033	9.0458	0.1111	44
17	8.8338	0.0682	8732	0747	8.9108	0814	9468	0885	8.9813	0958	0145	1034	9.0463	0.1112	43
18	8.8345	0.0683	8738	0748	8.9114	0816	9474	0886	8.9819	0959	0150	1035	9.0468	0.1114	42
19	8.8351	0.0684	8745	0749	8.9121	0817	9480	0887	8.9825	0960	0155	1036	9.0473	0.1115	41
20	8.8358	0.0685	8751	0750	8.9127	0818	9486	0888	8.9830	0962	0161	1038	9.0479	0.1116	40
21	8.8365	0.0686	8758	0751	8.9133	0819	9492	0890	8.9836	0963	0166	1039	9.0484	0.1118	39
22	8.8372	0.0687	8764	0752	8.9139	0820	9498	0891	8.9841	0964	0172	1040	9.0489	0.1119	38
23	8.8378	0.0688	8770	0753	8.9145	0821	9503	0892	8.9847	0965	0177	1042	9.0494	0.1121	37
24	8.8385	0.0689	8777	0755	8.9151	0822	9509	0893	8.9853	0967	0182	1043	9.0499	0.1122	36
25	8.8392	0.0691	8783	0756	8.9157	0824	9515	0894	8.9858	0968	0188	1044	9.0505	0.1123	35
26	8.8398	0.0692	8790	0757	8.9163	0825	9521	0896	8.9864	0969	0193	1045	9.0510	0.1125	34
27	8.8405	0.0693	8796	0758	8.9169	0826	9527	0897	8.9869	0970	0198	1047	9.0515	0.1126	33
28	8.8412	0.0694	8802	0759	8.9175	0827	9533	0898	8.9875	0972	0204	1048	9.0520	0.1127	32
29	8.8418	0.0695	8809	0760	8.9182	0828	9538	0899	8.9881	0973	0209	1049	9.0525	0.1129	31
30	8.8425	0.0696	8815	0761	8.9188	0829	9544	0900	8.9886	0974	0215	1051	9.0530	0.1130	30
31	8.8432	0.0697	8821	0762	8.9194	0831	9550	0902	8.9892	0975	0220	1052	9.0536	0.1131	29
32	8.8438	0.0698	8828	0763	8.9200	0832	9556	0903	8.9897	0977	0225	1053	9.0541	0.1133	28
33	8.8445	0.0699	8834	0765	8.9206	0833	9562	0904	8.9903	0978	0231	1055	9.0546	0.1134	27
34	8.8452	0.0700	8840	0766	8.9212	0834	9568	0905	8.9909	0979	0236	1056	9.0551	0.1135	26
35	8.8458	0.0701	8847	0767	8.9218	0835	9573	0906	8.9914	0980	0241	1057	9.0556	0.1137	25
36	8.8465	0.0702	8853	0768	8.9224	0836	9579	0908	8.9920	0982	0247	1058	9.0561	0.1138	24
37	8.8471	0.0703	8859	0769	8.9230	0838	9585	0909	8.9925	0983	0252	1060	9.0566	0.1139	23
38	8.8478	0.0704	8866	0770	8.9236	0839	9591	0910	8.9931	0984	0257	1061	9.0572	0.1141	22
39	8.8485	0.0705	8872	0771	8.9242	0840	9596	0911	8.9936	0985	0263	1062	9.0577	0.1142	21
40	8.8491	0.0707	8878	0772	8.9248	0841	9602	0912	8.9942	0987	0268	1064	9.0582	0.1143	20
41	8.8498	0.0708	8885	0773	8.9254	0842	9608	0914	8.9947	0988	0273	1065	9.0587	0.1145	19
42	8.8504	0.0709	8891	0775	8.9260	0843	9614	0915	8.9953	0989	0279	1066	9.0592	0.1146	18
43	8.8511	0.0710	8897	0776	8.9266	0845	9620	0916	8.9959	0990	0284	1068	9.0597	0.1147	17
44	8.8518	0.0711	8903	0777	8.9272	0846	9625	0917	8.9964	0992	0289	1069	9.0602	0.1149	16
45	8.8524	0.0712	8910	0778	8.9278	0847	9631	0919	8.9970	0993	0295	1070	9.0607	0.1150	15
46	8.8531	0.0713	8916	0779	8.9284	0848	9637	0920	8.9975	0994	0300	1072	9.0613	0.1151	14
47	8.8537	0.0714	8922	0780	8.9290	0849	9643	0921	8.9981	0996	0305	1073	9.0618	0.1153	13
48	8.8544	0.0715	8929	0781	8.9296	0850	9648	0922	8.9986	0997	0311	1074	9.0623	0.1154	12
49	8.8550	0.0716	8935	0782	8.9302	0852	9654	0923	8.9992	0998	0316	1075	9.0628	0.1156	11
50	8.8557	0.0717	8941	0784	8.9308	0853	9660	0925	8.9997	0999	0321	1077	9.0633	0.1157	10
51	8.8564	0.0718	8947	0785	8.9314	0854	9666	0926	9.0003	1001	0327	1078	9.0638	0.1158	9
52	8.8570	0.0719	8954	0786	8.9320	0855	9671	0927	9.0008	1002	0332	1079	9.0643	0.1160	8
53	8.8577	0.0721	8960	0787	8.9326	0856	9677	0928	9.0014	1003	0337	1081	9.0648	0.1161	7
54	8.8583	0.0722	8966	0788	8.9332	0857	9683	0930	9.0019	1004	0342	1082	9.0653	0.1162	6
55	8.8590	0.0723	8972	0789	8.9338	0859	9688	0931	9.0025	1006	0348	1083	9.0658	0.1164	5
56	8.8596	0.0724	8979	0790	8.9344	0860	9694	0932	9.0030	1007	0353	1085	9.0664	0.1165	4
57	8.8603	0.0725	8985	0792	8.9350	0861	9700	0933	9.0036	1008	0358	1086	9.0669	0.1166	3
58	8.8609	0.0726	8991	0793	8.9356	0862	9706	0934	9.0041	1010	0363	1087	9.0674	0.1168	2
59	8.8616	0.0727	8997	0794	8.9362	0863	9711	0936	9.0047	1011	0369	1089	9.0679	0.1169	1
60	8.8622	0.0728	9003	0795	8.9368	0865	9717	0937	9.0052	1012	0374	1090	9.0684	0.1171	0
′	Log 338°	Nat 338°	Log 337°	Nat 337°	Log 336°	Nat 336°	Log 335°	Nat 335°	Log 334°	Nat 334°	Log 333°	Nat 333°	Log 332°	Nat 332°	′

NOTE: For compactness. the leading digit is not displayed in every column. Scan other columns for necessary digit.

′	28° Log	28° Nat	29° Log	29° Nat	30° Log	30° Nat	31° Log	31° Nat	32° Log	32° Nat	33° Log	33° Nat	34° Log	34° Nat	′
0	9.0684	0.1171	0982	1254	9.1270	1340	1548	1428	9.1817	1520	2077	1613	9.2329	0.1710	60
1	9.0689	0.1172	0987	1255	9.1275	1341	1553	1430	9.1821	1521	2081	1615	9.2333	0.1711	59
2	9.0694	0.1173	0992	1257	9.1280	1343	1557	1431	9.1826	1523	2086	1616	9.2337	0.1713	58
3	9.0699	0.1175	0997	1258	9.1284	1344	1562	1433	9.1830	1524	2090	1618	9.2341	0.1715	57
4	9.0704	0.1176	1002	1259	9.1289	1346	1566	1434	9.1835	1526	2094	1620	9.2346	0.1716	56
5	9.0709	0.1177	1007	1261	9.1294	1347	1571	1436	9.1839	1527	2098	1621	9.2350	0.1718	55
6	9.0714	0.1179	1012	1262	9.1298	1348	1576	1437	9.1843	1529	2103	1623	9.2354	0.1719	54
7	9.0719	0.1180	1016	1264	9.1303	1350	1580	1439	9.1848	1530	2107	1624	9.2358	0.1721	53
8	9.0724	0.1181	1021	1265	9.1308	1351	1585	1440	9.1852	1532	2111	1626	9.2362	0.1723	52
9	9.0729	0.1183	1026	1267	9.1313	1353	1589	1442	9.1857	1533	2115	1628	9.2366	0.1724	51
10	9.0734	0.1184	1031	1268	9.1317	1354	1594	1443	9.1861	1535	2120	1629	9.2370	0.1726	50
11	9.0739	0.1186	1036	1269	9.1322	1356	1598	1445	9.1865	1537	2124	1631	9.2374	0.1728	49
12	9.0744	0.1187	1041	1271	9.1327	1357	1603	1446	9.1870	1538	2128	1632	9.2378	0.1729	48
13	9.0749	0.1188	1046	1272	9.1331	1359	1607	1448	9.1874	1540	2132	1634	9.2383	0.1731	47
14	9.0754	0.1190	1050	1274	9.1336	1360	1612	1449	9.1879	1541	2137	1636	9.2387	0.1732	46
15	9.0759	0.1191	1055	1275	9.1341	1362	1616	1451	9.1883	1543	2141	1637	9.2391	0.1734	45
16	9.0764	0.1192	1060	1276	9.1345	1363	1621	1452	9.1887	1544	2145	1639	9.2395	0.1736	44
17	9.0769	0.1194	1065	1278	9.1350	1365	1625	1454	9.1892	1546	2149	1640	9.2399	0.1737	43
18	9.0775	0.1195	1070	1279	9.1355	1366	1630	1455	9.1896	1547	2154	1642	9.2403	0.1739	42
19	9.0780	0.1197	1075	1281	9.1359	1368	1634	1457	9.1900	1549	2158	1644	9.2407	0.1741	41
20	9.0785	0.1198	1079	1282	9.1364	1369	1639	1458	9.1905	1550	2162	1645	9.2411	0.1742	40
21	9.0790	0.1199	1084	1284	9.1369	1370	1643	1460	9.1909	1552	2166	1647	9.2415	0.1744	39
22	9.0795	0.1201	1089	1285	9.1373	1372	1648	1461	9.1913	1554	2170	1648	9.2419	0.1746	38
23	9.0800	0.1202	1094	1286	9.1378	1373	1652	1463	9.1918	1555	2175	1650	9.2423	0.1747	37
24	9.0805	0.1204	1099	1288	9.1383	1375	1657	1464	9.1922	1557	2179	1652	9.2428	0.1749	36
25	9.0810	0.1205	1104	1289	9.1387	1376	1661	1466	9.1926	1558	2183	1653	9.2432	0.1751	35
26	9.0814	0.1206	1108	1291	9.1392	1378	1666	1468	9.1931	1560	2187	1655	9.2436	0.1752	34
27	9.0819	0.1208	1113	1292	9.1397	1379	1670	1469	9.1935	1561	2191	1656	9.2440	0.1754	33
28	9.0824	0.1209	1118	1294	9.1401	1381	1675	1471	9.1939	1563	2196	1658	9.2444	0.1755	32
29	9.0829	0.1210	1123	1295	9.1406	1382	1679	1472	9.1944	1565	2200	1660	9.2448	0.1757	31
30	9.0834	0.1212	1128	1296	9.1410	1384	1684	1474	9.1948	1566	2204	1661	9.2452	0.1759	30
31	9.0839	0.1213	1132	1298	9.1415	1385	1688	1475	9.1952	1568	2208	1663	9.2456	0.1760	29
32	9.0844	0.1215	1137	1299	9.1420	1387	1693	1477	9.1957	1569	2212	1664	9.2460	0.1762	28
33	9.0849	0.1216	1142	1301	9.1424	1388	1697	1478	9.1961	1571	2217	1666	9.2464	0.1764	27
34	9.0854	0.1217	1147	1302	9.1429	1390	1702	1480	9.1965	1572	2221	1668	9.2468	0.1765	26
35	9.0859	0.1219	1151	1304	9.1434	1391	1706	1481	9.1970	1574	2225	1669	9.2472	0.1767	25
36	9.0864	0.1220	1156	1305	9.1438	1393	1711	1483	9.1974	1575	2229	1671	9.2476	0.1769	24
37	9.0869	0.1222	1161	1306	9.1443	1394	1715	1484	9.1978	1577	2233	1672	9.2480	0.1770	23
38	9.0874	0.1223	1166	1308	9.1447	1396	1720	1486	9.1983	1579	2238	1674	9.2484	0.1772	22
39	9.0879	0.1224	1171	1309	9.1452	1397	1724	1487	9.1987	1580	2242	1676	9.2489	0.1774	21
40	9.0884	0.1226	1175	1311	9.1457	1399	1728	1489	9.1991	1582	2246	1677	9.2493	0.1775	20
41	9.0889	0.1227	1180	1312	9.1461	1400	1733	1490	9.1996	1583	2250	1679	9.2497	0.1777	19
42	9.0894	0.1229	1185	1314	9.1466	1401	1737	1492	9.2000	1585	2254	1680	9.2501	0.1779	18
43	9.0899	0.1230	1190	1315	9.1470	1403	1742	1493	9.2004	1586	2258	1682	9.2505	0.1780	17
44	9.0904	0.1231	1194	1317	9.1475	1404	1746	1495	9.2009	1588	2263	1684	9.2509	0.1782	16
45	9.0909	0.1233	1199	1318	9.1480	1406	1751	1496	9.2013	1590	2267	1685	9.2513	0.1784	15
46	9.0914	0.1234	1204	1319	9.1484	1407	1755	1498	9.2017	1591	2271	1687	9.2517	0.1785	14
47	9.0919	0.1236	1209	1321	9.1489	1409	1760	1500	9.2021	1593	2275	1689	9.2521	0.1787	13
48	9.0923	0.1237	1213	1322	9.1493	1410	1764	1501	9.2026	1594	2279	1690	9.2525	0.1789	12
49	9.0928	0.1238	1218	1324	9.1498	1412	1768	1503	9.2030	1596	2283	1692	9.2529	0.1790	11
50	9.0933	0.1240	1223	1325	9.1503	1413	1773	1504	9.2034	1597	2288	1693	9.2533	0.1792	10
51	9.0938	0.1241	1228	1327	9.1507	1415	1777	1506	9.2039	1599	2292	1695	9.2537	0.1793	9
52	9.0943	0.1243	1232	1328	9.1512	1416	1782	1507	9.2043	1601	2296	1697	9.2541	0.1795	8
53	9.0948	0.1244	1237	1330	9.1516	1418	1786	1509	9.2047	1602	2300	1698	9.2545	0.1797	7
54	9.0953	0.1245	1242	1331	9.1521	1419	1791	1510	9.2052	1604	2304	1700	9.2549	0.1798	6
55	9.0958	0.1247	1247	1332	9.1525	1421	1795	1512	9.2056	1605	2308	1701	9.2553	0.1800	5
56	9.0963	0.1248	1251	1334	9.1530	1422	1799	1513	9.2060	1607	2312	1703	9.2557	0.1802	4
57	9.0968	0.1250	1256	1335	9.1535	1424	1804	1515	9.2064	1609	2317	1705	9.2561	0.1803	3
58	9.0973	0.1251	1261	1337	9.1539	1425	1808	1516	9.2069	1610	2321	1706	9.2565	0.1805	2
59	9.0977	0.1252	1266	1338	9.1544	1427	1813	1518	9.2073	1612	2325	1708	9.2569	0.1807	1
60	9.0982	0.1254	1270	1340	9.1548	1428	1817	1520	9.2077	1613	2329	1710	9.2573	0.1808	0
′	Log	Nat	Log	Nat	Log	Nat	Log	Nat	Log	Nat	Log	Nat	Log	Nat	′
	331°		330°		329°		328°		327°		327°		325°		

NOTE: For compactness. the leading digit is not displayed in every column. Scan other columns for necessary digit.

VERSINES

′	35° Log	35° Nat	36° Log	36° Nat	37° Log	37° Nat	38° Log	38° Nat	39° Log	39° Nat	40° Log	40° Nat	41° Log	41° Nat	′
0	9.2573	0.1808	2810	1910	9.3040	2014	3263	2120	9.3480	2229	3691	2340	9.3897	0.2453	60
1	9.2577	0.1810	2814	1912	9.3044	2015	3267	2122	9.3484	2230	3695	2341	9.3900	0.2455	59
2	9.2581	0.1812	2818	1913	9.3047	2017	3270	2123	9.3487	2232	3698	2343	9.3904	0.2457	58
3	9.2585	0.1813	2822	1915	9.3051	2019	3274	2125	9.3491	2234	3702	2345	9.3907	0.2459	57
4	9.2589	0.1815	2825	1917	9.3055	2021	3278	2127	9.3494	2236	3705	2347	9.3910	0.2461	56
5	9.2593	0.1817	2829	1918	9.3059	2022	3281	2129	9.3498	2238	3709	2349	9.3914	0.2462	55
6	9.2597	0.1819	2833	1920	9.3062	2024	3285	2131	9.3502	2240	3712	2351	9.3917	0.2464	54
7	9.2601	0.1820	2837	1922	9.3066	2026	3289	2132	9.3505	2241	3716	2353	9.3920	0.2466	53
8	9.2605	0.1822	2841	1924	9.3070	2028	3292	2134	9.3509	2243	3719	2355	9.3924	0.2468	52
9	9.2609	0.1824	2845	1925	9.3074	2029	3296	2136	9.3512	2245	3723	2356	9.3927	0.2470	51
10	9.2613	0.1825	2849	1927	9.3077	2031	3300	2138	9.3516	2247	3726	2358	9.3931	0.2472	50
11	9.2617	0.1827	2853	1929	9.3081	2033	3303	2140	9.3519	2249	3729	2360	9.3934	0.2474	49
12	9.2621	0.1829	2856	1930	9.3085	2035	3307	2141	9.3523	2251	3733	2362	9.3937	0.2476	48
13	9.2625	0.1830	2860	1932	9.3089	2036	3311	2143	9.3526	2252	3736	2364	9.3941	0.2478	47
14	9.2629	0.1832	2864	1934	9.3093	2038	3314	2145	9.3530	2254	3740	2366	9.3944	0.2480	46
15	9.2633	0.1834	2868	1936	9.3096	2040	3318	2147	9.3534	2256	3743	2368	9.3947	0.2482	45
16	9.2637	0.1835	2872	1937	9.3100	2042	3322	2149	9.3537	2258	3747	2370	9.3951	0.2484	44
17	9.2641	0.1837	2876	1939	9.3104	2044	3325	2150	9.3541	2260	3750	2371	9.3954	0.2485	43
18	9.2645	0.1839	2880	1941	9.3107	2045	3329	2152	9.3544	2262	3754	2373	9.3957	0.2487	42
19	9.2649	0.1840	2883	1942	9.3111	2047	3333	2154	9.3548	2263	3757	2375	9.3961	0.2489	41
20	9.2653	0.1842	2887	1944	9.3115	2049	3336	2156	9.3551	2265	3760	2377	9.3964	0.2491	40
21	9.2657	0.1844	2891	1946	9.3119	2051	3340	2158	9.3555	2267	3764	2379	9.3967	0.2493	39
22	9.2661	0.1845	2895	1948	9.3122	2052	3343	2159	9.3558	2269	3767	2381	9.3971	0.2495	38
23	9.2665	0.1847	2899	1949	9.3126	2054	3347	2161	9.3562	2271	3771	2383	9.3974	0.2497	37
24	9.2669	0.1849	2903	1951	9.3130	2056	3351	2163	9.3565	2273	3774	2385	9.3977	0.2499	36
25	9.2673	0.1850	2907	1953	9.3134	2058	3354	2165	9.3569	2275	3778	2387	9.3981	0.2501	35
26	9.2677	0.1852	2910	1955	9.3137	2059	3358	2167	9.3572	2276	3781	2388	9.3984	0.2503	34
27	9.2681	0.1854	2914	1956	9.3141	2061	3362	2168	9.3576	2278	3784	2390	9.3987	0.2505	33
28	9.2685	0.1855	2918	1958	9.3145	2063	3365	2170	9.3579	2280	3788	2392	9.3991	0.2507	32
29	9.2688	0.1857	2922	1960	9.3149	2065	3369	2172	9.3583	2282	3791	2394	9.3994	0.2509	31
30	9.2692	0.1859	2926	1961	9.3152	2066	3372	2174	9.3586	2284	3795	2396	9.3998	0.2510	30
31	9.2696	0.1861	2930	1963	9.3156	2068	3376	2176	9.3590	2286	3798	2398	9.4001	0.2512	29
32	9.2700	0.1862	2933	1965	9.3160	2070	3380	2178	9.3594	2287	3802	2400	9.4004	0.2514	28
33	9.2704	0.1864	2937	1967	9.3163	2072	3383	2179	9.3597	2289	3805	2402	9.4008	0.2516	27
34	9.2708	0.1866	2941	1968	9.3167	2074	3387	2181	9.3601	2291	3808	2404	9.4011	0.2518	26
35	9.2712	0.1867	2945	1970	9.3171	2075	3390	2183	9.3604	2293	3812	2405	9.4014	0.2520	25
36	9.2716	0.1869	2949	1972	9.3175	2077	3394	2185	9.3608	2295	3815	2407	9.4017	0.2522	24
37	9.2720	0.1871	2953	1974	9.3178	2079	3398	2187	9.3611	2297	3819	2409	9.4021	0.2524	23
38	9.2724	0.1872	2956	1975	9.3182	2081	3401	2188	9.3615	2299	3822	2411	9.4024	0.2526	22
39	9.2728	0.1874	2960	1977	9.3186	2082	3405	2190	9.3618	2300	3826	2413	9.4027	0.2528	21
40	9.2732	0.1876	2964	1979	9.3189	2084	3409	2192	9.3622	2302	3829	2415	9.4031	0.2530	20
41	9.2736	0.1877	2968	1981	9.3193	2086	3412	2194	9.3625	2304	3832	2417	9.4034	0.2532	19
42	9.2740	0.1879	2972	1982	9.3197	2088	3416	2196	9.3629	2306	3836	2419	9.4037	0.2534	18
43	9.2744	0.1881	2975	1984	9.3201	2090	3419	2198	9.3632	2308	3839	2421	9.4041	0.2536	17
44	9.2747	0.1883	2979	1986	9.3204	2091	3423	2199	9.3636	2310	3843	2422	9.4044	0.2537	16
45	9.2751	0.1884	2983	1987	9.3208	2093	3427	2201	9.3639	2312	3846	2424	9.4047	0.2539	15
46	9.2755	0.1886	2987	1989	9.3212	2095	3430	2203	9.3643	2313	3849	2426	9.4051	0.2541	14
47	9.2759	0.1888	2991	1991	9.3215	2097	3434	2205	9.3646	2315	3853	2428	9.4054	0.2543	13
48	9.2763	0.1889	2994	1993	9.3219	2098	3437	2207	9.3650	2317	3856	2430	9.4057	0.2545	12
49	9.2767	0.1891	2998	1994	9.3223	2100	3441	2208	9.3653	2319	3860	2432	9.4061	0.2547	11
50	9.2771	0.1893	3002	1996	9.3226	2102	3444	2210	9.3657	2321	3863	2434	9.4064	0.2549	10
51	9.2775	0.1894	3006	1998	9.3230	2104	3448	2212	9.3660	2323	3866	2436	9.4067	0.2551	9
52	9.2779	0.1896	3010	2000	9.3234	2106	3452	2214	9.3664	2325	3870	2438	9.4071	0.2553	8
53	9.2783	0.1898	3013	2001	9.3237	2107	3455	2216	9.3667	2326	3873	2440	9.4074	0.2555	7
54	9.2787	0.1900	3017	2003	9.3241	2109	3459	2218	9.3670	2328	3877	2441	9.4077	0.2557	6
55	9.2790	0.1901	3021	2005	9.3245	2111	3462	2219	9.3674	2330	3880	2443	9.4080	0.2559	5
56	9.2794	0.1903	3025	2007	9.3248	2113	3466	2221	9.3677	2332	3883	2445	9.4084	0.2561	4
57	9.2798	0.1905	3028	2008	9.3252	2115	3469	2223	9.3681	2334	3887	2447	9.4087	0.2563	3
58	9.2802	0.1906	3032	2010	9.3256	2116	3473	2225	9.3684	2336	3890	2449	9.4090	0.2565	2
59	9.2806	0.1908	3036	2012	9.3259	2118	3477	2227	9.3688	2338	3893	2451	9.4094	0.2567	1
60	9.2810	0.1910	3040	2014	9.3263	2120	3480	2229	9.3691	2340	3897	2453	9.4097	0.2569	0
′	Log	Nat	Log	Nat	Log	Nat	Log	Nat	Log	Nat	Log	Nat	Log	Nat	′
	324°		323°		322°		321°		320°		319°		318°		

NOTE: For compactness. the leading digit is not displayed in every column. Scan other columns for necessary digit.

VERSINES

57

VERSINES

′	42° Log	Nat	43° Log	Nat	44° Log	Nat	45° Log	Nat	46° Log	Nat	47° Log	Nat	48° Log	Nat	′
0	9.4097	0.2569	4292	2686	9.4482	2807	4667	2929	9.4848	3053	5024	3180	9.5197	0.3309	60
1	9.4100	0.2570	4295	2688	9.4485	2809	4670	2931	9.4851	3056	5027	3182	9.5199	0.3311	59
2	9.4103	0.2572	4298	2690	9.4488	2811	4673	2933	9.4854	3058	5030	3184	9.5202	0.3313	58
3	9.4107	0.2574	4301	2692	9.4491	2813	4676	2935	9.4857	3060	5033	3186	9.5205	0.3315	57
4	9.4110	0.2576	4305	2694	9.4494	2815	4679	2937	9.4860	3062	5036	3189	9.5208	0.3317	56
5	9.4113	0.2578	4308	2696	9.4497	2817	4682	2939	9.4863	3064	5039	3191	9.5211	0.3320	55
6	9.4117	0.2580	4311	2698	9.4501	2819	4685	2941	9.4866	3066	5042	3193	9.5214	0.3322	54
7	9.4120	0.2582	4314	2700	9.4504	2821	4688	2943	9.4869	3068	5045	3195	9.5216	0.3324	53
8	9.4123	0.2584	4317	2702	9.4507	2823	4691	2945	9.4872	3070	5048	3197	9.5219	0.3326	52
9	9.4126	0.2586	4321	2704	9.4510	2825	4694	2947	9.4875	3072	5050	3199	9.5222	0.3328	51
10	9.4130	0.2588	4324	2706	9.4513	2827	4698	2950	9.4878	3074	5053	3201	9.5225	0.3330	50
11	9.4133	0.2590	4327	2708	9.4516	2829	4701	2952	9.4881	3076	5056	3203	9.5228	0.3333	49
12	9.4136	0.2592	4330	2710	9.4519	2831	4704	2954	9.4883	3079	5059	3206	9.5231	0.3335	48
13	9.4140	0.2594	4333	2712	9.4522	2833	4707	2956	9.4886	3081	5062	3208	9.5233	0.3337	47
14	9.4143	0.2596	4337	2714	9.4525	2835	4710	2958	9.4889	3083	5065	3210	9.5236	0.3339	46
15	9.4146	0.2598	4340	2716	9.4529	2837	4713	2960	9.4892	3085	5068	3212	9.5239	0.3341	45
16	9.4149	0.2600	4343	2718	9.4532	2839	4716	2962	9.4895	3087	5071	3214	9.5242	0.3343	44
17	9.4153	0.2602	4346	2720	9.4535	2841	4719	2964	9.4898	3089	5074	3216	9.5245	0.3346	43
18	9.4156	0.2604	4349	2722	9.4538	2843	4722	2966	9.4901	3091	5076	3218	9.5247	0.3348	42
19	9.4159	0.2606	4352	2724	9.4541	2845	4725	2968	9.4904	3093	5079	3221	9.5250	0.3350	41
20	9.4162	0.2608	4356	2726	9.4544	2847	4728	2970	9.4907	3095	5082	3223	9.5253	0.3352	40
21	9.4166	0.2610	4359	2728	9.4547	2849	4731	2972	9.4910	3097	5085	3225	9.5256	0.3354	39
22	9.4169	0.2612	4362	2730	9.4550	2851	4734	2974	9.4913	3100	5088	3227	9.5259	0.3356	38
23	9.4172	0.2613	4365	2732	9.4553	2853	4737	2976	9.4916	3102	5091	3229	9.5262	0.3359	37
24	9.4175	0.2615	4368	2734	9.4556	2855	4740	2978	9.4919	3104	5094	3231	9.5264	0.3361	36
25	9.4179	0.2617	4372	2736	9.4560	2857	4743	2981	9.4922	3106	5097	3233	9.5267	0.3363	35
26	9.4182	0.2619	4375	2738	9.4563	2859	4746	2983	9.4925	3108	5099	3236	9.5270	0.3365	34
27	9.4185	0.2621	4378	2740	9.4566	2861	4749	2985	9.4928	3110	5102	3238	9.5273	0.3367	33
28	9.4188	0.2623	4381	2742	9.4569	2863	4752	2987	9.4931	3112	5105	3240	9.5276	0.3369	32
29	9.4192	0.2625	4384	2744	9.4572	2865	4755	2989	9.4934	3114	5108	3242	9.5278	0.3372	31
30	9.4195	0.2627	4387	2746	9.4575	2867	4758	2991	9.4937	3116	5111	3244	9.5281	0.3374	30
31	9.4198	0.2629	4391	2748	9.4578	2870	4761	2993	9.4940	3119	5114	3246	9.5284	0.3376	29
32	9.4201	0.2631	4394	2750	9.4581	2872	4764	2995	9.4942	3121	5117	3248	9.5287	0.3378	28
33	9.4205	0.2633	4397	2752	9.4584	2874	4767	2997	9.4945	3123	5120	3251	9.5290	0.3380	27
34	9.4208	0.2635	4400	2754	9.4587	2876	4770	2999	9.4948	3125	5122	3253	9.5292	0.3383	26
35	9.4211	0.2637	4403	2756	9.4590	2878	4773	3001	9.4951	3127	5125	3255	9.5295	0.3385	25
36	9.4214	0.2639	4406	2758	9.4594	2880	4776	3003	9.4954	3129	5128	3257	9.5298	0.3387	24
37	9.4218	0.2641	4410	2760	9.4597	2882	4779	3005	9.4957	3131	5131	3259	9.5301	0.3389	23
38	9.4221	0.2643	4413	2762	9.4600	2884	4782	3008	9.4960	3133	5134	3261	9.5304	0.3391	22
39	9.4224	0.2645	4416	2764	9.4603	2886	4785	3010	9.4963	3135	5137	3263	9.5306	0.3393	21
40	9.4227	0.2647	4419	2766	9.4606	2888	4788	3012	9.4966	3138	5140	3266	9.5309	0.3396	20
41	9.4231	0.2649	4422	2768	9.4609	2890	4791	3014	9.4969	3140	5142	3268	9.5312	0.3398	19
42	9.4234	0.2651	4425	2770	9.4612	2892	4794	3016	9.4972	3142	5145	3270	9.5315	0.3400	18
43	9.4237	0.2653	4428	2772	9.4615	2894	4797	3018	9.4975	3144	5148	3272	9.5318	0.3402	17
44	9.4240	0.2655	4432	2774	9.4618	2896	4800	3020	9.4978	3146	5151	3274	9.5320	0.3404	16
45	9.4244	0.2657	4435	2776	9.4621	2898	4803	3022	9.4981	3148	5154	3276	9.5323	0.3407	15
46	9.4247	0.2659	4438	2778	9.4624	2900	4806	3024	9.4984	3150	5157	3278	9.5326	0.3409	14
47	9.4250	0.2661	4441	2780	9.4627	2902	4809	3026	9.4986	3152	5160	3281	9.5329	0.3411	13
48	9.4253	0.2663	4444	2782	9.4630	2904	4812	3028	9.4989	3155	5162	3283	9.5331	0.3413	12
49	9.4256	0.2665	4447	2784	9.4633	2906	4815	3030	9.4992	3157	5165	3285	9.5334	0.3415	11
50	9.4260	0.2667	4450	2786	9.4637	2908	4818	3033	9.4995	3159	5168	3287	9.5337	0.3417	10
51	9.4263	0.2669	4454	2788	9.4640	2910	4821	3035	9.4998	3161	5171	3289	9.5340	0.3420	9
52	9.4266	0.2671	4457	2790	9.4643	2912	4824	3037	9.5001	3163	5174	3291	9.5343	0.3422	8
53	9.4269	0.2673	4460	2792	9.4646	2915	4827	3039	9.5004	3165	5177	3294	9.5345	0.3424	7
54	9.4273	0.2675	4463	2794	9.4649	2917	4830	3041	9.5007	3167	5180	3296	9.5348	0.3426	6
55	9.4276	0.2677	4466	2797	9.4652	2919	4833	3043	9.5010	3169	5182	3298	9.5351	0.3428	5
56	9.4279	0.2679	4469	2799	9.4655	2921	4836	3045	9.5013	3172	5185	3300	9.5354	0.3431	4
57	9.4282	0.2681	4472	2801	9.4658	2923	4839	3047	9.5016	3174	5188	3302	9.5357	0.3433	3
58	9.4285	0.2682	4476	2803	9.4661	2925	4842	3049	9.5018	3176	5191	3304	9.5359	0.3435	2
59	9.4289	0.2684	4479	2805	9.4664	2927	4845	3051	9.5021	3178	5194	3307	9.5362	0.3437	1
60	9.4292	0.2686	4482	2807	9.4667	2929	4848	3053	9.5024	3180	5197	3309	9.5365	0.3439	0
′	Log	Nat 317°	Log	Nat 316°	Log	Nat 315°	Log	Nat 314°	Log	Nat 313°	Log	Nat 312°	Log	Nat 311°	′

NOTE: For compactness. the leading digit is not displayed in every column. Scan other columns for necessary digit.

VERSINES

′	49° Log	Nat	50° Log	Nat	51° Log	Nat	52° Log	Nat	53° Log	Nat	54° Log	Nat	55° Log	Nat	′
0	9.5365	0.3439	5529	3572	9.5690	3707	5847	3843	9.6001	3982	6151	4122	9.6298	0.4264	60
1	9.5368	0.3442	5532	3574	9.5693	3709	5850	3846	9.6003	3984	6154	4125	9.6301	0.4267	59
2	9.5370	0.3444	5535	3577	9.5695	3711	5852	3848	9.6006	3986	6156	4127	9.6303	0.4269	58
3	9.5373	0.3446	5537	3579	9.5698	3714	5855	3850	9.6008	3989	6159	4129	9.6306	0.4271	57
4	9.5376	0.3448	5540	3581	9.5701	3716	5857	3853	9.6011	3991	6161	4132	9.6308	0.4274	56
5	9.5379	0.3450	5543	3583	9.5703	3718	5860	3855	9.6014	3993	6164	4134	9.6311	0.4276	55
6	9.5381	0.3453	5546	3586	9.5706	3720	5863	3857	9.6016	3996	6166	4136	9.6313	0.4279	54
7	9.5384	0.3455	5548	3588	9.5709	3723	5865	3859	9.6019	3998	6169	4139	9.6315	0.4281	53
8	9.5387	0.3457	5551	3590	9.5711	3725	5868	3862	9.6021	4000	6171	4141	9.6318	0.4283	52
9	9.5390	0.3459	5554	3592	9.5714	3727	5870	3864	9.6024	4003	6174	4143	9.6320	0.4286	51
10	9.5393	0.3461	5556	3594	9.5716	3729	5873	3866	9.6026	4005	6176	4146	9.6323	0.4288	50
11	9.5395	0.3464	5559	3597	9.5719	3732	5876	3869	9.6029	4007	6178	4148	9.6325	0.4290	49
12	9.5398	0.3466	5562	3599	9.5722	3734	5878	3871	9.6031	4010	6181	4150	9.6327	0.4293	48
13	9.5401	0.3468	5564	3601	9.5724	3736	5881	3873	9.6034	4012	6183	4153	9.6330	0.4295	47
14	9.5404	0.3470	5567	3603	9.5727	3738	5883	3876	9.6036	4014	6186	4155	9.6332	0.4298	46
15	9.5406	0.3472	5570	3606	9.5730	3741	5886	3878	9.6039	4017	6188	4158	9.6335	0.4300	45
16	9.5409	0.3475	5572	3608	9.5732	3743	5888	3880	96041	4019	6191	4160	9.6337	0.4302	44
17	9.5412	0.3477	5575	3610	9.5735	3745	5891	3882	9.6044	4021	6193	4162	9.6340	0.4305	43
18	9.5415	0.3479	5578	3612	9.5738	3748	5894	3885	9.6046	4024	6196	4165	9.6342	0.4307	42
19	9.5417	0.3481	5581	3615	9.5740	3750	5896	3887	9.6049	4026	6198	4167	9.6344	0.4310	41
20	9.5420	0.3483	5583	3617	9.5743	3752	5899	3889	9.6051	4028	6201	4169	9.6347	0.4312	40
21	9.5423	0.3486	5586	3619	9.5745	3754	5901	3892	9.6054	4031	6203	4172	9.6349	0.4314	39
22	9.5426	0.3488	5589	3621	9.5748	3757	5904	3894	9.6056	4033	6206	4174	9.6352	0.4317	38
23	9.5428	0.3490	5591	3624	9.5751	3759	5906	3896	9.6059	4035	6208	4176	9.6354	0.4319	37
24	9.5431	0.3492	5594	3626	9.5753	3761	5909	3899	9.6061	4038	6210	4179	9.6356	0.4322	36
25	9.5434	0.3494	5597	3628	9.5756	3763	5912	3901	9.6064	4040	6213	4181	9.6359	0.4324	35
26	9.5437	0.3497	5599	3630	9.5759	3766	5914	3903	9.6066	4042	6215	4184	9.6361	0.4326	34
27	9.5439	0.3499	5602	3632	9.5761	3768	5917	3905	9.6069	4045	6218	4186	9.6364	0.4329	33
28	9.5442	0.3501	5605	3635	9.5764	3770	5919	3908	9.6071	4047	6220	4188	9.6366	0.4331	32
29	9.5445	0.3503	5607	3637	9.5766	3773	5922	3910	9.6074	4049	6223	4191	9.6368	0.4334	31
30	9.5448	0.3506	5610	3639	9.5769	3775	5924	3912	9.6076	4052	6225	4193	9.6371	0.4336	30
31	9.5450	0.3508	5613	3641	9.5772	3777	5927	3915	9.6079	4054	6228	4195	9.6373	0.4338	29
32	9.5453	0.3510	5615	3644	9.5774	3779	5930	3917	9.6081	4056	6230	4198	9.6376	0.4341	28
33	9.5456	0.3512	5618	3646	9.5777	3782	5932	3919	9.6084	4059	6233	4200	9.6378	0.4343	27
34	9.5458	0.3514	5621	3648	9.5779	3784	5935	3922	9.6086	4061	6235	4202	9.6380	0.4346	26
35	9.5461	0.3517	5623	3650	9.5782	3786	5937	3924	9.6089	4063	6237	4205	9.6383	0.4348	25
36	9.5464	0.3519	5626	3653	9.5785	3789	5940	3926	9.6091	4066	6240	4207	9.6385	0.4350	24
37	9.5467	0.3521	5629	3655	9.5787	3791	5942	3929	9.6094	4068	6242	4210	9.6388	0.4353	23
38	9.5469	0.3523	5631	3657	9.5790	3793	5945	3931	9.6096	4070	6245	4212	9.6390	0.4355	22
39	9.5472	0.3525	5634	3659	9.5793	3795	5947	3933	9.6099	4073	6247	4214	9.6392	0.4358	21
40	9.5475	0.3528	5637	3662	9.5795	3798	5950	3935	9.6101	4075	6250	4217	9.6395	0.4360	20
41	9.5478	0.3530	5639	3664	9.5798	3800	5953	3938	9.6104	4078	6252	4219	9.6397	0.4362	19
42	9.5480	0.3532	5642	3666	9.5800	3802	5955	3940	9.6106	4080	6255	4221	9.6400	0.4365	18
43	9.5483	0.3534	5645	3668	9.5803	3804	5958	3942	9.6109	4082	6257	4224	9.6402	0.4367	17
44	9.5486	0.3537	5647	3671	9.5806	3807	5960	3945	9.6111	4085	6259	4226	9.6404	0.4370	16
45	9.5489	0.3539	5650	3673	9.5808	3809	5963	3947	9.6114	4087	6262	4229	9.6407	0.4372	15
46	9.5491	0.3541	5653	3675	9.5811	3811	5965	3949	9.6116	4089	6264	4231	9.6409	0.4374	14
47	9.5494	0.3543	5655	3677	9.5813	3814	5968	3952	9.6119	4092	6267	4233	9.6412	0.4377	13
48	9.5497	0.3545	5658	3680	9.5816	3816	5970	3954	9.6121	4094	6269	4236	9.6414	0.4379	12
49	9.5499	0.3548	5661	3682	9.5819	3818	5973	3956	9.6124	4096	6272	4238	9.6416	0.4382	11
50	9.5502	0.3550	5663	3684	9.5821	3820	5975	3959	9.6126	4099	6274	4240	9.6419	0.4384	10
51	9.5505	0.3552	5666	3686	9.5824	3823	5978	3961	9.6129	4101	6277	4243	9.6421	0.4386	9
52	9.5508	0.3554	5669	3689	9.5826	3825	5981	3963	9.6131	4103	6279	4245	9.6423	0.4389	8
53	9.5510	0.3557	5671	3691	9.5829	3827	5983	3966	9.6134	4106	6281	4248	9.6426	0.4391	7
54	9.5513	0.3559	5674	3693	9.5832	3830	5986	3968	9.6136	4108	6284	4250	9.6428	0.4394	6
55	9.5516	0.3561	5677	3695	9.5834	3832	5988	3970	9.6139	4110	6286	4252	9.6431	0.4396	5
56	9.5518	0.3563	5679	3698	9.5837	3834	5991	3973	9.6141	4113	6289	4255	9.6433	0.4398	4
57	9.5521	0.3565	5682	3700	9.5839	3837	5993	3975	9.6144	4115	6291	4257	9.6435	0.4401	3
58	9.5524	0.3568	5685	3702	9.5842	3839	5996	3977	9.6146	4117	6294	4259	9.6438	0.4403	2
59	9.5527	0.3570	5687	3705	9.5845	3841	5998	3980	9.6149	4120	6296	4262	9.6440	0.4406	1
60	9.5529	0.3572	5690	3707	9.5847	3843	6001	3982	9.6151	4122	6298	4264	9.6442	0.4408	0
′	Log	Nat 310°	Log	Nat 309°	Log	Nat 308°	Log	Nat 307°	Log	Nat 306°	Log	Nat 305°	Log	Nat 304°	′

NOTE: For compactness. the leading digit is not displayed in every column. Scan other columns for necessary digit.

VERSINES

′	56° Log	56° Nat	57° Log	57° Nat	58° Log	58° Nat	59° Log	59° Nat	60° Log	60° Nat	61° Log	61° Nat	62° Log	62° Nat	′
0	9.6442	0.4408	6584	4554	9.6722	4701	6857	4850	9.6990	5000	7120	5152	9.7247	0.5305	60
1	9.6445	0.4410	6586	4556	9.6724	4703	6859	4852	9.6992	5003	7122	5154	9.7249	0.5308	59
2	9.6447	0.4413	6588	4558	9.6726	4706	6862	4855	9.6994	5005	7124	5157	9.7251	0.5310	58
3	9.6450	0.4415	6591	4561	9.6729	4708	6864	4857	9.6996	5008	7126	5160	9.7253	0.5313	57
4	9.6452	0.4418	6593	4563	9.6731	4711	6866	4860	9.6998	5010	7128	5162	9.7255	0.5316	56
5	9.6454	0.4420	6595	4566	9.6733	4713	6868	4862	9.7001	5013	7130	5165	9.7258	0.5318	55
6	9.6457	0.4423	6598	4568	9.6735	4716	6870	4865	9.7003	5015	7133	5167	9.7260	0.5321	54
7	9.6459	0.4425	6600	4571	9.6738	4718	6873	4867	9.7005	5018	7135	5170	9.7262	0.5323	53
8	9.6461	0.4427	6602	4573	9.6740	4721	6875	4870	9.7007	5020	7137	5172	9.7264	0.5326	52
9	9.6464	0.4430	6604	4576	9.6742	4723	6877	4872	9.7009	5023	7139	5175	9.7266	0.5328	51
10	9.6466	0.4432	6607	4578	9.6744	4725	6879	4875	9.7012	5025	7141	5177	9.7268	0.5331	50
11	9.6469	0.4435	6609	4580	9.6747	4728	6882	4877	9.7014	5028	7143	5180	9.7270	0.5334	49
12	9.6471	0.4437	6611	4583	9.6749	4730	6884	4880	9.7016	5030	7145	5182	9.7272	0.5336	48
13	9.6473	0.4439	6614	4585	9.6751	4733	6886	4882	9.7018	5033	7147	5185	9.7274	0.5339	47
14	9.6476	0.4442	6616	4588	9.6754	4735	6888	4885	9.7020	5035	7150	5188	9.7276	0.5341	46
15	9.6478	0.4444	6618	4590	9.6756	4738	6890	4887	9.7022	5038	7152	5190	9.7279	0.5344	45
16	9.6480	0.4447	6621	4593	9.6758	4740	6893	4890	9.7025	5040	7154	5193	9.7281	0.5346	44
17	9.6483	0.4449	6623	4595	9.6760	4743	6895	4892	9.7027	5043	7156	5195	9.7283	0.5349	43
18	9.6485	0.4452	6625	4598	9.6763	4745	6897	4895	9.7029	5045	7158	5198	9.7285	0.5352	42
19	9.6487	0.4454	6628	4600	9.6765	4748	6899	4897	9.7031	5048	7160	5200	9.7287	0.5354	41
20	9.6490	0.4456	6630	4602	9.6767	4750	6902	4900	9.7033	5050	7162	5203	9.7289	0.5357	40
21	9.6492	0.4459	6632	4605	9.6769	4753	6904	4902	9.7035	5053	7165	5205	9.7291	0.5359	39
22	9.6495	0.4461	6635	4607	9.6772	4755	6906	4905	9.7038	5056	7167	5208	9.7293	0.5362	38
23	9.6497	0.4464	6637	4610	9.6774	4758	6908	4907	9.7040	5058	7169	5211	9.7295	0.5364	37
24	9.6499	0.4466	6639	4612	9.6776	4760	6910	4910	9.7042	5061	7171	5213	9.7297	0.5367	36
25	9.6502	0.4469	6641	4615	9.6778	4763	6913	4912	9.7044	5063	7173	5216	9.7299	0.5370	35
26	9.6504	0.4471	6644	4617	9.6781	4765	6915	4915	9.7046	5066	7175	5218	9.7302	0.5372	34
27	9.6506	0.4473	6646	4620	9.6783	4768	6917	4917	9.7049	5068	7177	5221	9.7304	0.5375	33
28	9.6509	0.4476	6648	4622	9.6785	4770	6919	4920	9.7051	5071	7179	5223	9.7306	0.5377	32
29	9.6511	0.4478	6651	4625	9.6787	4773	6922	4922	9.7053	5073	7182	5226	9.7308	0.5380	31
30	9.6513	0.4481	6653	4627	9.6790	4775	6924	4925	9.7055	5076	7184	5228	9.7310	0.5383	30
31	9.6516	0.4483	6655	4629	9.6792	4777	6926	4927	9.7057	5078	7186	5231	9.7312	0.5385	29
32	9.6518	0.4485	6658	4632	9.6794	4780	6928	4930	9.7059	5081	7188	5234	9.7314	0.5388	28
33	9.6520	0.4488	6660	4634	9.6797	4782	6930	4932	9.7062	5083	7190	5236	9.7316	0.5390	27
34	9.6523	0.4490	6662	4637	9.6799	4785	6933	4935	9.7064	5086	7192	5239	9.7318	0.5393	26
35	9.6525	0.4493	6665	4639	9.6801	4787	6935	4937	9.7066	5088	7194	5241	9.7320	0.5395	25
36	9.6527	0.4495	6667	4642	9.6803	4790	6937	4940	9.7068	5091	7196	5244	9.7322	0.5398	24
37	9.6530	0.4498	6669	4644	9.6806	4792	6939	4942	9.7070	5093	7199	5246	9.7324	0.5401	23
38	9.6532	0.4500	6671	4647	9.6808	4795	6941	4945	9.7072	5096	7201	5249	9.7326	0.5403	22
39	9.6535	0.4502	6674	4649	9.6810	4797	6944	4947	9.7074	5099	7203	5251	9.7329	0.5406	21
40	9.6537	0.4505	6676	4652	9.6812	4800	6946	4950	9.7077	5101	7205	5254	9.7331	0.5408	20
41	9.6539	0.4507	6678	4654	9.6815	4802	6948	4952	9.7079	5104	7207	5257	9.7333	0.5411	19
42	9.6542	0.4510	6681	4656	9.6817	4805	6950	4955	9.7081	5106	7209	5259	9.7335	0.5414	18
43	9.6544	0.4512	6683	4659	9.6819	4807	6952	4957	9.7083	5109	7211	5262	9.7337	0.5416	17
44	9.6546	0.4515	6685	4661	9.6821	4810	6955	4960	9.7085	5111	7213	5264	9.7339	0.5419	16
45	9.6549	0.4517	6687	4664	9.6823	4812	6957	4962	9.7087	5114	7215	5267	9.7341	0.5421	15
46	9.6551	0.4520	6690	4666	9.6826	4815	6959	4965	9.7090	5116	7218	5269	9.7343	0.5424	14
47	9.6553	0.4522	6692	4669	9.6828	4817	6961	4967	9.7092	5119	7220	5272	9.7345	0.5426	13
48	9.6556	0.4524	6694	4671	9.6830	4820	6963	4970	9.7094	5121	7222	5274	9.7347	0.5429	12
49	9.6558	0.4527	6697	4674	9.6832	4822	6966	4972	9.7096	5124	7224	5277	9.7349	0.5432	11
50	9.6560	0.4529	6699	4676	9.6835	4825	6968	4975	9.7098	5126	7226	5280	9.7351	0.5434	10
51	9.6563	0.4532	6701	4679	9.6837	4827	6970	4977	9.7100	5129	7228	5282	9.7353	0.5437	9
52	9.6565	0.4534	6703	4681	9.6839	4830	6972	4980	9.7102	5132	7230	5285	9.7355	0.5439	8
53	9.6567	0.4537	6706	4684	9.6841	4832	6974	4982	9.7105	5134	7232	5287	9.7358	0.5442	7
54	9.6570	0.4539	6708	4686	9.6844	4835	6977	4985	9.7107	5137	7234	5290	9.7360	0.5445	6
55	9.6572	0.4541	6710	4688	9.6846	4837	6979	4987	9.7109	5139	7237	5292	9.7362	0.5447	5
56	9.6574	0.4544	6713	4691	9.6848	4840	6981	4990	9.7111	5142	7239	5295	9.7364	0.5450	4
57	9.6577	0.4546	6715	4693	9.6850	4842	6983	4992	9.7113	5144	7241	5298	9.7366	0.5452	3
58	9.6579	0.4549	6717	4696	9.6853	4845	6985	4995	9.7115	5147	7243	5300	9.7368	0.5455	2
59	9.6581	0.4551	6719	4698	9.6855	4847	6988	4997	9.7118	5149	7245	5303	9.7370	0.5458	1
60	9.6584	0.4554	6722	4701	9.6857	4850	6990	5000	9.7120	5152	7247	5305	9.7372	0.5460	0
′	Log 303°	Nat	Log 302°	Nat	Log 301°	Nat	Log 300°	Nat	Log 299°	Nat	Log 298°	Nat	Log 297°	Nat	′

NOTE: For compactness. the leading digit is not displayed in every column. Scan other columns for necessary digit.

VERSINES

′	63° Log	63° Nat	64° Log	64° Nat	65° Log	65° Nat	66° Log	66° Nat	67° Log	67° Nat	68° Log	68° Nat	69° Log	69° Nat	′
0	9.7372	0.5460	7494	5616	9.7615	5774	7732	5933	9.7848	6093	7962	6254	9.8073	0.6416	60
4	9.7380	0.5470	7503	5627	9.7623	5784	7740	5943	9.7856	6103	7969	6265	9.8080	0.6427	56
8	9.7388	0.5481	7511	5637	9.7630	5795	7748	5954	9.7863	6114	7976	6276	9.8088	0.6438	52
12	9.7397	0.5491	7519	5648	9.7638	5805	7756	5965	9.7871	6125	7984	6286	9.8095	0.6449	48
16	9.7405	0.5502	7527	5658	9.7646	5816	7764	5975	9.7879	6136	7991	6297	9.8102	0.6460	44
20	9.7413	0.5512	7535	5669	9.7654	5827	7771	5986	9.7886	6146	7999	6308	9.8110	0.6471	40
24	9.7421	0.5522	7543	5679	9.7662	5837	7779	5997	9.7894	6157	8006	6319	9.8117	0.6482	36
28	9.7429	0.5533	7551	5690	9.7670	5848	7787	6007	9.7901	6168	8014	6330	9.8124	0.6492	32
32	9.7438	0.5543	7559	5700	9.7678	5858	7794	6018	9.7909	6179	8021	6340	9.8131	0.6503	28
36	9.7446	0.5554	7567	5711	9.7686	5869	7802	6029	9.7916	6189	8029	6351	9.8139	0.6514	24
40	9.7454	0.5564	7575	5721	9.7693	5880	7810	6039	9.7924	6200	8036	6362	9.8146	0.6525	20
44	9.7462	0.5575	7583	5732	9.7701	5890	7817	6050	9.7931	6211	8043	6373	9.8153	0.6536	16
48	9.7470	0.5585	7591	5742	9.7709	5901	7825	6061	9.7939	6222	8051	6384	9.8160	0.6547	12
52	9.7478	0.5595	7599	5753	9.7717	5911	7833	6071	9.7947	6232	8058	6395	9.8168	0.6558	8
56	9.7486	0.5606	7607	5763	9.7725	5922	7840	6082	9.7954	6243	8066	6405	9.8175	0.6569	4
60	9.7494	0.5616	7615	5774	9.7732	5933	7848	6093	9.7962	6254	8073	6416	9.8182	0.6580	0
	Log 296°	Nat	Log 295°	Nat	Log 294°	Nat	Log 293°	Nat	Log 292°	Nat	Log 291°	Nat	Log 290°	Nat	
	70° Log	70° Nat	71° Log	71° Nat	72° Log	72° Nat	73° Log	73° Nat	74° Log	74° Nat	75° Log	75° Nat	76° Log	76° Nat	
0	9.8182	0.6580	8289	6744	9.8395	6910	8498	7076	9.8600	7244	8699	7412	9.8797	0.7581	60
4	9.8189	0.6591	8296	6755	9.8402	6921	8505	7087	9.8606	7255	8706	7423	9.8804	0.7592	56
8	9.8197	0.6602	8304	6766	9.8409	6932	8512	7099	9.8613	7266	8712	7434	9.8810	0.7603	52
12	9.8204	0.6613	8311	6777	9.8416	6943	8519	7110	9.8620	7277	8719	7446	9.8817	0.7615	48
16	9.8211	0.6624	8318	6788	9.8422	6954	8525	7121	9.8626	7288	8726	7457	9.8823	0.7626	44
20	9.8218	0.6635	8325	6799	9.8429	6965	8532	7132	9.8633	7300	8732	7468	9.8829	0.7637	40
24	9.8225	0.6645	8332	6810	9.8436	6976	8539	7143	9.8640	7311	8739	7479	9.8836	0.7649	36
28	9.8232	0.6656	8339	6821	9.8443	6987	8546	7154	9.8646	7322	8745	7491	9.8842	0.7660	32
32	9.8240	0.6667	8346	6832	9.8450	6998	8552	7165	9.8653	7333	8752	7502	9.8849	0.7671	28
36	9.8247	0.6678	8353	6844	9.8457	7010	8559	7177	9.8660	7344	8758	7513	9.8855	0.7683	24
40	9.8254	0.6689	8360	6855	9.8464	7021	8566	7188	9.8666	7356	8765	7524	9.8861	0.7694	20
44	9.8261	0.6700	8367	6866	9.8471	7032	8573	7199	9.8673	7367	8771	7536	9.8868	0.7705	16
48	9.8268	0.6711	8374	6877	9.8478	7043	8579	7210	9.8679	7378	8778	7547	9.8874	0.7716	12
52	9.8275	0.6722	8381	6888	9.8484	7054	8586	7221	9.8686	7389	8784	7558	9.8881	0.7728	8
56	9.8282	0.6733	8388	6899	9.8491	7065	8593	7232	9.8693	7401	8791	7569	9.8887	0.7739	4
60	9.8289	0.6744	8395	6910	9.8498	7076	8600	7244	9.8699	7412	8797	7581	9.8893	0.7750	0
	Log 289°	Nat	Log 288°	Nat	Log 287°	Nat	Log 286°	Nat	Log 285°	Nat	Log 284°	Nat	Log 283°	Nat	
	77° Log	77° Nat	78° Log	78° Nat	79° Log	79° Nat	80° Log	80° Nat	81° Log	81° Nat	82° Log	82° Nat	83° Log	83° Nat	
0	9.8893	0.7750	8988	7921	9.9081	8092	9172	8264	9.9261	8436	9349	8608	9.9436	0.8781	60
4	9.8900	0.7762	8994	7932	9.9087	8103	9178	8275	9.9267	8447	9355	8620	9.9441	0.8793	56
8	9.8906	0.7773	9000	7944	9.9093	8115	9184	8286	9.9273	8459	9361	8631	9.9447	0.8804	52
12	9.8912	0.7785	9006	7955	9.9099	8126	9190	8298	9.9279	8470	9367	8643	9.9453	0.8816	48
16	9.8919	0.7796	9013	7966	9.9105	8138	9196	8309	9.9285	8482	9372	8654	9.9458	0.8828	44
20	9.8925	0.7807	9019	7978	9.9111	8149	9202	8321	9.9291	8493	9378	8666	9.9464	0.8839	40
24	9.8931	0.7819	9025	7989	9.9117	8160	9208	8332	9.9297	8505	9384	8677	9.9470	0.8851	36
28	9.8938	0.7830	9031	8001	9.9123	8172	9214	8344	9.9302	8516	9390	8689	9.9475	0.8862	32
32	9.8944	0.7841	9037	8012	9.9129	8183	9220	8355	9.9308	8528	9395	8701	9.9481	0.8874	28
36	9.8950	0.7853	9044	8023	9.9135	8195	9226	8367	9.9314	8539	9401	8712	9.9487	0.8885	24
40	9.8956	0.7864	9050	8035	9.9141	8206	9232	8378	9.9320	8551	9407	8724	9.9492	0.8897	20
44	9.8963	0.7875	9056	8046	9.9148	8218	9237	8390	9.9326	8562	9413	8735	9.9498	0.8908	16
48	9.8969	0.7887	9062	8058	9.9154	8229	9243	8401	9.9332	8574	9418	8747	9.9504	0.8920	12
52	9.8975	0.7898	9068	8069	9.9160	8241	9249	8413	9.9338	8585	9424	8758	9.9509	0.8932	8
56	9.8981	0.7910	9074	8080	9.9166	8252	9255	8424	9.9343	8597	9430	8770	9.9515	0.8943	4
60	9.8988	0.7921	9081	8092	9.9172	8264	9261	8436	9.9349	8608	9436	8781	9.9521	0.8955	0
′	Log 282°	Nat	Log 281°	Nat	Log 280°	Nat	Log 279°	Nat	Log 278°	Nat	Log 277°	Nat	Log 276°	Nat	′

NOTE: For compactness. the leading digit is not displayed in every column. Scan other columns for necessary digit.

VERSINES

′	84° Log	84° Nat	85° Log	85° Nat	86° Log	86° Nat	87° Log	87° Nat	88° Log	88° Nat	89° Log	89° Nat	90° Log	90° Nat	′
0	9.9521	0.8955	9604	9128	9.9686	9302	9767	9477	9.9846	9651	9924	9825	0.0000	1.0000	60
4	9.9526	0.8966	9609	9140	9.9691	9314	9772	9488	9.9851	9663	9929	9837	0.0005	1.0012	56
8	9.9532	0.8978	9615	9152	9.9697	9326	9777	9500	9.9856	9674	9934	9849	0.0010	1.0023	52
12	9.9537	0.8989	9620	9163	9.9702	9337	9782	9512	9.9861	9686	9939	9860	0.0015	1.0035	48
16	9.9543	0.9001	9626	9175	9.9708	9349	9788	9523	9.9867	9698	9944	9872	0.0020	1.0047	44
20	9.9548	0.9013	9631	9186	9.9713	9360	9793	9535	9.9872	9709	9949	9884	0.0025	1.0058	40
24	9.9554	0.9024	9637	9198	9.9718	9372	9798	9546	9.9877	9721	9954	9895	0.0030	1.0070	36
28	9.9560	0.9036	9642	9210	9.9724	9384	9804	9558	9.9882	9732	9959	9907	0.0035	1.0081	32
32	9.9565	0.9047	9648	9221	9.9729	9395	9809	9570	9.9887	9744	9964	9919	0.0040	1.0093	28
36	9.9571	0.9059	9653	9233	9.9734	9407	9814	9581	9.9893	9756	9970	9930	0.0045	1.0105	24
40	9.9576	0.9071	9659	9244	9.9740	9419	9819	9593	9.9898	9767	9975	9942	0.0050	1.0116	20
44	9.9582	0.9082	9664	9256	9.9745	9430	9825	9604	9.9903	9779	9980	9953	0.0055	1.0128	16
48	9.9587	0.9094	9670	9268	9.9751	9442	9830	9616	9.9908	9791	9985	9965	0.0060	1.0140	12
52	9.9593	0.9105	9675	9279	9.9756	9453	9835	9628	9.9913	9802	9990	9977	0.0065	1.0151	8
56	9.9598	0.9117	9681	9291	9.9761	9465	9840	9639	9.9918	9814	9995	9988	0.0070	1.0163	4
60	9.9604	0.9128	9686	9302	9.9767	9477	9846	9651	9.9924	9825	0000	0000	0.0075	1.0175	0
	Log 275°	Nat	Log 274°	Nat	Log 273°	Nat	Log 272°	Nat	Log 271°	Nat	Log 270°	Nat	Log 269°	Nat	

′	91° Log	91° Nat	92° Log	92° Nat	93° Log	93° Nat	94° Log	94° Nat	95° Log	95° Nat	96° Log	96° Nat	97° Log	97° Nat	′
0	0.0075	1.0175	0149	0349	0.0222	0523	0293	0698	0.0363	0872	0432	1045	0.0499	1.1219	60
4	0.0080	1.0186	0154	0361	0.0226	0535	0298	0709	0.0368	0883	0436	1057	0.0504	1.1230	56
8	0.0085	1.0198	0159	0372	0.0231	0547	0302	0721	0.0372	0895	0441	1068	0.0508	1.1242	52
12	0.0090	1.0209	0164	0384	0.0236	0558	0307	0732	0.0377	0906	0445	1080	0.0513	1.1253	48
16	0.0095	1.0221	0168	0396	0.0241	0570	0312	0744	0.0381	0918	0450	1092	0.0517	1.1265	44
20	0.0100	1.0233	0173	0407	0.0245	0581	0316	0756	0.0386	0929	0454	1103	0.0522	1.1276	40
24	0.0105	1.0244	0178	0419	0.0250	0593	0321	0767	0.0391	0941	0459	1115	0.0526	1.1288	36
28	0.0110	1.0256	0183	0430	0.0255	0605	0326	0779	0.0395	0953	0463	1126	0.0531	1.1299	32
32	0.0115	1.0268	0188	0442	0.0260	0616	0330	0790	0.0400	0964	0468	1138	0.0535	1.1311	28
36	0.0120	1.0279	0193	0454	0.0264	0628	0335	0802	0.0404	0976	0473	1149	0.0539	1.1323	24
40	0.0125	1.0291	0197	0465	0.0269	0640	0340	0814	0.0409	0987	0477	1161	0.0544	1.1334	20
44	0.0129	1.0302	0202	0477	0.0274	0651	0344	0825	0.0414	0999	0481	1172	0.0548	1.1346	16
48	0.0134	1.0314	0207	0488	0.0279	0663	0349	0837	0.0418	1011	0486	1184	0.0553	1.1357	12
52	0.0139	1.0326	0212	0500	0.0283	0674	0354	0848	0.0423	1022	0490	1196	0.0557	1.1369	8
56	0.0144	1.0337	0217	0512	0.0288	0686	0358	0860	0.0427	1034	0495	1207	0.0562	1.1380	4
60	0.0149	1.0349	0222	0523	0.0293	0698	0363	0872	0.0432	1045	0499	1219	0.0566	1.1392	0
	Log 268°	Nat	Log 267°	Nat	Log 266°	Nat	Log 265°	Nat	Log 264°	Nat	Log 263°	Nat	Log 262°	Nat	

′	98° Log	98° Nat	99° Log	99° Nat	100° Log	100° Nat	101° Log	101° Nat	102° Log	102° Nat	103° Log	103° Nat	104° Log	104° Nat	′
0	0.0566	1.1392	0631	1564	0.0695	1736	0758	1908	0.0820	2079	0881	2250	0.0941	1.2419	60
4	0.0570	1.1403	0636	1576	0.0700	1748	0763	1920	0.0824	2090	0885	2261	0.0945	1.2431	56
8	0.0575	1.1415	0640	1587	0.0704	1759	0767	1931	0.0829	2102	0889	2272	0.0949	1.2442	52
12	0.0579	1.1426	0644	1599	0.0708	1771	0771	1942	0.0833	2113	0893	2284	0.0953	1.2453	48
16	0.0583	1.1438	0648	1610	0.0712	1782	0775	1954	0.0837	2125	0897	2295	0.0957	1.2464	44
20	0.0588	1.1449	0653	1622	0.0717	1794	0779	1965	0.0841	2136	0901	2306	0.0961	1.2476	40
24	0.0592	1.1461	0657	1633	0.0721	1805	0783	1977	0.0845	2147	0905	2317	0.0965	1.2487	36
28	0.0597	1.1472	0661	1645	0.0725	1817	0787	1988	0.0849	2159	0909	2329	0.0968	1.2498	32
32	0.0601	1.1484	0666	1656	0.0729	1828	0792	1999	0.0853	2170	0913	2340	0.0972	1.2509	28
36	0.0605	1.1495	0670	1668	0.0733	1840	0796	2011	0.0857	2181	0917	2351	0.0976	1.2521	24
40	0.0610	1.1507	0674	1679	0.0738	1851	0800	2022	0.0861	2193	0921	2363	0.0980	1.2532	20
44	0.0614	1.1518	0678	1691	0.0742	1862	0804	2034	0.0865	2204	0925	2374	0.0984	1.2543	16
48	0.0618	1.1530	0683	1702	0.0746	1874	0808	2045	0.0869	2215	0929	2385	0.0988	1.2554	12
52	0.0623	1.1541	0687	1714	0.0750	1885	0812	2056	0.0873	2227	0933	2397	0.0992	1.2566	8
56	0.0627	1.1553	0691	1725	0.0754	1897	0816	2068	0.0877	2238	0937	2408	0.0996	1.2577	4
60	0.0631	1.1564	0695	1736	0.0758	1908	0820	2079	0.0881	2250	0941	2419	0.1000	1.2588	0
′	Log 261°	Nat	Log 260°	Nat	Log 259°	Nat	Log 258°	Nat	Log 257°	Nat	Log 256°	Nat	Log 255°	Nat	′

NOTE: For compactness. the leading digit is not displayed in every column. Scan other columns for necessary digit.

VERSINES

′	105° Log	105° Nat	106° Log	106° Nat	107° Log	107° Nat	108° Log	108° Nat	109° Log	109° Nat	110° Log	110° Nat	111° Log	111° Nat	′
0	0.1000	1.2588	1057	2756	0.1114	2924	1169	3090	0.1224	3256	1278	3420	0.1330	1.3584	60
4	0.1004	1.2599	1061	2768	0.1118	2935	1173	3101	0.1228	3267	1281	3431	0.1334	1.3595	56
8	0.1007	1.2611	1065	2779	0.1121	2946	1177	3112	0.1231	3278	1285	3442	0.1337	1.3605	52
12	0.1011	1.2622	1069	2790	0.1125	2957	1180	3123	0.1235	3289	1288	3453	0.1341	1.3616	48
16	0.1015	1.2633	1072	2801	0.1129	2968	1184	3134	0.1238	3300	1292	3464	0.1344	1.3627	44
20	0.1019	1.2644	1076	2812	0.1133	2979	1188	3145	0.1242	3311	1295	3475	0.1347	1.3638	40
24	0.1023	1.2656	1080	2823	0.1136	2990	1191	3156	0.1246	3322	1299	3486	0.1351	1.3649	36
28	0.1027	1.2667	1084	2835	0.1140	3002	1195	3168	0.1249	3333	1302	3497	0.1354	1.3660	32
32	0.1031	1.2678	1088	2846	0.1144	3013	1199	3179	0.1253	3344	1306	3508	0.1358	1.3670	28
36	0.1034	1.2689	1091	2857	0.1147	3024	1202	3190	0.1256	3355	1309	3518	0.1361	1.3681	24
40	0.1038	1.2700	1095	2868	0.1151	3035	1206	3201	0.1260	3365	1313	3529	0.1365	1.3692	20
44	0.1042	1.2712	1099	2879	0.1155	3046	1210	3212	0.1263	3376	1316	3540	0.1368	1.3703	16
48	0.1046	1.2723	1103	2890	0.1158	3057	1213	3223	0.1267	3387	1320	3551	0.1372	1.3714	12
52	0.1050	1.2734	1106	2901	0.1162	3068	1217	3234	0.1271	3398	1323	3562	0.1375	1.3724	8
56	0.1053	1.2745	1110	2913	0.1166	3079	1220	3245	0.1274	3409	1327	3573	0.1378	1.3735	4
60	0.1057	1.2756	1114	2924	0.1169	3090	1224	3256	0.1278	3420	1330	3584	0.1382	1.3746	0
	Log 254°	Nat	Log 253°	Nat	Log 252°	Nat	Log 251°	Nat	Log 250°	Nat	Log 249°	Nat	Log 248°	Nat	

′	112° Log	112° Nat	113° Log	113° Nat	114° Log	114° Nat	115° Log	115° Nat	116° Log	116° Nat	117° Log	117° Nat	118° Log	118° Nat	′
0	0.1382	1.3746	1432	3907	0.1482	4067	1531	4226	0.1579	4384	1626	4540	0.1672	1.4695	60
4	0.1385	1.3757	1436	3918	0.1485	4078	1534	4237	0.1582	4394	1629	4550	0.1675	1.4705	56
8	0.1389	1.3768	1439	3929	0.1489	4089	1537	4247	0.1585	4405	1632	4561	0.1678	1.4715	52
12	0.1392	1.3778	1442	3939	0.1492	4099	1541	4258	0.1588	4415	1635	4571	0.1681	1.4726	48
16	0.1395	1.3789	1446	3950	0.1495	4110	1544	4268	0.1591	4425	1638	4581	0.1684	1.4736	44
20	0.1399	1.3800	1449	3961	0.1498	4120	1547	4279	0.1594	4436	1641	4592	0.1687	1.4746	40
24	0.1402	1.3811	1452	3971	0.1502	4131	1550	4289	0.1598	4446	1644	4602	0.1690	1.4756	36
28	0.1406	1.3821	1456	3982	0.1505	4142	1553	4300	0.1601	4457	1647	4612	0.1693	1.4766	32
32	0.1409	1.3832	1459	3993	0.1508	4152	1557	4310	0.1604	4467	1650	4623	0.1696	1.4777	28
36	0.1412	1.3843	1462	4003	0.1511	4163	1560	4321	0.1607	4478	1653	4633	0.1699	1.4787	24
40	0.1416	1.3854	1466	4014	0.1515	4173	1563	4331	0.1610	4488	1656	4643	0.1702	1.4797	20
44	0.1419	1.3864	1469	4025	0.1518	4184	1566	4342	0.1613	4498	1659	4654	0.1705	1.4807	16
48	0.1422	1.3875	1472	4035	0.1521	4195	1569	4352	0.1616	4509	1662	4664	0.1708	1.4818	12
52	0.1426	1.3886	1476	4046	0.1524	4205	1572	4363	0.1619	4519	1666	4674	0.1711	1.4828	8
56	0.1429	1.3897	1479	4057	0.1528	4216	1576	4373	0.1623	4530	1669	4684	0.1714	1.4838	4
60	0.1432	1.3907	1482	4067	0.1531	4226	1579	4384	0.1626	4540	1672	4695	0.1717	1.4848	0
	Log 247°	Nat	Log 246°	Nat	Log 245°	Nat	Log 244°	Nat	Log 243°	Nat	Log 242°	Nat	Log 241°	Nat	

′	119° Log	119° Nat	120° Log	120° Nat	121° Log	121° Nat	122° Log	122° Nat	123° Log	123° Nat	124° Log	124° Nat	125° Log	125° Nat	′
0	0.1717	1.4848	1761	5000	0.1804	5150	1847	5299	0.1888	5446	1929	5592	0.1969	1.5736	60
4	0.1720	1.4858	1764	5010	0.1807	5160	1849	5309	0.1891	5456	1932	5602	0.1972	1.5745	56
8	0.1723	1.4868	1767	5020	0.1810	5170	1852	5319	0.1894	5466	1934	5611	0.1974	1.5755	52
12	0.1726	1.4879	1770	5030	0.1813	5180	1855	5329	0.1896	5476	1937	5621	0.1977	1.5764	48
16	0.1729	1.4889	1773	5040	0.1816	5190	1858	5339	0.1899	5485	1940	5630	0.1979	1.5774	44
20	0.1732	1.4899	1775	5050	0.1818	5200	1861	5348	0.1902	5495	1942	5640	0.1982	1.5783	40
24	0.1734	1.4909	1778	5060	0.1821	5210	1863	5358	0.1905	5505	1945	5650	0.1985	1.5793	36
28	0.1737	1.4919	1781	5070	0.1824	5220	1866	5368	0.1907	5515	1948	5659	0.1987	1.5802	32
32	0.1740	1.4929	1784	5080	0.1827	5230	1869	5378	0.1910	5524	1950	5669	0.1990	1.5812	28
36	0.1743	1.4939	1787	5090	0.1830	5240	1872	5388	0.1913	5534	1953	5678	0.1992	1.5821	24
40	0.1746	1.4950	1790	5100	0.1833	5250	1875	5398	0.1916	5544	1956	5688	0.1995	1.5831	20
44	0.1749	1.4960	1793	5110	0.1835	5260	1877	5407	0.1918	5553	1958	5698	0.1998	1.5840	16
48	0.1752	1.4970	1796	5120	0.1838	5270	1880	5417	0.1921	5563	1961	5707	0.2000	1.5850	12
52	0.1755	1.4980	1799	5130	0.1841	5279	1883	5427	0.1924	5573	1964	5717	0.2003	1.5859	8
56	0.1758	1.4990	1801	5140	0.1844	5289	1886	5437	0.1926	5582	1966	5726	0.2005	1.5868	4
60	0.1761	1.5000	1804	5150	0.1847	5299	1888	5446	0.1929	5592	1969	5736	0.2008	1.5878	0
′	Log 240°	Nat	Log 239°	Nat	Log 238°	Nat	Log 237°	Nat	Log 236°	Nat	Log 235°	Nat	Log 234°	Nat	′

NOTE: For compactness. the leading digit is not displayed in every column. Scan other columns for necessary digit.

′	126°		127°		128°		129°		130°		131°		132°		′
	Log	Nat	Log	Nat	Log	Nat	Log	Nat	Log	Nat	Log	Nat	Log	Nat	
0	0.2008	1.5878	2046	6018	0.2084	6157	2120	6293	0.2156	6428	2191	6561	0.2225	1.6691	60
6	0.2012	1.5892	2050	6032	0.2087	6170	2124	6307	0.2159	6441	2194	6574	0.2228	1.6704	54
12	0.2016	1.5906	2054	6046	0.2091	6184	2127	6320	0.2163	6455	2198	6587	0.2232	1.6717	48
18	0.2019	1.5920	2057	6060	0.2095	6198	2131	6334	0.2166	6468	2201	6600	0.2235	1.6730	42
24	0.2023	1.5934	2061	6074	0.2098	6211	2134	6347	0.2170	6481	2205	6613	0.2238	1.6743	36
30	0.2027	1.5948	2065	6088	0.2102	6225	2138	6361	0.2173	6494	2208	6626	0.2242	1.6756	30
36	0.2031	1.5962	2069	6101	0.2106	6239	2142	6374	0.2177	6508	2211	6639	0.2245	1.6769	24
42	0.2035	1.5976	2072	6115	0.2109	6252	2145	6388	0.2180	6521	2215	6652	0.2248	1.6782	18
48	0.2039	1.5990	2076	6129	0.2113	6266	2149	6401	0.2184	6534	2218	6665	0.2252	1.6794	12
54	0.2042	1.6004	2080	6143	0.2116	6280	2152	6414	0.2187	6547	2222	6678	0.2255	1.6807	6
60	0.2046	1.6018	2084	6157	0.2120	6293	2156	6428	0.2191	6561	2225	6691	0.2258	1.6820	0
	Log	Nat	Log	Nat	Log	Nat	Log	Nat	Log	Nat	Log	Nat	Log	Nat	
	233°		232°		231°		230°		229°		228°		227°		
	133°		134°		134°		136°		137°		138°		139°		
	Log	Nat	Log	Nat	Log	Nat	Log	Nat	Log	Nat	Log	Nat	Log	Nat	
0	0.2258	1.6820	2291	6947	0.2323	7071	2354	7193	0.2384	7314	2413	7431	0.2442	1.7547	60
6	0.2262	1.6833	2294	6959	0.2326	7083	2357	7206	0.2387	7325	2416	7443	0.2445	1.7559	54
12	0.2265	1.6845	2297	6972	0.2329	7096	2360	7218	0.2390	7337	2419	7455	0.2448	1.7570	48
18	0.2268	1.6858	2300	6984	0.2332	7108	2363	7230	0.2393	7349	2422	7466	0.2451	1.7581	42
24	0.2271	1.6871	2304	6997	0.2335	7120	2366	7242	0.2396	7361	2425	7478	0.2453	1.7593	36
30	0.2275	1.6864	2307	7009	0.2338	7133	2369	7254	0.2399	7373	2428	7490	0.2456	1.7604	30
36	0.2278	1.6896	2310	7022	0.2341	7145	2372	7266	0.2402	7385	2431	7501	0.2459	1.7615	24
42	0.2281	1.6909	2313	7034	0.2344	7157	2375	7278	0.2405	7396	2434	7513	0.2462	1.7627	18
48	0.2284	1.6921	2316	7046	0.2347	7169	2378	7290	0.2408	7408	2436	7524	0.2464	1.7638	12
54	0.2288	1.6934	2319	7059	0.2351	7181	2381	7302	0.2410	7420	2439	7536	0.2467	1.7649	6
60	0.2291	1.6947	2323	7071	0.2354	7193	2384	7314	0.2413	7431	2442	7547	0.2470	1.7660	0
	Log	Nat	Log	Nat	Log	Nat	Log	Nat	Log	Nat	Log	Nat	Log	Nat	
	226°		225°		224°		223°		222°		221°		220°		
	140°		141°		142°		143°		144°		145°		146°		
	Log	Nat	Log	Nat	Log	Nat	Log	Nat	Log	Nat	Log	Nat	Log	Nat	
0	0.2470	1.7660	2497	7771	0.2524	7880	2549	7986	0.2574	8090	2599	8192	0.2622	1.8290	60
6	0.2473	1.7672	2500	7782	0.2526	7891	2552	7997	0.2577	8100	2601	8202	0.2625	1.8300	54
12	0.2476	1.7683	2503	7793	0.2529	7902	2554	8007	0.2579	8111	2603	8211	0.2627	1.8310	48
18	0.2478	1.7694	2505	7804	0.2531	7912	2557	8018	0.2582	8121	2606	8221	0.2629	1.8320	42
24	0.2481	1.7705	2508	7815	0.2534	7923	2560	8028	0.2584	8131	2608	8231	0.2631	1.8329	36
30	0.2484	1.7716	2511	7826	0.2537	7934	2562	8039	0.2587	8141	2611	8241	0.2634	1.8339	30
36	0.2486	1.7727	2513	7837	0.2539	7944	2565	8049	0.2589	8151	2613	8251	0.2636	1.8348	24
42	0.2489	1.7738	2516	7848	0.2542	7955	2567	8059	0.2591	8161	2615	8261	0.2638	1.8358	18
48	0.2492	1.7749	2518	7859	0.2544	7965	2569	8070	0.2594	8171	2618	8271	0.2641	1.8368	12
54	0.2495	1.7760	2521	7869	0.2547	7976	2572	8080	0.2596	8181	2620	8281	0.2643	1.8377	6
60	0.2497	1.7771	2524	7880	0.2549	7986	2574	8090	0.2599	8192	2622	8290	0.2645	1.8387	0
	Log	Nat	Log	Nat	Log	Nat	Log	Nat	Log	Nat	Log	Nat	Log	Nat	
	219°		218°		217°		216°		215°		214°		213°		
	147°		148°		149°		150°		151°		152°		153°		
	Log	Nat	Log	Nat	Log	Nat	Log	Nat	Log	Nat	Log	Nat	Log	Nat	
0	0.2645	1.8387	2667	8480	0.2689	8572	2709	8660	0.2729	8746	2748	8829	0.2767	1.8910	60
6	0.2647	1.8396	2669	8490	0.2691	8581	2711	8669	0.2731	8755	2750	8838	0.2769	1.8918	54
12	0.2650	1.8406	2671	8499	0.2693	8590	2713	8678	0.2733	8763	2752	8846	0.2771	1.8926	48
18	0.2652	1.8415	2674	8508	0.2695	8599	2715	8686	0.2735	8771	2754	8854	0.2772	1.8934	42
24	0.2654	1.8425	2676	8517	0.2697	8607	2717	8695	0.2737	8780	2756	8862	0.2774	1.8942	36
30	0.2656	1.8434	2678	8526	0.2699	8616	2719	8704	0.2739	8788	2758	8870	0.2776	1.8949	30
36	0.2658	1.8443	2680	8536	0.2701	8625	2721	8712	0.2741	8796	2760	8878	0.2778	1.8957	24
42	0.2661	1.8453	2682	8545	0.2703	8634	2723	8721	0.2743	8805	2761	8886	0.2779	1.8965	18
48	0.2663	1.8462	2684	8554	0.2705	8643	2725	8729	0.2745	8813	2763	8894	0.2781	1.8973	12
54	0.2665	1.8471	2686	8563	0.2707	8652	2727	8738	0.2746	8821	2765	8902	0.2783	1.8980	6
60	0.2667	1.8480	2689	8572	0.2709	8660	2729	8746	0.2748	8829	2767	8910	0.2785	1.8988	0
′	Log	Nat	Log	Nat	Log	Nat	Log	Nat	Log	Nat	Log	Nat	Log	Nat	′
	212°		211°		210°		209°		208°		207°		206°		

NOTE: For compactness. the leading digit is not displayed in every column. Scan other columns for necessary digit.

VERSINES

′	154°		155°		156°		157°		158°		159°		160°		′
	Log	Nat	Log	Nat	Log	Nat	Log	Nat	Log	Nat	Log	Nat	Log	Nat	
0	0.2785	1.8988	2802	9063	0.2818	9135	2834	9205	0.2849	9272	2864	9336	0.2877	1.9397	60
6	0.2787	1.8996	2804	9070	0.2820	9143	2836	9212	0.2851	9278	2865	9342	0.2879	1.9403	54
12	0.2788	1.9003	2805	9078	0.2822	9150	2837	9219	0.2852	9285	2866	9348	0.2880	1.9409	48
18	0.2790	1.9011	2807	9085	0.2823	9157	2839	9225	0.2854	9291	2868	9354	0.2881	1.9415	42
24	0.2792	1.9018	2809	9092	0.2825	9164	2840	9232	0.2855	9298	2869	9361	0.2883	1.9421	36
30	0.2793	1.9026	2810	9100	0.2826	9171	2842	9239	0.2857	9304	2871	9367	0.2884	1.9426	30
36	0.2795	1.9033	2812	9107	0.2828	9178	2843	9245	0.2858	9311	2872	9373	0.2885	1.9432	24
42	0.2797	1.9041	2814	9114	0.2829	9184	2845	9252	0.2859	9317	2873	9379	0.2887	1.9438	18
48	0.2799	1.9048	2815	9121	0.2831	9191	2846	9259	0.2861	9323	2875	9385	0.2888	1.9444	12
54	0.2800	1.9056	2817	9128	0.2833	9198	2848	9265	0.2862	9330	2876	9391	0.2889	1.9449	6
60	0.2802	1.9063	2818	9135	0.2834	9205	2849	9272	0.2864	9336	2877	9397	0.2890	1.9455	0
	Log	Nat	Log	Nat	Log	Nat	Log	Nat	Log	Nat	Log	Nat	Log	Nat	
	205°		204°		203°		202°		201°		200°		199°		
	161°		162°		163°		164°		165°		166°		167°		
	Log	Nat	Log	Nat	Log	Nat	Log	Nat	Log	Nat	Log	Nat	Log	Nat	
0	0.2890	1.9455	2903	9511	0.2914	9563	2925	9613	0.2936	9659	2945	9703	0.2954	1.9744	60
6	0.2892	1.9461	2904	9516	0.2915	9568	2926	9617	0.2937	9664	2946	9707	0.2955	1.9748	54
12	0.2893	1.9466	2905	9521	0.2917	9573	2927	9622	0.2938	9668	2947	9711	0.2956	1.9751	48
18	0.2894	1.9472	2906	9527	0.2918	9578	2929	9627	0.2939	9673	2948	9715	0.2957	1.9755	42
24	0.2895	1.9478	2907	9532	0.2919	9583	2930	9632	0.2940	9677	2949	9720	0.2958	1.9759	36
30	0.2897	1.9483	2909	9537	0.2920	9588	2931	9636	0.2941	9681	2950	9724	0.2959	1.9763	30
36	0.2898	1.9489	2910	9542	0.2921	9593	2932	9641	0.2942	9686	2951	9728	0.2959	1.9767	24
42	0.2899	1.9494	2911	9548	0.2922	9598	2933	9646	0.2942	9690	2952	9732	0.2960	1.9770	18
48	0.2900	1.9500	2912	9553	0.2923	9603	2934	9650	0.2943	9694	2953	9736	0.2961	1.9774	12
54	0.2901	1.9505	2913	9558	0.2924	9608	2935	9655	0.2944	9699	2953	9740	0.2962	1.9778	6
60	0.2903	1.9511	2914	9563	0.2925	9613	2936	9659	0.2945	9703	2954	9744	0.2963	1.9781	0
	Log	Nat	Log	Nat	Log	Nat	Log	Nat	Log	Nat	Log	Nat	Log	Nat	
	198°		197°		196°		195°		194°		193°		192°		
	168°		169°		170°		171°		172°		173°		174°		
	Log	Nat	Log	Nat	Log	Nat	Log	Nat	Log	Nat	Log	Nat	Log	Nat	
0	0.2963	1.9781	2970	9816	0.2977	9848	2983	9877	0.2989	9903	2994	9925	0.2998	1.9945	60
6	0.2963	1.9785	2971	9820	0.2978	9851	2984	9880	0.2990	9905	2995	9928	0.2999	1.9947	54
12	0.2964	1.9789	2972	9823	0.2978	9854	2985	9882	0.2990	9907	2995	9930	0.2999	1.9949	48
18	0.2965	1.9792	2972	9826	0.2979	9857	2985	9885	0.2991	9910	2995	9932	0.3000	1.9951	42
24	0.2966	1.9796	2973	9829	0.2980	9860	2986	9888	0.2991	9912	2996	9934	0.3000	1.9952	36
30	0.2966	1.9799	2974	9833	0.2980	9863	2986	9890	0.2992	9914	2996	9936	0.3000	1.9954	30
36	0.2967	1.9803	2974	9836	0.2981	9866	2987	9893	0.2992	9917	2997	9938	0.3001	1.9956	24
42	0.2968	1 9806	2975	9839	0.2982	9869	2987	9895	0.2993	9919	2997	9940	0.3001	1.9957	18
48	0.2969	1.9810	2976	9842	0.2982	9871	2988	9898	0.2993	9921	2998	9942	0.3001	1.9959	12
54	0.2969	1.9813	2977	9845	0.2983	9874	2989	9900	0.2994	9923	2998	9943	0.3002	1.9960	6
60	0.2970	1.9816	2977	9848	0.2983	9877	2989	9903	0.2994	9925	2998	9945	0.3002	1.9962	0
	Log	Nat	Log	Nat	Log	Nat	Log	Nat	Log	Nat	Log	Nat	Log	Nat	
	191°		190°		189°		188°		187°		186°		185°		
	175°		176°		177°		178°		179°						
	Log	Nat	Log	Nat	Log	Nat	Log	Nat	Log	Nat					
0	0.3002	1.9962	3005	9976	0.3007	9986	3009	9994	0.3010	9998	60				
6	0.3002	1.9963	3005	9977	0.3008	9987	3009	9995	0.3010	9999	54				
12	0.3003	1.9965	3006	9978	0.3008	9988	3009	9995	0.3010	9999	48				
18	0.3003	1.9966	3006	9979	0.3008	9989	3009	9996	0.3010	9999	42				
24	0.3003	1.9968	3006	9980	0.3008	9990	3009	9996	0.3010	9999	36				
30	0.3004	1.9969	3006	9981	0.3008	9990	3010	9997	0.3010	0000	30				
36	0.3004	1.9971	3006	9982	0.3008	9991	3010	9997	0.3010	0000	24				
42	0.3004	1.9972	3007	9983	0.3009	9992	3010	9997	0.3010	0000	18				
48	0.3004	1.9973	3007	9984	0.3009	9993	3010	9998	0.3010	0000	12				
54	0.3005	1.9974	3007	9985	0.3009	9993	3010	9998	0.3010	0000	6				
60	0.3005	1.9976	3007	9986	0.3009	9994	3010	9998	0.3010	0000	0				
′	Log	Nat	Log	Nat	Log	Nat	Log	Nat	Log	Nat	′				
	184°		183°		182°		181°		180°						

NOTE: For compactness. the leading digit is not displayed in every column. Scan other columns for necessary digit.

VERSINES

LOG COSINES

	0°	1°	2°	3°	4°	5°	6°	7°	8°	9°	10°	11°	12°	13°	14°	
0	0.0000	9999	9997	9994	9989	9.9983	9976	9968	9958	9.9946	9934	9919	9904	9887	9.9869	60
1	0.0000	9999	9997	9994	9989	9.9983	9976	9967	9957	9.9946	9933	9919	9904	9887	9.9869	59
2	0.0000	9999	9997	9994	9989	9.9983	9976	9967	9957	9.9946	9933	9919	9904	9887	9.9868	58
3	0.0000	9999	9997	9994	9989	9.9983	9976	9967	9957	9.9946	9933	9919	9903	9886	9.9868	57
4	0.0000	9999	9997	9994	9989	9.9983	9976	9967	9957	9.9945	9933	9918	9903	9886	9.9868	56
5	0.0000	9999	9997	9994	9989	9.9983	9975	9967	9957	9.9945	9932	9918	9903	9886	9.9867	55
6	0.0000	9999	9997	9994	9989	9.9983	9975	9967	9956	9.9945	9932	9918	9902	9885	9.9867	54
7	0.0000	9999	9997	9994	9989	9.9983	9975	9966	9956	9.9945	9932	9918	9902	9885	9.9867	53
8	0.0000	9999	9997	9994	9989	9.9983	9975	9966	9956	9.9945	9932	9917	9902	9885	9.9867	52
9	0.0000	9999	9997	9993	9989	9.9982	9975	9966	9956	9.9944	9931	9917	9902	9885	9.9866	51
10	0.0000	9999	9997	9993	9989	9.9982	9975	9966	9956	9.9944	9931	9917	9901	9884	9.9866	50
11	0.0000	9999	9997	9993	9988	9.9982	9975	9966	9956	9.9944	9931	9917	9901	9884	9.9866	49
12	0.0000	9999	9997	9993	9988	9.9982	9975	9966	9955	9.9944	9931	9916	9901	9884	9.9865	48
13	0.0000	9999	9997	9993	9988	9.9982	9974	9965	9955	9.9944	9931	9916	9901	9883	9.9865	47
14	0.0000	9999	9997	9993	9988	9.9982	9974	9965	9955	9.9943	9930	9916	9900	9883	9.9865	46
15	0.0000	9999	9997	9993	9988	9.9982	9974	9965	9955	9.9943	9930	9916	9900	9883	9.9864	45
16	0.0000	9999	9997	9993	9988	9.9982	9974	9965	9955	9.9943	9930	9915	9900	9883	9.9864	44
17	0.0000	9999	9997	9993	9988	9.9982	9974	9965	9954	9.9943	9930	9915	9899	9882	9.9864	43
18	0.0000	9999	9996	9993	9988	9.9981	9974	9965	9954	9.9943	9929	9915	9899	9882	9.9863	42
19	0.0000	9999	9996	9993	9988	9.9981	9974	9964	9954	9.9942	9929	9915	9899	9882	9.9863	41
20	0.0000	9999	9996	9993	9988	9.9981	9973	9964	9954	9.9942	9929	9914	9899	9881	9.9863	40
21	0.0000	9999	9996	9993	9987	9.9981	9973	9964	9954	9.9942	9929	9914	9898	9881	9.9862	39
22	0.0000	9999	9996	9992	9987	9.9981	9973	9964	9954	9.9942	9929	9914	9898	9881	9.9862	38
23	0.0000	9999	9996	9992	9987	9.9981	9973	9964	9953	9.9941	9928	9914	9898	9880	9.9862	37
24	0.0000	9999	9996	9992	9987	9.9981	9973	9964	9953	9.9941	9928	9913	9897	9880	9.9861	36
25	0.0000	9999	9996	9992	9987	9.9981	9973	9964	9953	9.9941	9928	9913	9897	9880	9.9861	35
26	0.0000	9999	9996	9992	9987	9.9980	9973	9963	9953	9.9941	9928	9913	9897	9880	9.9861	34
27	0.0000	9999	9996	9992	9987	9.9980	9972	9963	9953	9.9941	9927	9913	9897	9879	9.9860	33
28	0.0000	9999	9996	9992	9987	9.9980	9972	9963	9952	9.9940	9927	9912	9896	9879	9.9860	32
29	0.0000	9999	9996	9992	9987	9.9980	9972	9963	9952	9.9940	9927	9912	9896	9879	9.9860	31
30	0.0000	9999	9996	9992	9987	9.9980	9972	9963	9952	9.9940	9927	9912	9896	9878	9.9859	30
31	0.0000	9998	9996	9992	9986	9.9980	9972	9963	9952	9.9940	9926	9912	9896	9878	9.9859	29
32	0.0000	9998	9996	9992	9986	9.9980	9972	9962	9952	9.9940	9926	9911	9895	9878	9.9859	28
33	0.0000	9998	9996	9992	9986	9.9980	9972	9962	9951	9.9939	9926	9911	9895	9877	9.9858	27
34	0.0000	9998	9996	9992	9986	9.9979	9971	9962	9951	9.9939	9926	9911	9895	9877	9.9858	26
35	0.0000	9998	9996	9992	9986	9.9979	9971	9962	9951	9.9939	9925	9911	9894	9877	9.9858	25
36	0.0000	9998	9996	9991	9986	9.9979	9971	9962	9951	9.9939	9925	9910	9894	9876	9.9857	24
37	0.0000	9998	9995	9991	9986	9.9979	9971	9962	9951	9.9939	9925	9910	9894	9876	9.9857	23
38	0.0000	9998	9995	9991	9986	9.9979	9971	9961	9951	9.9938	9925	9910	9894	9876	9.9857	22
39	0.0000	9998	9995	9991	9986	9.9979	9971	9961	9950	9.9938	9925	9910	9893	9876	9.9856	21
40	0.0000	9998	9995	9991	9986	9.9979	9971	9961	9950	9.9938	9924	9909	9893	9875	9.9856	20
41	0.0000	9998	9995	9991	9985	9.9979	9970	9961	9950	9.9938	9924	9909	9893	9875	9.9856	19
42	0.0000	9998	9995	9991	9985	9.9978	9970	9961	9950	9.9937	9924	9909	9892	9875	9.9855	18
43	0.0000	9998	9995	9991	9985	9.9978	9970	9960	9950	9.9937	9924	9909	9892	9874	9.9855	17
44	0.0000	9998	9995	9991	9985	9.9978	9970	9960	9949	9.9937	9923	9908	9892	9874	9.9855	16
45	0.0000	9998	9995	9991	9985	9.9978	9970	9960	9949	9.9937	9923	9908	9892	9874	9.9854	15
46	0.0000	9998	9995	9991	9985	9.9978	9970	9960	9949	9.9937	9923	9908	9891	9873	9.9854	14
47	0.0000	9998	9995	9991	9985	9.9978	9969	9960	9949	9.9936	9923	9908	9891	9873	9.9854	13
48	0.0000	9998	9995	9990	9985	9.9978	9969	9960	9949	9.9936	9922	9907	9891	9873	9.9853	12
49	0.0000	9998	9995	9990	9985	9.9978	9969	9959	9948	9.9936	9922	9907	9890	9872	9.9853	11
50	0.0000	9998	9995	9990	9985	9.9977	9969	9959	9948	9.9936	9922	9907	9890	9872	9.9853	10
51	0.0000	9998	9995	9990	9984	9.9977	9969	9959	9948	9.9936	9922	9906	9890	9872	9.9852	9
52	0.0000	9998	9995	9990	9984	9.9977	9969	9959	9948	9.9935	9921	9906	9890	9872	9.9852	8
53	9.9999	9998	9994	9990	9984	9.9977	9969	9959	9948	9.9935	9921	9906	9889	9871	9.9852	7
54	9.9999	9998	9994	9990	9984	9.9977	9968	9959	9947	9.9935	9921	9906	9889	9871	9.9851	6
55	9.9999	9998	9994	9990	9984	9.9977	9968	9958	9947	9.9935	9921	9905	9889	9871	9.9851	5
56	9.9999	9998	9994	9990	9984	9.9977	9968	9958	9947	9.9934	9920	9905	9888	9870	9.9851	4
57	9.9999	9997	9994	9990	9984	9.9977	9968	9958	9947	9.9934	9920	9905	9888	9870	9.9850	3
58	9.9999	9997	9994	9990	9984	9.9976	9968	9958	9947	9.9934	9920	9905	9888	9870	9.9850	2
59	9.9999	9997	9994	9989	9984	9.9976	9968	9958	9946	9.9934	9920	9904	9888	9869	9.9850	1
60	9.9999	9997	9994	9989	9983	9.9976	9968	9958	9945	9.9934	9919	9904	9887	9869	9.9849	0
	89°	88°	87°	86°	85°	84°	83°	82°	81°	80°	79°	78°	77°	76°	75°	

NOTE: For compactness, the leading digit is not displayed in every column. Scan other columns for necessary digit.

LOG SINES

LOG COSINES

	15°	16°	17°	18°	19°	20°	21°	22°	23°	24°	25°	26°	27°	28°	29°	
0	9.9849	9828	9806	9782	9757	9.9730	9702	9672	9640	9.9607	9573	9537	9499	9459	9.9418	60
1	9.9849	9828	9806	9782	9756	9.9729	9701	9671	9640	9.9607	9572	9536	9498	9459	9.9417	59
2	9.9849	9828	9805	9781	9756	9.9729	9701	9671	9639	9.9606	9572	9535	9498	9458	9.9417	58
3	9.9848	9827	9805	9781	9755	9.9728	9700	9670	9639	9.9606	9571	9535	9497	9457	9.9416	57
4	9.9848	9827	9804	9780	9755	9.9728	9700	9670	9638	9.9605	9570	9534	9496	9457	9.9415	56
5	9.9848	9827	9804	9780	9755	9.9728	9699	9669	9638	9.9604	9570	9534	9496	9456	9.9415	55
6	9.9847	9826	9804	9780	9754	9.9727	9699	9669	9637	9.9604	9569	9533	9495	9455	9.9414	54
7	9.9847	9826	9803	9779	9754	9.9727	9698	9668	9636	9.9603	9569	9532	9494	9455	9.9413	53
8	9.9847	9826	9803	9779	9753	9.9726	9698	9668	9636	9.9603	9568	9532	9494	9454	9.9413	52
9	9.9846	9825	9802	9778	9753	9.9726	9697	9667	9635	9.9602	9567	9531	9493	9453	9.9412	51
10	9.9846	9825	9802	9778	9752	9.9725	9697	9667	9635	9.9602	9567	9530	9492	9453	9.9411	50
11	9.9846	9824	9802	9778	9752	9.9725	9696	9666	9634	9.9601	9566	9530	9492	9452	9.9410	49
12	9.9845	9824	9801	9777	9751	9.9724	9696	9666	9634	9.9601	9566	9529	9491	9451	9.9410	48
13	9.9845	9824	9801	9777	9751	9.9724	9695	9665	9633	9.9600	9565	9529	9490	9451	9.9409	47
14	9.9845	9823	9801	9776	9751	9.9723	9695	9664	9633	9.9599	9564	9528	9490	9450	9.9408	46
15	9.9844	9823	9800	9776	9750	9.9723	9694	9664	9632	9.9599	9564	9527	9489	9449	9.9408	45
16	9.9844	9823	9800	9775	9750	9.9722	9694	9663	9632	9.9598	9563	9527	9488	9449	9.9407	44
17	9.9844	9822	9799	9775	9749	9.9722	9693	9663	9631	9.9598	9563	9526	9488	9448	9.9406	43
18	9.9843	9822	9799	9775	9749	9.9722	9693	9662	9631	9.9597	9562	9525	9487	9447	9.9406	42
19	9.9843	9821	9799	9774	9748	9.9721	9692	9662	9630	9.9597	9561	9525	9486	9447	9.9405	41
20	9.9843	9821	9798	9774	9748	9.9721	9692	9661	9629	9.9596	9561	9524	9486	9446	9.9404	40
21	9.9842	9821	9798	9773	9747	9.9720	9691	9661	9629	9.9595	9560	9524	9485	9445	9.9403	39
22	9.9842	9820	9797	9773	9747	9.9720	9691	9660	9628	9.9595	9560	9523	9485	9444	9.9403	38
23	9.9842	9820	9797	9773	9747	9.9719	9690	9660	9628	9.9594	9559	9522	9484	9444	9.9402	37
24	9.9841	9820	9797	9772	9746	9.9719	9690	9659	9627	9.9594	9558	9522	9483	9443	9.9401	36
25	9.9841	9819	9796	9772	9746	9.9718	9689	9659	9627	9.9593	9558	9521	9483	9442	9.9401	35
26	9.9841	9819	9796	9771	9745	9.9718	9689	9658	9626	9.9593	9557	9520	9482	9442	9.9400	34
27	9.9840	9818	9795	9771	9745	9.9717	9688	9658	9626	9.9592	9557	9520	9481	9441	9.9399	33
28	9.9840	9818	9795	9770	9744	9.9717	9688	9657	9625	9.9591	9556	9519	9481	9440	9.9398	32
29	9.9839	9818	9795	9770	9744	9.9716	9687	9657	9625	9.9591	9555	9519	9480	9440	9.9398	31
30	9.9839	9817	9794	9770	9743	9.9716	9687	9656	9624	9.9590	9555	9518	9479	9439	9.9397	30
31	9.9839	9817	9794	9769	9743	9.9715	9686	9656	9623	9.9590	9554	9517	9479	9438	9.9396	29
32	9.9838	9817	9793	9769	9743	9.9715	9686	9655	9623	9.9589	9554	9517	9478	9438	9.9396	28
33	9.9838	9816	9793	9768	9742	9.9714	9685	9655	9622	9.9589	9553	9516	9477	9437	9.9395	27
34	9.9838	9816	9793	9768	9742	9.9714	9685	9654	9622	9.9588	9552	9515	9477	9436	9.9394	26
35	9.9837	9815	9792	9767	9741	9.9714	9684	9654	9621	9.9587	9552	9515	9476	9436	9.9393	25
36	9.9837	9815	9792	9767	9741	9.9713	9684	9653	9621	9.9587	9551	9514	9475	9435	9.9393	24
37	9.9837	9815	9791	9767	9740	9.9713	9683	9652	9620	9.9586	9551	9513	9475	9434	9.9392	23
38	9.9836	9814	9791	9766	9740	9.9712	9683	9652	9620	9.9586	9550	9513	9474	9433	9.9391	22
39	9.9836	9814	9791	9766	9739	9.9712	9682	9651	9619	9.9585	9549	9512	9473	9433	9.9391	21
40	9.9836	9814	9790	9765	9739	9.9711	9682	9651	9618	9.9584	9549	9512	9473	9432	9.9390	20
41	9.9835	9813	9790	9765	9739	9.9711	9681	9650	9618	9.9584	9548	9511	9472	9431	9.9389	19
42	9.9835	9813	9789	9764	9738	9.9710	9681	9650	9617	9.9583	9548	9510	9471	9431	9.9388	18
43	9.9835	9812	9789	9764	9738	9.9710	9680	9649	9617	9.9583	9547	9510	9471	9430	9.9388	17
44	9.9834	9812	9789	9764	9737	9.9709	9680	9649	9616	9.9582	9546	9509	9470	9429	9.9387	16
45	9.9834	9812	9788	9763	9737	9.9709	9679	9648	9616	9.9582	9546	9508	9469	9429	9.9386	15
46	9.9833	9811	9788	9763	9736	9.9708	9679	9648	9615	9.9581	9545	9508	9469	9428	9.9385	14
47	9.9833	9811	9787	9762	9736	9.9708	9678	9647	9615	9.9580	9545	9507	9468	9427	9.9385	13
48	9.9833	9811	9787	9762	9735	9.9707	9678	9647	9614	9.9580	9544	9506	9467	9427	9.9384	12
49	9.9832	9810	9787	9761	9735	9.9707	9677	9646	9613	9.9579	9543	9506	9467	9426	9.9383	11
50	9.9832	9810	9786	9761	9734	9.9706	9677	9646	9613	9.9579	9543	9505	9466	9425	9.9383	10
51	9.9832	9809	9786	9761	9734	9.9706	9676	9645	9612	9.9578	9542	9505	9465	9424	9.9382	9
52	9.9831	9809	9785	9760	9734	9.9705	9676	9645	9612	9.9577	9542	9504	9465	9424	9.9381	8
53	9.9831	9809	9785	9760	9733	9.9705	9675	9644	9611	9.9577	9541	9503	9464	9423	9.9380	7
54	9.9831	9808	9785	9759	9733	9.9704	9675	9643	9611	9.9576	9540	9503	9463	9422	9.9380	6
55	9.9830	9808	9784	9759	9732	9.9704	9674	9643	9610	9.9576	9540	9502	9463	9422	9.9379	5
56	9.9830	9808	9784	9758	9732	9.9703	9674	9642	9610	9.9575	9539	9501	9462	9421	9.9378	4
57	9.9830	9807	9783	9758	9731	9.9703	9673	9642	9609	9.9575	9538	9501	9461	9420	9.9377	3
58	9.9829	9807	9783	9758	9731	9.9702	9673	9641	9608	9.9574	9538	9500	9461	9420	9.9377	2
59	9.9829	9806	9782	9757	9730	9.9702	9672	9641	9608	9.9573	9537	9499	9460	9419	9.9376	1
60	9.9828	9806	9782	9757	9730	9.9702	9672	9640	9607	9.9573	9537	9499	9459	9418	9.9375	0
	74°	73°	72°	71°	70°	69°	68°	67°	66°	65°	64°	63°	62°	61°	60°	

NOTE: For compactness, the leading digit is not displayed in every column. Scan other columns for necessary digit.

LOG SINES

LOG COSINES

	30°	31°	32°	33°	34°	35°	36°	37°	38°	39°	40°	41°	42°	43°	44°	
0	9.9375	9331	9284	9236	9186	9.9134	9080	9023	8965	9.8905	8843	8778	8711	8641	9.8569	60
1	9.9375	9330	9283	9235	9185	9.9133	9079	9023	8964	9.8904	8841	8777	8710	8640	9.8568	59
2	9.9374	9329	9283	9234	9184	9.9132	9078	9022	8963	9.8903	8840	8776	8708	8639	9.8567	58
3	9.9373	9328	9282	9233	9183	9.9131	9077	9021	8962	9.8902	8839	8775	8707	8638	9.8566	57
4	9.9372	9328	9281	9233	9182	9.9130	9076	9020	8961	9.8901	8838	8773	8706	8637	9.8564	56
5	9.9372	9327	9280	9232	9181	9.9129	9075	9019	8960	9.8900	8837	8772	8705	8635	9.8563	55
6	9.9371	9326	9279	9231	9181	9.9128	9074	9018	8959	9.8899	8836	8771	8704	8634	9.8562	54
7	9.9370	9325	9279	9230	9180	9.9127	9073	9017	8958	9.8898	8835	8770	8703	8633	9.8561	53
8	9.9369	9325	9278	9229	9179	9.9127	9072	9016	8957	9.8897	8834	8769	8702	8632	9.8560	52
9	9.9369	9324	9277	9229	9178	9.9126	9071	9015	8956	9.8896	8833	8768	8700	8631	9.8558	51
10	9.9368	9323	9276	9228	9177	9.9125	9070	9014	8955	9.8895	8832	8767	8699	8629	9.8557	50
11	9.9367	9322	9275	9227	9176	9.9124	9069	9013	8954	9.8894	8831	8766	8698	8628	9.8556	49
12	9.9367	9322	9275	9226	9175	9.9123	9069	9012	8953	9.8893	8830	8765	8697	8627	9.8555	48
13	9.9366	9321	9274	9225	9175	9.9122	9068	9011	8952	9.8892	8829	8763	8696	8626	9.8553	47
14	9.9365	9320	9273	9224	9174	9.9121	9067	9010	8951	9.8891	8828	8762	8695	8625	9.8552	46
15	9.9364	9319	9272	9224	9173	9.9120	9066	9009	8950	9.8890	8827	8761	8694	8624	9.8551	45
16	9.9364	9318	9272	9223	9172	9.9119	9065	9008	8949	9.8889	8825	8760	8692	8622	9.8550	44
17	9.9363	9318	9271	9222	9171	9.9119	9064	9007	8948	9.8888	8824	8759	8691	8621	9.8549	43
18	9.9362	9317	9270	9221	9170	9.9118	9063	9006	8947	9.8887	8823	8758	8690	8620	9.8547	42
19	9.9361	9316	9269	9220	9169	9.9117	9062	9005	8946	9.8885	8822	8757	8689	8619	9.8546	41
20	9.9361	9315	9268	9219	9169	9.9116	9061	9004	8945	9.8884	8821	8756	8688	8618	9.8545	40
21	9.9360	9315	9268	9219	9168	9.9115	9060	9003	8944	9.8883	8820	8755	8687	8616	9.8544	39
22	9.9359	9314	9267	9218	9167	9.9114	9059	9002	8943	9.8882	8819	8753	8686	8615	9.8542	38
23	9.9358	9313	9266	9217	9166	9.9113	9058	9001	8942	9.8881	8818	8752	8684	8614	9.8541	37
24	9.9358	9312	9265	9216	9165	9.9112	9057	9000	8941	9.8880	8817	8751	8683	8613	9.8540	36
25	9.9357	9312	9264	9215	9164	9.9111	9056	9000	8940	9.8879	8816	8750	8682	8612	9.8539	35
26	9.9356	9311	9264	9214	9163	9.9110	9056	8999	8939	9.8878	8815	8749	8681	8610	9.8537	34
27	9.9355	9310	9263	9214	9163	9.9110	9055	8998	8938	9.8877	8814	8748	8680	8609	9.8536	33
28	9.9355	9309	9262	9213	9162	9.9109	9054	8997	8937	9.8876	8813	8747	8679	8608	9.8535	32
29	9.9354	9308	9261	9212	9161	9.9108	9053	8996	8936	9.8875	8812	8746	8677	8607	9.8534	31
30	9.9353	9308	9260	9211	9160	9.9107	9052	8995	8935	9.8874	8810	8745	8676	8606	9.8532	30
31	9.9352	9307	9259	9210	9159	9.9106	9051	8994	8934	9.8873	8809	8743	8675	8604	9.8531	29
32	9.9352	9306	9259	9209	9158	9.9105	9050	8993	8933	9.8872	8808	8742	8674	8603	9.8530	28
33	9.9351	9305	9258	9209	9157	9.9104	9049	8992	8932	9.8871	8807	8741	8673	8602	9.8529	27
34	9.9350	9305	9257	9208	9156	9.9103	9048	8991	8931	9.8870	8806	8740	8672	8601	9.8527	26
35	9.9349	9304	9256	9207	9156	9.9102	9047	8990	8930	9.8869	8805	8739	8671	8600	9.8526	25
36	9.9349	9303	9255	9206	9155	9.9101	9046	8989	8929	9.8868	8804	8738	8669	8598	9.8525	24
37	9.9348	9302	9255	9205	9154	9.9101	9045	8988	8928	9.8867	8803	8737	8668	8597	9.8524	23
38	9.9347	9301	9254	9204	9153	9.9100	9044	8987	8927	9.8866	8802	8736	8667	8596	9.8522	22
39	9.9346	9301	9253	9204	9152	9.9099	9043	8986	8926	9.8865	8801	8734	8666	8595	9.8521	21
40	9.9346	9300	9252	9203	9151	9.9098	9042	8985	8925	9.8864	8800	8733	8665	8594	9.8520	20
41	9.9345	9299	9251	9202	9150	9.9097	9041	8984	8924	9.8863	8799	8732	8664	8592	9.8519	19
42	9.9344	9298	9251	9201	9149	9.9096	9041	8983	8923	9.8862	8797	8731	8662	8591	9.8517	18
43	9.9343	9298	9250	9200	9149	9.9095	9040	8982	8922	9.8860	8796	8730	8661	8590	9.8516	17
44	99343	9297	9249	9199	9148	9.9094	9039	8981	8921	9.8859	8795	8729	8660	8589	9.8515	16
45	9.9342	9296	9248	9198	9147	9.9093	9038	8980	8920	9.8858	8794	8728	8659	8588	9.8514	15
46	9.9341	9295	9247	9198	9146	9.9092	9037	8979	8919	9.8857	8793	8727	8658	8586	9.8512	14
47	9.9340	9294	9247	9197	9145	9.9091	9036	8978	8918	9.8856	8792	8725	8657	8585	9.8511	13
48	9.9340	9294	9246	9196	9144	9.9091	9035	8977	8917	9.8855	8791	8724	8655	8584	9.8510	12
49	9.9339	9293	9245	9195	9143	9.9090	9034	8976	8916	9.8854	8790	8723	8654	8583	9.8509	11
50	9.9338	9292	9244	9194	9142	9.9089	9033	8975	8915	9.8853	8789	8722	8653	8582	9.8507	10
51	9.9337	9291	9243	9193	9142	9.9088	9032	8974	8914	9.8852	8788	8721	8652	8580	9.8506	9
52	9.9337	9291	9242	9193	9141	9.9087	9031	8973	8913	9.8851	8787	8720	8651	8579	9.8505	8
53	9.9336	9290	9242	9192	9140	9.9086	9030	8972	8912	9.8850	8785	8719	8650	8578	9.8504	7
54	9.9335	9289	9241	9191	9139	9.9085	9029	8971	8911	9.8849	8784	8718	8648	8577	9.8502	6
55	9.9334	9288	9240	9190	9138	9.9084	9028	8970	8910	9.8848	8783	8716	8647	8575	9.8501	5
56	9.9334	9287	9239	9189	9137	9.9083	9027	8969	8909	9.8847	8782	8715	8646	8574	9.8500	4
57	9.9333	9287	9238	9188	9136	9.9082	9026	8968	8908	9.8846	8781	8714	8645	8573	9.8499	3
58	9.9332	9286	9238	9187	9135	9.9081	9025	8967	8907	9.8845	8780	8713	8644	8572	9.8497	2
59	9.9331	9285	9237	9187	9135	9.9080	9024	8966	8906	9.8844	8779	8712	8642	8571	9.8496	1
60	9.9331	9284	9236	9186	9134	9.9080	9023	8965	8905	9.8843	8778	8711	8641	8569	9.8495	0
	59°	58°	57°	56°	55°	54°	53°	52°	51°	50°	49°	48°	47°	46°	45°	

NOTE: For compactness, the leading digit is not displayed in every column. Scan other columns for necessary digit.

LOG SINES

LOG COSINES

	45°	46°	47°	48°	49°	50°	51°	52°	53°	54°	55°	56°	57°	58°	59°	
0	9.8495	8418	8338	8255	8169	9.8081	7989	7893	7795	9.7692	7586	7476	7361	7242	9.7118	60
1	9.8494	8416	8336	8254	8168	9.8079	7987	7892	7793	9.7690	7584	7474	7359	7240	9.7116	59
2	9.8492	8415	8335	8252	8167	9.8078	7986	7890	7791	9.7689	7582	7472	7357	7238	9.7114	58
3	9.8491	8414	8334	8251	8165	9.8076	7984	7889	7790	9.7687	7580	7470	7355	7236	9.7112	57
4	9.8490	8412	8332	8249	8164	9.8075	7982	7887	7788	9.7685	7579	7468	7353	7234	9.7110	56
5	9.8489	8411	8331	8248	8162	9.8073	7981	7885	7786	9.7683	7577	7466	7351	7232	9.7108	55
6	9.8487	8410	8330	8247	8161	9.8072	7979	7884	7785	9.7682	7575	7464	7349	7230	9.7106	54
7	9.8486	8409	8328	8245	8159	9.8070	7978	7882	7783	9.7680	7573	7462	7347	7228	9.7104	53
8	9.8485	8407	8327	8244	8158	9.8069	7976	7880	7781	9.7678	7571	7461	7345	7226	9.7102	52
9	9.8483	8406	8326	8242	8156	9.8067	7975	7879	7780	9.7676	7570	7459	7344	7224	9.7099	51
10	9.8482	8405	8324	8241	8155	9.8066	7973	7877	7778	9.7675	7568	7457	7342	7222	9.7097	50
11	9.8481	8403	8323	8240	8153	9.8064	7972	7876	7776	9.7673	7566	7455	7340	7220	9.7095	49
12	9.8480	8402	8322	8238	8152	9.8063	7970	7874	7774	9.7671	7564	7453	7338	7218	9.7093	48
13	9.8478	8401	8320	8237	8150	9.8061	7968	7872	7773	9.7669	7562	7451	7336	7216	9.7091	47
14	9.8477	8399	8319	8235	8149	9.8060	7967	7871	7771	9.7668	7561	7449	7334	7214	9.7089	46
15	9.8476	8398	8317	8234	8148	9.8058	7965	7869	7769	9.7666	7559	7447	7332	7212	9.7087	45
16	9.8475	8397	8316	8233	8146	9.8056	7964	7867	7768	9.7664	7557	7445	7330	7210	9.7085	44
17	9.8473	8395	8315	8231	8145	9.8055	7962	7866	7766	9.7662	7555	7444	7328	7208	9.7082	43
18	9.8472	8394	8313	8230	8143	9.8053	7960	7864	7764	9.7661	7553	7442	7326	7205	9.7080	42
19	9.8471	8393	8312	8228	8142	9.8052	7959	7863	7763	9.7659	7551	7440	7324	7203	9.7078	41
20	9.8469	8391	8311	8227	8140	9.8050	7957	7861	7761	9.7657	7550	7438	7322	7201	9.7076	40
21	9.8468	8390	8309	8225	8139	9.8049	7956	7859	7759	9.7655	7548	7436	7320	7199	9.7074	39
22	9.8467	8389	8308	8224	8137	9.8047	7954	7858	7758	9.7654	7546	7434	7318	7197	9.7072	38
23	9.8466	8387	8306	8223	8136	9.8046	7953	7856	7756	9.7652	7544	7432	7316	7195	9.7070	37
24	9.8464	8386	8305	8221	8134	9.8044	7951	7854	7754	9.7650	7542	7430	7314	7193	9.7068	36
25	9.8463	8385	8304	8220	8133	9.8043	7949	7853	7752	9.7648	7540	7428	7312	7191	9.7065	35
26	9.8462	8383	8302	8218	8131	9.8041	7948	7851	7751	9.7647	7539	7427	7310	7189	9.7063	34
27	9.8460	8382	8301	8217	8130	9.8040	7946	7849	7749	9.7645	7537	7425	7308	7187	9.7061	33
28	9.8459	8381	8300	8215	8128	9.8038	7945	7848	7747	9.7643	7535	7423	7306	7185	9.7059	32
29	9.8458	8379	8298	8214	8127	9.8037	7943	7846	7746	9.7641	7533	7421	7304	7183	9.7057	31
30	9.8457	8378	8297	8213	8125	9.8035	7941	7844	7744	9.7640	7531	7419	7302	7181	9.7055	30
31	9.8455	8377	8295	8211	8124	9.8034	7940	7843	7742	9.7638	7529	7417	7300	7179	9.7053	29
32	9.8454	8375	8294	8210	8122	9.8032	7938	7841	7740	9.7636	7528	7415	7298	7177	9.7050	28
33	9.8453	8374	8293	8208	8121	9.8031	7937	7840	7739	9.7634	7526	7413	7296	7175	9.7048	27
34	9.8451	8373	8291	8207	8120	9.8029	7935	7838	7737	9.7632	7524	7411	7294	7173	9.7046	26
35	9.8450	8371	8290	8205	8118	9.8027	7934	7836	7735	9.7631	7522	7409	7292	7171	9.7044	25
36	9.8449	8370	8289	8204	8117	9.8026	7932	7835	7734	9.7629	7520	7407	7290	7168	9.7042	24
37	9.8448	8369	8287	8203	8115	9.8024	7930	7833	7732	9.7627	7518	7406	7288	7166	9.7040	23
38	9.8446	8367	8286	8201	8114	9.8023	7929	7831	7730	9.7625	7517	7404	7286	7164	9.7037	22
39	9.8445	8366	8284	8200	8112	9.8021	7927	7830	7728	9.7624	7515	7402	7284	7162	9.7035	21
40	9.8444	8365	8283	8198	8111	9.8020	7926	7828	7727	9.7622	7513	7400	7282	7160	9.7033	20
41	9.8442	8363	8282	8197	8109	9.8018	7924	7826	7725	9.7620	7511	7398	7280	7158	9.7031	19
42	9.8441	8362	8280	8195	8108	9.8017	7922	7825	7723	9.7618	7509	7396	7278	7156	9.7029	18
43	9.8440	8361	8279	8194	8106	9.8015	7921	7823	7722	9.7616	7507	7394	7276	7154	9.7027	17
44	9.8439	8359	8277	8193	8105	9.8014	7919	7821	7720	9.7615	7505	7392	7274	7152	9.7025	16
45	9.8437	8358	8276	8191	8103	9.8012	7918	7820	7718	9.7613	7504	7390	7272	7150	9.7022	15
46	9.8436	8357	8275	8190	8102	9.8010	7916	7818	7716	9.7611	7502	7388	7270	7148	9.7020	14
47	9.8435	8355	8273	8188	8100	9.8009	7914	7816	7715	9.7609	7500	7386	7268	7146	9.7018	13
48	9.8433	8354	8272	8187	8099	9.8007	7913	7815	7713	9.7607	7498	7384	7266	7144	9.7016	12
49	9.8432	8353	8270	8185	8097	9.8006	7911	7813	7711	9.7606	7496	7382	7264	7141	9.7014	11
50	9.8431	8351	8269	8184	8096	9.8004	7910	7811	7710	9.7604	7494	7380	7262	7139	9.7012	10
51	9.8429	8350	8268	8182	8094	9.8003	7908	7810	7708	9.7602	7492	7379	7260	7137	9.7009	9
52	9.8428	8349	8266	8181	8093	9.8001	7906	7808	7706	9.7600	7491	7377	7258	7135	9.7007	8
53	9.8427	8347	8265	8180	8091	9.8000	7905	7806	7704	9.7599	7489	7375	7256	7133	9.7005	7
54	9.8426	8346	8264	8178	8090	9.7998	7903	7805	7703	9.7597	7487	7373	7254	7131	9.7003	6
55	9.8424	8345	8262	8177	8088	9.7997	7901	7803	7701	9.7595	7485	7371	7252	7129	9.7001	5
56	9.8423	8343	8261	8175	8087	97995	7900	7801	7699	9.7593	7483	7369	7250	7127	9.6998	4
57	9.8422	8342	8259	8174	8085	9.7993	7898	7800	7697	9.7591	7481	7367	7248	7125	9.6996	3
58	9.8420	8341	8258	8172	8084	9.7992	7897	7798	7696	9.7590	7479	7365	7246	7123	9.6994	2
59	9.8419	8339	8257	8171	8082	9.7990	7895	7796	7694	9.7588	7477	7363	7244	7120	9.6992	1
60	9.8418	8338	8255	8169	8081	9.7989	7893	7795	7692	9.7586	7476	7361	7242	7118	9.6990	0
	44°	43°	42°	41°	40°	39°	38°	37°	36°	35°	34°	33°	32°	31°	30°	

NOTE: For compactness, the leading digit is not displayed in every column. Scan other columns for necessary digit.

LOG SINES

63

LOG COSINES

	60°	61°	62°	63°	64°	65°	66°	67°	68°	69°	70°	71°	72°	73°	74°	
0	9.6990	6856	6716	6570	6418	9.6259	6093	5919	5736	9.5543	5341	5126	4900	4659	9.4403	60
1	9.6988	6853	6714	6568	6416	9.6257	6090	5916	5733	9.5540	5337	5123	4896	4655	9.4399	59
2	9.6985	6851	6711	6566	6413	9.6254	6087	5913	5729	9.5537	5334	5119	4892	4651	9.4395	58
3	9.6983	6849	6709	6563	6411	9.6251	6085	5910	5726	9.5533	5330	5115	4888	4647	9.4390	57
4	9.6981	6847	6707	6561	6408	9.6249	6082	5907	5723	9.5530	5327	5112	4884	4643	9.4386	56
5	9.6979	6844	6704	6558	6405	9.6246	6079	5904	5720	9.5527	5323	5108	4880	4639	9.4381	55
6	9.6977	6842	6702	6556	6403	9.6243	6076	5901	5717	9.5524	5320	5104	4876	4634	9.4377	54
7	9.6974	6840	6699	6553	6400	9.6240	6073	5898	5714	9.5520	5316	5101	4873	4630	9.4372	53
8	9.6972	6837	6697	6551	6398	9.6238	6070	5895	5711	9.5517	5313	5097	4869	4626	9.4368	52
9	9.6970	6835	6695	6548	6395	9.6235	6068	5892	5708	9.5514	5309	5093	4865	4622	9.4364	51
10	9.6968	6833	6692	6546	6392	9.6232	6065	5889	5704	9.5510	5306	5090	4861	4618	9.4359	50
11	9.6966	6831	6690	6543	6390	9.6230	6062	5886	5701	9.5507	5302	5086	4857	4614	9.4355	49
12	9.6963	6828	6687	6541	6387	9.6227	6059	5883	5698	9.5504	5299	5082	4853	4609	9.4350	48
13	9.6961	6826	6685	6538	6385	9.6224	6056	5880	5695	9.5500	5295	5078	4849	4605	9.4346	47
14	9.6959	6824	6683	6536	6382	9.6221	6053	5877	5692	9.5497	5292	5075	4845	4601	9.4341	46
15	9.6957	6821	6680	6533	6379	9.6219	6050	5874	5689	9.5494	5288	5071	4841	4597	9.4337	45
16	9.6955	6819	6678	6531	6377	9.6216	6047	5871	5685	9.5490	5285	5067	4837	4593	9.4332	44
17	9.6952	6817	6675	6528	6374	9.6213	6045	5868	5682	9.5487	5281	5064	4833	4588	9.4328	43
18	9.6950	6814	6673	6526	6371	9.6210	6042	5865	5679	9.5484	5278	5060	4829	4584	9.4323	42
19	9.6948	6812	6671	6523	6369	9.6208	6039	5862	5676	9.5480	5274	5056	4825	4580	9.4319	41
20	9.6946	6810	6668	6521	6366	9.6205	6036	5859	5673	9.5477	5270	5052	4821	4576	9.4314	40
21	9.6943	6808	6666	6518	6364	9.6202	6033	5856	5670	9.5474	5267	5049	4817	4572	9.4310	39
22	9.6941	6805	6663	6515	6361	9.6199	6030	5853	5666	9.5470	5263	5045	4813	4567	9.4305	38
23	9.6939	6803	6661	6513	6358	9.6197	6027	5850	5663	9.5467	5260	5041	4809	4563	9.4301	37
24	9.6937	6801	6659	6510	6356	9.6194	6024	5847	5660	9.5463	5256	5037	4805	4559	9.4296	36
25	9.6935	6798	6656	6508	6353	9.6191	6021	5844	5657	9.5460	5253	5034	4801	4555	9.4292	35
26	9.6932	6796	6654	6505	6350	9.6188	6019	5841	5654	9.5457	5249	5030	4797	4550	9.4287	34
27	9.6930	6794	6651	6503	6348	9.6186	6016	5838	5650	9.5453	5246	5026	4793	4546	9.4283	33
28	9.6928	6791	6649	6500	6345	9.6183	6013	5834	5647	9.5450	5242	5022	4789	4542	9.4278	32
29	9.6926	6789	6646	6498	6342	9.6180	6010	5831	5644	9.5447	5239	5019	4785	4538	9.4274	31
30	9.6923	6787	6644	6495	6340	9.6177	6007	5828	5641	9.5443	5235	5015	4781	4533	9.4269	30
31	9.6921	6784	6642	6493	6337	9.6175	6004	5825	5638	M440	5231	5011	4777	4529	9.4264	29
32	9.6919	6782	6639	6490	6335	9.6172	6001	5822	5634	9.5437	5228	5007	4773	4525	9.4260	28
33	9.6917	6780	6637	6488	6332	9.6169	5998	5819	5631	9.5433	5224	5003	4769	4521	9.4255	27
34	9.6914	6777	6634	6485	6329	9.6166	5995	5816	5628	9.5430	5221	5000	4765	4516	9.4251	26
35	9.6912	6775	6632	6483	6327	9.6163	5992	5813	5625	9.5426	5217	4996	4761	4512	9.4246	25
36	9.6910	6773	6629	6480	6324	9.6161	5990	5810	5621	9.5423	5213	4992	4757	4508	9.4242	24
37	9.6908	6770	6627	6477	6321	9.6158	5987	5807	5618	9.5420	5210	4988	4753	4503	9.4237	23
38	9.6905	6768	6625	6475	6319	9.6155	5984	5804	5615	9.5416	5206	4984	4749	4499	9.4232	22
39	9.6903	6766	6622	6472	6316	9.6152	5981	5801	5612	9.5413	5203	4981	4745	4495	9.4228	21
40	9.6901	6763	6620	6470	6313	9.6149	5978	5798	5609	9.5409	5199	4977	4741	4491	9.4223	20
41	9.6899	6761	6617	6467	6311	9.6147	5975	5795	5605	9.5406	5196	4973	4737	4486	9.4219	19
42	9.6896	6759	6615	6465	6308	9.6144	5972	5792	5602	9.5403	5192	4969	4733	4482	9.4214	18
43	9.6894	6756	6612	6462	6305	9.6141	5969	5789	5599	9.5399	5188	4965	4729	4478	9.4209	17
44	9.6892	6754	6610	6460	6303	9.6138	5966	5785	5596	9.5396	5185	4962	4725	4473	9.4205	16
45	9.6890	6752	6607	6457	6300	9.6135	5963	5782	5592	9.5392	5181	4958	4721	4469	9.4200	15
46	9.6887	6749	6605	6454	6297	9.6133	5960	5779	5589	9.5389	5177	4954	4717	4465	9.4195	14
47	9.6885	6747	6603	6452	6295	9.6130	5957	5776	5586	9.5385	5174	4950	4713	4460	9.4191	13
48	9.6883	6744	6600	6449	6292	9.6127	5954	5773	5583	9.5382	5170	4946	4709	4456	9.4186	12
49	9.6881	6742	6598	6447	6289	9.6124	5951	5770	5579	9.5379	5167	4942	4705	4452	9.4182	11
50	9.6878	6740	6595	6444	6286	9.6121	5948	5767	5576	9.5375	5163	4939	4700	4447	9.4177	10
51	9.6876	6737	6593	6442	6284	9.6119	5945	5764	5573	9.5372	5159	4935	4696	4443	9.4172	9
52	9.6874	6735	6590	6439	6281	9.6116	5943	5761	5570	9.5368	5156	4931	4692	4438	9.4168	8
53	9.6872	6733	6588	6437	6278	9.6113	5940	5758	5566	9.5365	5152	4927	4688	4434	9.4163	7
54	9.6869	6730	6585	6434	6276	9.6110	5937	5754	5563	9.5361	5148	4923	4684	4430	9.4158	6
55	9.6867	6728	6583	6431	6273	9.6107	5934	5751	5560	9.5358	5145	4919	4680	4425	9.4153	5
56	9.6865	6726	6580	6429	6270	9.6104	5931	5748	5556	9.5354	5141	4915	4676	4421	9.4149	4
57	9.6863	6723	6578	6426	6268	9.6102	5928	5745	5553	9.5351	5137	4911	4672	4417	9.4144	3
58	9.6860	6721	6575	6424	6265	9.6099	5925	5742	5550	9.5347	5134	4908	4668	4412	9.4139	2
59	9.6858	6718	6573	6421	6262	9.6096	5922	5739	5547	9.5344	5130	4904	4663	4408	9.4135	1
60	9.6856	6716	6570	6418	6259	9.6093	5919	5736	5543	9.5341	5126	4900	4659	4403	9.4130	0
	29°	28°	27°	26°	25°	24°	23°	22°	21°	20°	19°	18°	17°	16°	15°	

NOTE: For compactness, the leading digit is not displayed in every column. Scan other columns for necessary digit.

LOG SINES

LOG COSINES

	75°	76°	77°	78°	79°	80°	81°	82°	83°	84°	85°	86°	87°	88°	89°	
0	9.4130	3837	3521	3179	2806	9.2397	1943	1436	0859	9.0192	9403	8436	7188	5428	8.2419	60
1	9.4125	3832	3515	3173	2799	9.2390	1935	1427	0849	9.0180	9388	8418	7164	5392	8.2346	59
2	9.4121	3827	3510	3167	2793	9.2382	1927	1418	0838	9.0168	9374	8400	7140	5355	8.2271	58
3	9.4116	3822	3504	3161	2786	9.2375	1919	1409	0828	9.0156	9359	8381	7115	5318	8.2196	57
4	9.4111	3816	3499	3155	2780	9.2368	1911	1399	0818	9.0144	9345	8363	7090	5281	8.2119	56
5	9.4106	3811	3493	3149	2773	9.2361	1903	1390	0807	9.0132	9330	8345	7066	5243	8.2041	55
6	9.4102	3806	3488	3143	2767	9.2354	1895	1381	0797	9.0120	9315	8326	7041	5206	8.1961	54
7	9.4097	3801	3482	3137	2760	9.2346	1887	1372	0786	9.0107	9301	8307	7016	5167	8.1880	53
8	9.4092	3796	3477	3131	2754	9.2339	1879	1363	0776	9.0095	9286	8289	6991	5129	8.1797	52
9	9.4087	3791	3471	3125	2747	9.2332	1871	1354	0765	9.0083	9271	8270	6965	5090	8.1713	51
10	9.4083	3786	3466	3119	2740	9.2324	1863	1345	0755	9.0070	9256	8251	6940	5050	8.1627	50
11	9.4078	3781	3460	3113	2734	9.2317	1855	1336	0744	9.0058	9241	8232	6914	5011	8.1539	49
12	9.4073	3775	3455	3107	2727	9.2310	1847	1326	0734	9.0046	9226	8213	6889	4971	8.1450	48
13	9.4068	3770	3449	3101	2721	9.2303	1838	1317	0723	9.0033	9211	8194	6863	4930	8.1358	47
14	9.4063	3765	3444	3095	2714	9.2295	1830	1308	0712	9.0021	9196	8175	6837	4890	8.1265	46
15	9.4059	3760	3438	3089	2707	9.2288	1822	1299	0702	9.0008	9181	8156	6810	4848	8.1169	45
16	9.4054	3755	3432	3083	2701	9.2281	1814	1289	0691	8.9996	9166	8137	6784	4807	8.1072	44
17	9.4049	3750	3427	3077	2694	9.2273	1806	1280	0680	8.9983	9150	8117	6758	4765	8.0972	43
18	9.4044	3745	3421	3070	2687	9.2266	1797	1271	0670	8.9970	9135	8098	6731	4723	8.0870	42
19	9.4039	3739	3416	3064	2681	9.2258	1789	1261	0659	8.9958	9119	8078	6704	4680	8.0765	41
20	9.4035	3734	3410	3058	2674	9.2251	1781	1252	0648	8.9945	9104	8059	6677	4637	8.0658	40
21	9.4030	3729	3404	3052	2667	9.2244	1772	1242	0637	8.9932	9089	8039	6650	4593	8.0548	39
22	9.4025	3724	3399	3046	2661	9.2236	1764	1233	0626	8.9919	9073	8019	6622	4549	8.0435	38
23	9.4020	3719	3393	3040	2654	9.2229	1756	1224	0616	8.9907	9057	7999	6595	4504	8.0319	37
24	9.4015	3713	3387	3034	2647	9.2221	1747	1214	0605	8.9894	9042	7979	6567	4459	8.0200	36
25	9.4010	3708	3382	3027	2640	9.2214	1739	1205	0594	8.9881	9026	7959	6539	4414	8.0078	35
26	9.4006	3703	3376	3021	2634	9.2206	1731	1195	0583	8.9868	9010	7939	6511	4368	7.9952	34
27	9.4001	3698	3370	3015	2627	9.2199	1722	1186	0572	8.9855	8994	7918	6483	4322	7.9822	33
28	9.3996	3692	3365	3009	2620	9.2191	1714	1176	0561	8.9842	8978	7898	6454	4275	7.9689	32
29	9.3991	3687	3359	3003	2613	9.2184	1705	1167	0550	8.9829	8962	7877	6426	4227	7.9551	31
30	9.3986	3682	3353	2997	2606	9.2176	1697	1157	0539	8.9816	8946	7857	6397	4179	7.9408	30
31	9.3981	3677	3348	2990	2600	9.2169	1689	1147	0527	8.9803	8930	7836	6368	4131	7.9261	29
32	9.3976	3671	3342	2984	2593	9.2161	1680	1138	0516	8.9789	8914	7815	6339	4082	7.9109	28
33	9.3971	3666	3336	2978	2586	9.2153	1672	1128	0505	8.9776	8898	7794	6309	4032	7.8951	27
34	9.3966	3661	3331	2972	2579	9.2146	1663	1118	0494	8.9763	8882	7773	6279	3982	7.8787	26
35	9.3962	3655	3325	2965	2572	9.2138	1655	1109	0483	8.9750	8865	7752	6250	3931	7.8617	25
36	9.3957	3650	3319	2959	2565	9.2131	1646	1099	0472	8.9736	8849	7731	6220	3880	7.8439	24
37	9.3952	3645	3313	2953	2558	9.2123	1637	1089	0460	8.9723	8833	7710	6189	3828	7.8255	23
38	9.3947	3640	3308	2947	2551	9.2115	1629	1080	0449	8.9710	8816	7688	6159	3775	7.8061	22
39	9.3942	3634	3302	2940	2545	9.2108	1620	1070	0438	8.9696	8799	7667	6128	3722	7.7859	21
40	9.3937	3629	3296	2934	2538	9.2100	1612	1060	0426	8.9683	8783	7645	6097	3668	7.7648	20
41	9.3932	3624	3290	2928	2531	9.2092	1603	1050	0415	8.9669	8766	7623	6066	3613	7.7425	19
42	9.3927	3618	3284	2921	2524	9.2085	1594	1040	0403	8.9655	8749	7602	6035	3558	7.7190	18
43	9.3922	3613	3279	2915	2517	9.2077	1586	1030	0392	8.9642	8733	7580	6003	3502	7.6942	17
44	9.3917	3608	3273	2909	2510	9.2069	1577	1020	0380	8.9628	8716	7557	5972	3445	7.6678	16
45	9.3912	3602	3267	2902	2503	9.2061	1568	1011	0369	8.9614	8689	7535	5939	3388	7.6398	15
46	9.3907	3597	3261	2896	2496	9.2054	1560	1001	0357	8.9601	8682	7513	5907	3329	7.6099	14
47	9.3902	3591	3255	2890	2489	9.2046	1551	0991	0346	8.9587	8665	7491	5875	3270	7.5777	13
48	9.3897	3586	3250	2883	2482	9.2038	1542	0981	0334	8.9573	8647	7468	5842	3210	7.5429	12
49	9.3892	3581	3244	2877	2475	9.2030	1533	0971	0323	8.9559	8630	7445	5809	3150	7.5051	11
50	9.3887	3575	3238	2870	2468	9.2022	1525	0961	0311	8.9545	8613	7423	5776	3088	7.4637	10
51	9.3882	3570	3232	2864	2461	9.2015	1516	0951	0299	8.9531	8595	7400	5742	3025	7.4180	9
52	9.3877	3564	3226	2858	2454	9.2007	1507	0940	0287	8.9517	8578	7377	5708	2962	7.3668	8
53	9.3872	3559	3220	2851	2447	9.1999	1498	0930	0276	8.9503	8560	7354	5674	2898	7.3088	7
54	9.3867	3554	3214	2845	2439	9.1991	1489	0920	0264	8.9489	8543	7330	5640	2832	7.2419	6
55	9.3862	3548	3208	2838	2432	9.1983	1480	0910	0252	8.9475	8525	7307	5605	2766	7.1627	5
56	9.3857	3543	3202	2832	2425	9.1975	1471	0900	0240	8.9460	8508	7283	5571	2699	7.0658	4
57	9.3852	3537	3197	2825	2418	9.1967	1462	0890	0228	8.9446	8490	7260	5535	2630	6.9409	3
58	9.3847	3532	3191	2819	2411	9.1959	1453	0879	0216	8.9432	8472	7236	5500	2561	6.7648	2
59	9.3842	3526	3185	2812	2404	9.1951	1445	0869	0204	8.9417	8454	7212	5464	2490	6.4637	1
60	9.3837	3521	3179	2806	2397	9.1943	1436	0859	0192	8.9403	8436	7188	5428	2419	– ∞	0
	14°	13°	12°	11°	10°	9°	8°	7°	6°	5°	4°	3°	2°	1°	0°	

NOTE: For compactness, the leading digit is not displayed in every column. Scan other columns for necessary digit.

LOG SINES

A

\+ if hour angle is listed at top
− if hour angle is listed at bottom

LHA	1° 359°	2° 358°	3° 357°	4° 356°	5° 355°	6° 354°	7° 353°	8° 352°	9° 351°	10° 350°	11° 349°	12° 348°	13° 347°	14° 346°	15° 345°
0	0.00	.000	.000	.000	.000	.000	.000	.000	.000	.000	.000	.000	.000	.000	.000
3	3.00	1.50	1.00	.749	.599	.499	.427	.373	.331	.297	.270	.247	.227	.210	.196
6	6.02	3.01	2.01	1.50	1.20	1.00	.856	.748	.664	.596	.541	.494	.455	.422	.392
9	9.07	4.54	3.02	2.27	1.81	1.51	1.29	1.13	1.00	.898	.815	.745	.686	.635	.591
12	12.2	6.09	4.06	3.04	2.43	2.02	1.73	1.51	1.34	1.21	1.09	1.00	.921	.853	.793
15	15.4	7.67	5.11	3.83	3.06	2.55	2.18	1.91	1.69	1.52	1.38	1.26	1.16	1.07	1.00
18	18.6	9.30	6.20	4.65	3.71	3.09	2.65	2.31	2.05	1.84	1.67	1.53	1.41	1.30	1.21
21	22.0	11.0	7.32	5.49	4.39	3.65	3.13	2.73	2.42	2.18	1.97	1.81	1.66	1.54	1.43
24	25.5	12.7	8.50	6.37	5.09	4.24	3.63	3.17	2.81	2.53	2.29	2.09	1.93	1.79	1.66
27	29.2	14.6	9.72	7.29	5.82	4.85	4.15	3.63	3.22	2.89	2.62	2.40	2.21	2.04	1.90
30	33.1	16.5	11.0	8.26	6.60	5.49	4.70	4.11	3.65	3.27	2.97	2.72	2.50	2.32	2.15
33	37.2	18.6	12.4	9.29	7.42	6.18	5.29	4.62	4.10	3.68	3.34	3.06	2.81	2.61	2.42
36	41.6	20.8	13.9	10.4	8.30	6.91	5.92	5.17	4.59	4.12	3.74	3.42	3.15	2.91	2.71
38	44.8	22.4	14.9	11.2	8.93	7.43	6.36	5.56	4.93	4.43	4.02	3.68	3.38	3.13	2.92
40	48.1	24.0	16.0	12.0	9.59	7.98	6.83	5.97	5.30	4.76	4.32	3.95	3.63	3.37	3.13
42	51.6	25.8	17.2	12.9	10.3	8.57	7.33	6.41	5.69	5.11	4.63	4.24	3.90	3.61	3.36
44	55.3	27.7	18.4	13.8	11.0	9.19	7.86	6.87	6.10	5.48	4.97	4.54	4.18	3.87	3.60
46	59.3	29.7	19.8	14.8	11.8	9.85	8.43	7.37	6.54	5.87	5.33	4.87	4.49	4.15	3.86
48	63.6	31.8	21.2	15.9	12.7	10.6	9.05	7.90	7.01	6.30	5.71	5.23	4.81	4.45	4.14
50	68.3	34.1	22.7	17.0	13.6	11.3	9.71	8.48	7.52	6.76	6.13	5.61	5.16	4.78	4.45
52	73.3	36.7	24.4	18.3	14.6	12.2	10.4	9.11	8.08	7.26	6.58	6.02	5.55	5.13	4.78
54	78.9	39.4	26.3	19.7	15.7	13.1	11.2	9.79	8.69	7.81	7.08	6.48	5.96	5.52	5.14
56	84.9	42.5	28.3	21.2	16.9	14.1	12.1	10.5	9.36	8.41	7.63	6.97	6.42	5.95	5.53
58	91.7	45.8	30.5	22.9	18.3	15.2	13.0	11.4	10.1	9.08	8.23	7.53	6.93	6.42	5.97
60	99.2	49.6	33.0	24.8	19.8	16.5	14.1	12.3	10.9	9.82	8.91	8.15	7.50	6.95	6.46
62	108	53.9	35.9	26.9	21.5	17.9	15.3	13.4	11.9	10.7	9.68	8.85	8.15	7.54	7.02
64	117	58.7	39.1	29.3	23.4	19.5	16.7	14.6	12.9	11.6	10.5	9.65	8.88	8.22	7.65
66	129	64.3	42.9	32.1	25.7	21.4	18.3	16.0	14.2	12.7	11.6	10.6	9.72	9.01	8.38
LHA	179° 181°	178° 183°	177° 183°	176° 184°	175° 185°	174° 186°	173° 187°	172° 188°	171° 189°	170° 190°	169° 191°	168° 192°	167° 193°	166° 194°	165° 195°

B

− if latitude and declination have same name
\+ if latitude and declination have different names

LHA	1° 359°	2° 358°	3° 357°	4° 356°	5° 355°	6° 354°	7° 353°	8° 352°	9° 351°	10° 350°	11° 349°	12° 348°	13° 347°	14° 346°	15° 345°
0	0.00	.000	.000	.000	.000	.000	.000	.000	.000	.000	.000	.000	.000	.000	.000
3	3.00	1.50	1.00	.751	.601	.501	.430	.377	.335	.302	.275	.252	.233	.217	.202
6	6.02	3.01	2.01	1.51	1.21	1.01	.862	.755	.672	.605	.551	.506	.467	.434	.406
9	9.08	4.54	3.03	2.27	1.82	1.52	1.30	1.14	1.01	.912	.830	.762	.704	.655	.612
12	12.2	6.09	4.06	3.05	2.44	2.03	1.74	1.53	1.36	1.22	1.11	1.02	.945	.879	.821
15	15.4	7.68	5.12	3.84	3.07	2.56	2.20	1.93	1.71	1.54	1.40	1.29	1.19	1.11	1.04
18	18.6	9.31	6.21	4.66	3.73	3.11	2.67	2.33	2.08	1.87	1.70	1.56	1.44	1.34	1.26
21	22.0	11.0	7.33	5.50	4.40	3.67	3.15	2.76	2.45	2.21	2.01	1.85	1.71	1.59	1.48
24	25.5	12.8	8.51	6.38	5.11	4.26	3.65	3.20	2.85	2.56	2.33	2.14	1.98	1.84	1.72
27	29.2	14.6	9.74	7.30	5.85	4.87	4.18	3.66	3.26	2.93	2.67	2.45	2.27	2.11	1.97
30	33.1	16.5	11.0	8.28	6.62	5.52	4.74	4.15	3.69	3.32	3.03	2.78	2.57	2.39	2.23
33	37.2	18.6	12.4	9.31	7.45	6.21	5.33	4.67	4.15	3.74	3.40	3.12	2.89	2.68	2.51
36	41.6	20.8	13.9	10.4	8.34	6.95	5.96	5.22	4.64	4.18	3.81	3.49	3.23	3.00	2.81
38	44.8	22.4	14.9	11.2	8.96	7.47	6.41	5.61	4.99	4.50	4.09	3.76	3.47	3.23	3.02
40	48.1	24.0	16.0	12.0	9.63	8.03	6.89	6.03	5.36	4.83	4.40	4.04	3.73	3.47	3.24
42	51.6	25.8	17.2	12.9	10.3	8.61	7.39	6.47	5.76	5.19	4.72	4.33	4.00	3.72	3.48
44	55.3	27.7	18.5	13.8	11.1	9.24	7.92	6.94	6.17	5.56	5.06	4.64	4.29	3.99	3.73
46	59.3	29.7	19.8	14.8	11.9	9.91	8.50	7.44	6.62	5.96	5.43	4.98	4.60	4.28	4.00
48	63.6	31.8	21.2	15.9	12.7	10.6	9.11	7.98	7.10	6.40	5.82	5.34	4.94	4.59	4.29
50	68.3	34.1	22.8	17.1	13.7	11.4	9.78	8.56	7.62	6.86	6.25	5.73	5.30	4.93	4.60
52	73.3	36.7	24.5	18.3	14.7	12.2	10.5	9.20	8.18	7.37	6.71	6.16	5.69	5.29	4.95
54	78.9	39.4	26.3	19.7	15.8	13.2	11.3	9.89	8.80	7.93	7.21	6.62	6.12	5.69	5.32
56	84.9	42.5	28.3	21.3	17.0	14.2	12.2	10.7	9.48	8.54	7.77	7.13	6.59	6.13	5.73
58	91.7	45.9	30.6	22.9	18.4	15.3	13.1	11.5	10.2	9.22	8.39	7.70	7.11	6.62	6.18
60	99.2	49.6	33.1	24.8	19.9	16.6	14.2	12.5	11.1	9.97	9.08	8.33	7.70	7.16	6.69
62	108	53.9	35.9	27.0	21.6	18.0	15.4	13.5	12.0	10.8	9.86	9.05	8.36	7.77	7.27
LHA	179° 181°	178° 183°	177° 183°	176° 184°	175° 185°	174° 186°	173° 187°	172° 188°	171° 189°	170° 190°	169° 191°	168° 192°	167° 193°	166° 194°	165° 195°

A

\+ if hour angle is listed at top
− if hour angle is listed at bottom

LHA	16° 344°	17° 343°	18° 342°	19° 341°	20° 340°	21° 339°	22° 338°	23° 337°	24° 336°	25° 335°	26° 334°	27° 333°	28° 332°	29° 331°	30° 330°
0	.000	.000	.000	.000	.000	.000	.000	.000	.000	.000	.000	.000	.000	.000	.000
3	.183	.171	.161	.152	.144	.137	.130	.123	.118	.112	.107	.103	.099	.095	.091
6	.367	.344	.323	.305	.289	.274	.260	.248	.236	.225	.215	.206	.198	.190	.182
9	.552	.518	.487	.460	.435	.413	.392	.373	.356	.340	.325	.311	.298	.286	.274
12	.741	.695	.654	.617	.584	.554	.526	.501	.477	.456	.436	.417	.400	.383	.368
15	.934	.876	.825	.778	.736	.698	.663	.631	.602	.575	.549	.526	.504	.483	.464
18	1.13	1.06	1.00	.944	.893	.846	.804	.765	.730	.697	.666	.638	.611	.586	.563
21	1.34	1.26	1.18	1.11	1.05	1.00	.950	.904	.862	.823	.787	.753	.722	.693	.665
24	1.55	1.46	1.37	1.29	1.22	1.16	1.10	1.05	1.00	.955	.913	.874	.837	.803	.771
27	1.78	1.67	1.57	1.48	1.40	1.33	1.26	1.20	1.14	1.09	1.04	1.00	.958	.919	.883
30	2.01	1.89	1.78	1.68	1.59	1.50	1.43	1.36	1.30	1.24	1.18	1.13	1.09	1.04	1.00
33	2.26	2.12	2.00	1.89	1.78	1.69	1.61	1.53	1.46	1.39	1.33	1.27	1.22	1.17	1.12
36	2.53	2.38	2.24	2.11	2.00	1.89	1.80	1.71	1.63	1.56	1.49	1.43	1.37	1.31	1.26
38	2.72	2.56	2.40	2.27	2.15	2.04	1.93	1.84	1.75	1.68	1.60	1.53	1.47	1.41	1.35
40	2.93	2.74	2.58	2.44	2.31	2.19	2.08	1.98	1.88	1.80	1.72	1.65	1.58	1.51	1.45
42	3.14	2.95	2.77	2.61	2.47	2.35	2.23	2.12	2.02	1.93	1.85	1.77	1.69	1.62	1.56
44	3.37	3.16	2.97	2.80	2.65	2.52	2.39	2.28	2.17	2.07	1.98	1.90	1.82	1.74	1.67
46	3.61	3.39	3.19	3.01	2.85	2.70	2.56	2.44	2.33	2.22	2.12	2.03	1.95	1.87	1.79
48	3.87	3.63	3.42	3.23	3.05	2.89	2.75	2.62	2.49	2.38	2.28	2.18	2.09	2.00	1.92
50	4.16	3.90	3.67	3.46	3.27	3.10	2.95	2.81	2.68	2.56	2.44	2.34	2.24	2.15	2.06
52	4.46	4.19	3.94	3.72	3.52	3.33	3.17	3.02	2.87	2.74	2.62	2.51	2.41	2.31	2.22
54	4.80	4.50	4.24	4.00	3.78	3.59	3.41	3.24	3.09	2.95	2.82	2.70	2.59	2.48	2.38
56	5.17	4.85	4.56	4.31	4.07	3.86	3.67	3.49	3.33	3.18	3.04	2.91	2.79	2.67	2.57
58	5.58	5.23	4.93	4.65	4.40	4.17	3.96	3.77	3.59	3.43	3.28	3.14	3.01	2.89	2.77
60	6.04	5.67	5.33	5.03	4.76	4.51	4.29	4.08	3.89	3.71	3.55	3.40	3.26	3.12	3.00
62	6.56	6.15	5.79	5.46	5.17	4.90	4.65	4.43	4.22	4.03	3.86	3.69	3.54	3.39	3.26
64	7.15	6.71	6.31	5.95	5.63	5.34	5.07	4.83	4.61	4.40	4.20	4.02	3.86	3.70	3.55
66	7.83	7.35	6.91	6.52-	6.17	5.85	5.56	5.29	5.04	4.82	4.61	4.41	4.22	4.05	3.89
LHA	164° 196°	163° 197°	162° 198°	161° 199°	160° 200°	159° 201°	158° 202°	157° 203°	156° 204°	155° 205°	154° 206°	153° 207°	152° 208°	151° 209°	150° 210°

B

− if latitude and declination have same name
\+ if latitude and declination have different names

LHA	16° 344°	17° 343°	18° 342°	19° 341°	20° 340°	21° 339°	22° 338°	23° 337°	24° 336°	25° 335°	26° 334°	27° 333°	28° 332°	29° 331°	30° 330°
0	.000	.000	.000	.000	.000	.000	.000	.000	.000	.000	.000	.000	.000	.000	.000
3	.190	.179	.170	.161	.153	.146	.140	.134	.129	.124	.120	.115	.112	.108	.105
6	.381	.359	.340	.323	.307	.293	.281	.269	.258	.249	.240	.232	.224	.217	.210
9	.575	.542	.513	.486	.463	.442	.423	.405	.389	.375	.361	.349	.337	.327	.317
12	.771	.727	.688	.653	.621	.593	.567	.544	.523	.503	.485	.468	.453	.438	.425
15	.972	.916	.867	.823	.783	.748	.715	.686	.659	.634	.611	.590	.571	.553	.536
18	1.18	1.11	1.05	.998	.950	.907	.867	.832	.799	.769	.741	.716	.692	.670	.650
21	1.39	1.31	1.24	1.18	1.12	1.07	1.02	.982	.944	.908	.876	.846	.818	.792	.768
24	1.62	1.52	1.44	1.37	1.30	1.24	1.19	1.14	1.09	1.05	1.02	.981	.948	.918	.890
27	1.85	1.74	1.65	1.57	1.49	1.42	1.36	1.30	1.25	1.21	1.16	1.12	1.09	1.05	1.02
30	2.09	1.97	1.87	1.77	1.69	1.61	1.54	1 A8	1.42	1.37	1.32	1.27	1.23	1.19	1.15
33	2.36	2.22	2.10	1.99	1.90	1.81	1.73	1.66	1.60	1.54	1.48	1.43	1.38	1.34	1.30
36	2.64	2.48	2.32	2.23	2.12	2.03	1.94	1.86	1.79	1.72	1.66	1.60	1.55	1.50	1.45
38	2.83	2.67	2.53	2.40	2.28	2.18	2.09	2.00	1.92	1.85	1.78	1.72	1.66	1.61	1.56
40	3.04	2.87	2.72	2.58	2.45	2.34	2.24	2.15	2.06	1.99	1.91	1.85	1.79-	1.73	1.68
42	3.27	3.08	2.91	2.77	2.63	2.51	2.40	2.30	2.21	2.13	2.05	1.98	1.92	1.86	1.80
44	3.50	3.30	3.13	2.97	2.82	2.69	2.58	2.47	2.37	2.29	2.20	2.13	2.06	1.99	1.93
46	3.76	3.54	3.35	3.18	3.03	2.89	2.76	2.65	2.55	2.45	2.36	2.28	2.21	2.14	2.07
48	4.03	3.80	3.59	3.41	3.25	3.10	2.96	2.84	2.73	2.63	2.53	2.45	2.37	2.29	2.22
50	4.32	4.08	3.86	3.66	3.48	3.33	3.18	3.05	2.93	2.82	2.72	2.63	2.54	2.46	2.38
52	4.64	4.38	4.14	3.93	3.74	3.57	3.42	3.28	3.15	3.03	2.92	2.82	2.73	2.64	2.56
54	4.99	4.71	4.45	4.23	4.02	3.84	3.67	3.52	3.38	3.26	3.14	3.03	2.93	2.84	2.75
56	5.38	5.07	4.80	4.55	4.33	4.14	3.96	3.79	3.65	3.51	3.38	3.27	3.16	3.06	2.97
58	5.81	5.47	5.18	4.92	4.68	4.47	4.27	4.10	3.93	3.79	3.65	3.53	3.41	3.30	3.20
60	6.28	5.92	5.61	5.32	5.06	4.83	4.62	4.43	4.26	4.10	3.95	3.82	3.69	3.57	3.46
62	6.82	6.43	6.09	5.78	5.50	5.25	5.02	4.81	4.62	4.45	4.29	4.14	4.01	3.88	3.76
LHA	164° 196°	163° 197°	162° 198°	161° 199°	160° 200°	159° 201°	158° 202°	157° 203°	156° 204°	155° 205°	154° 206°	153° 207°	152° 208°	151° 209°	150° 210°

65
ABC TABLES

ABC TABLES

A
+ if hour angle is listed at top
− if hour angle is listed at bottom

LHA	32° 328°	34° 326°	36° 324°	38° 322°	40° 320°	42° 318°	44° 316°	46° 314°	48° 312°	50° 310°	52° 308°	54° 306°	56° 304°	58° 302°	60° 300°
0	.000	.000	.000	.000	.000	.000	.000	.000	.000	.000	.000	.000	.000	.000	.000
3	.084	.078	.072	.067	.062	.058	.054	.051	.047	.044	.041	.038	.035	.033	.030
6	.168	.156	.145	.135	.125	.117	.109	.101	.095	.088	.082	.076	.071	.066	.061
9	.253	.235	.218	.203	.189	.176	.164	.153	.143	.133	.124	.115	.107	.099	.091
12	.340	.315	.293	.272	.253	.236	.220	.205	.191	.178	.166	.154	.143	.133	.123
15	.429	.397	.369	.343	.319	.298	.277	.259	.241	.225	.209	.195	.181	.167	.155
18	.520	.482	.447	.416	.387	.361	.336	.314	.293	.273	.254	.236	.219	.203	.188
21	.614	.569	.528	.491	.457	.426	.398	.371	.346	.322	.300	.279	.259	.240	.222
24	.713	.660	.613	.570	.531	.494	.461	.430	.401	.374	.348	.323	.300	.278	.257
27	.815	.755	.701	.652	.607	.566	.528	.492	.459	.428	.398	.370	.344	.318	.294
30	.924	.856	.795	.739	.688	.641	.598	.558	.520	.484	.451	.419	.389	.361	.333
33	1.04	.963	.894	.831	.774	.721	.672	.627	.585	.545	.507	.472	.438	.406	.375
36	1.16	1.08	1.00	.930	.866	.807	.752	.702	.654	.610	.568	.528	.490	.454	.419
38	1.25	1.16	1.08	1.00	.931	.868	.809	.754	.703	.656	.610	.568	.527	.488	.451
40	1.34	1.24	1.15	1.07	1.00	.932	.869	.810	.756	.704	.656	.610	.566	.524	.484
42	1.44	1.33	1.24	1.15	1.07	1.00	.932	.870	.811	.756	.703	.654	.607	.563	.520
44	1.55	1.43	1.33	1.24	1.15	1.07	1.00	.933	.870	.810	.754	.702	.651	.603	.558
46	1.66	1.54	1.43	1.33	1.23	1.15	1.07	1.00	.932	.869	.809	.752	.698	.647	.598
48	1.78	1.65	1.53	1.42	1.32	1.23	1.15	1.07	1.00	.932	.868	.807	.749	.694	.641
50	1.91	1.77	1.64	1.53	1.42	1.32	1.23	1.15	1.07	1.00	.931	.866	.804	.745	.688
52	2.05	1.90	1.76	1.64	1.53	1.42	1.33	1.24	1.15	1.07	1.00	.930	.863	.800	.739
54	2.20	2.04	1.89	1.76	1.64	1.53	1.43	1.33	1.24	1.15	1.08	1.00	.928	.860	.795
56	2.37	2.20	2.04	1.90	1.77	1.65	1.54	1.43	1.33	1.24	1.16	1.08	1.00	.926	.856
58	2.56	2.37	2.20	2.05	1.91	1.78	1.66	1.55	1.44	1.34	1.25	1.16	1.08	1.00	.924
60	2.77	2.57	2.38	2.22	2.06	1.92	1.79	1.67	1.56	1.45	1.35	1.26	1.17	1.08	1.00
62	3.01	2.79	2.59	2.41	2.24	2.09	1.95	1.82	1.69	1.58	1.47	1.37	1.27	1.18	1.09
64	3.28	3.04	2.82	2.62	2.44	2.28	2.12	1.98	1.85	1.72	1.60	1.49	1.38	1.28	1.18
66	3.59	3.33	3.09	2.87	2.68	2.49	2.33	2.17	2.02	1.88	1.75	1.63	1.52	1.40	1.30
LHA	148° 212°	146° 214°	144° 216°	142° 218°	140° 220°	138° 222°	136° 224°	134° 226°	132° 228°	130° 230°	128° 232°	126° 234°	124° 236°	122° 238°	120° 240°

B
− if if latitude and declination have same name
+ if latitude and declination have different names

LHA	32° 328°	34° 326°	36° 324°	38° 322°	40° 320°	42° 318°	44° 316°	46° 314°	48° 312°	50° 310°	52° 308°	54° 306°	56° 304°	58° 302°	60° 300°
0	.000	.000	.000	.000	.000	.000	.000	.000	.000	.000	.000	.000	.000	.000	.000
3	.099	.094	.089	.085	.082	.078	.075	.073	.071	.068	.067	.065	.063	.062	.061
6	.198	.188	.179	.171	.164	.157	.151	.146	.141	.137	.133	.130	.127	.124	.121
9	.299	.283	.269	.257	.246	.237	.228	.220	.213	.207	.201	.196	.191	.187	.183
12	.401	.380	.362	.345	.331	.318	.306	.295	.286	.277	.270	.263	.256	.251	.245
15	.506	.479	.456	.435	.417	.400	.386	.372	.361	.350	.340	.331	.323	.316	.309
18	.613	.581	.553	.528	.505	.486	.468	.452	.437	.424	.412	.402	.392	.383	.375
21	.724	.686	.653	.623	.597	.574	.553	.534	.517	.501	.487	.474	.463	.453	.443
24	.840	.796	.757	.723	.693	.665	.641	.619	.599	.581	.565	.550	.537	.525	.514
27	.962	.911	.867	.828	.793	.761	.733	.708	.686	.665	.647	.630	.615	.601	.588
30	1.09	1.03	.982	.938	.898	.863	.831	.803	.777	.754	.733	.714	.696	.681	.667
33	1.23	1.16	1.11	1.05	1.01	.971	.935	.903	.874	.848	.824	.803	.783	.766	.750
36	1.37	1.30	1.24	1.18	1.13	1.09	1.05	1.01	.978	.948	.922	.898	.876	.857	.839
38	1.47	1.40	1.33	1.27	1.22	1.17	1.12	1.09	1.05	1.02	.991	.966	.942	.921	.902
40	1.58	1.50	1.43	1.36	1.31	1.25	1.21	1.17	1.13	1.10	1.06	1.04	1.01	.989	.969
42	1.70	1.61	1.53	1.46	1.40	1.35	1.30	1.25	1.21	1.18	1.14	1.11	1.09	1.06	1.04
44	1.82	1.73	1.64	1.57	1.50	1.44	1.39	1.34	1.30	1.26	1.23	1.19	1.16	1.14	1.12
46	1.95	1.85	1.76	1.68	1.61	1.55	1.49	1.44	1.39	1.35	1.31	1.28	1.25	1.22	1.20
48	2.10	1.99	1.89	1.80	1.73	1.66	1.60	1.54	1.49	1.45	1.41	1.37	1.34	1.31	1.28
50	2.25	2.13	2.03	1.94	1.85	1.78	1.72	1.66	1.60	1.56	1.51	1.47	1.44	1.41	1.38
52	2.42	2.29	2.18	2.08	1.99	1.91	1.84	1.78	1.72	1.67	1.62	1.58	1.54	1.51	1.48
54	2.60	2.46	2.34	2.24	2.14	2.06	1.98	1.91	1.85	1.80	1.75	1.70	1.66	1.62	1.59
56	2.80	2.65	2.52	2.41	2.31	2.22	2.13	2.06	2.00	1.94	1.88	1.83	1.79	1.75	1.71
58	3.02	2.86	2.72	2.60	2.49	2.39	2.30	2.22	2.15	2.09	2.03	1.98	1.93	1.89	1.85
60	3.27	3.10	2.95	2.81	2.69	2.59	2.49	2.41	2.33	2.26	2.20	2.14	2.09	2.04	2.00
62	3.55	3.36	3.20	3.05	2.93	2.81	2.71	2.61	2.53	2.46	2.39	2.32	2.27	2.22	2.17
LHA	148° 212°	146° 214°	144° 216°	142° 218°	140° 220°	138° 222°	136° 224°	134° 226°	132° 228°	130° 230°	128° 232°	126° 234°	124° 236°	122° 238°	120° 240°

ABC TABLES

A
+ if hour angle is listed at top
− if hour angle is listed at bottom

LHA	62° 298°	64° 296°	66° 294°	68° 292°	70° 290°	72° 288°	74° 286°	76° 284°	78° 282°	80° 280°	82° 278°	84° 276°	86° 274°	88° 272°	90° 270°
0	.000	.000	.000	.000	.000	.000	.000	.000	.000	.000	.000	.000	.000	.000	.000
3	.028	.026	.023	.021	.019	.017	.015	.013	.011	.009	.007	.006	.004	.002	.000
6	.056	.051	.047	.043	.038	.034	.030	.026	.022	.019	.015	.011	.007	.004	.000
9	.084	.077	.071	.064	.058	.051	.045	.039	.034	.028	.022	.017	.011	.006	.000
12	.113	.104	.065	.086	.077	.069	.061	.053	.045	.037	.030	.022	.015	.007	.000
15	.142	.131	.119	.108	.098	.087	.077	.067	.057	.047	.038	.028	.019	.009	.000
18	.173	.158	.145	.131	.118	.106	.093	.081	.069	.057	.046	.034	.023	.012	.000
21	.204	.187	.171	.155	.140	.125	.110	.096	.082	.068	.054	.040	.027	.013	.000
24	.237	.217	.198	.180	.162	.145	.128	.111	.095	.079	.063	.047	.031	.016	.000
27	.271	.249	.227	.206	.185	.166	.146	.127	.108	.090	.072	.054	.036	.018	.000
30	.307	.282	.257	.233	.210	.188	.166	.144	.123	.102	.081	.061	.040	.020	.000
33	.345	.317	.289	.262	.236	.211	.186	.162	.138	.115	.091	.068	.045	.023	.000
36	.386	.354	.323	.294	.264	.236	.208	.181	.154	.128	.102	.076	.051	.025	.000
38	.415	.381	.348	.316	.284	.254	.224	.195	.166	.138	.110	.082	.055	.027	.000
40	.446	.409	.374	.339	.305	.273	.241	.209	.178	.148	.118	.088	.059	.029	.000
42	.479	.439	.401	.364	.328	.293	.258	.224	.191	.159	.127	.095	.063	.031	.000
44	.513	.471	.430	.390	.351	.314	.277	.241	.205	.170	.136	.101	.068	.034	.000
46	.551	.505	.461	.418	.377	.336	.297	.258	.220	.183	.146	.109	.072	.036	.000
48	.591	.542	.494	.449	.404	.361	.318	.277	.236	.196	.156	.117	.078	.039	.000
50	.634	.581	.531	.481	.434	.387	.342	.297	.253	.210	.167	.125	.083	.042	.000
52	.681	.624	.570	.517	.466	.416	.367	.319	.272	.226	.180	.135	.090	.045	.000
54	.732	.671	.613	.556	.501	.447	.395	.343	.293	.243	.193	.145	.096	.048	.000
56	.788	.723	.660	.559	.540	.482	.425	.370	.315	.261	.208	.156	.104	.052	.000
58	.851	.781	.713	.647	.582	.520	.459	.399	.340	.282	.225	.168	.112	.056	.000
60	.921	.845	.771	.700	.630	.563	.497	.432	.368	.305	.243	.182	.121	.060	.000
62	1.00	.917	.837	.760	.685	.611	.539	.469	.400	.332	.264	.198	.132	.066	.000
64	1.09	1.00	.913	.828	.746	.666	.588	.511	.436	.362	.288	.215	.143	.072	.000
66	1.19	1.10	1.00	.907	.817	.730	.644	.560	.477	.396	.316	.236	.157	.078	.000
LHA	118° 242°	116° 244°	114° 246°	112° 248°	110° 250°	108° 252°	106° 254°	104° 256°	102° 258°	100° 260°	98° 262°	96° 264°	94° 266°	92° 268°	90° 270°

B
− if if latitude and declination have same name
+ if latitude and declination have different names

LHA	62° 298°	64° 296°	66° 294°	68° 292°	70° 290°	72° 288°	74° 286°	76° 284°	78° 282°	80° 280°	82° 278°	84° 276°	86° 274°	88° 272°	90° 270°
0	.000	.000	.000	.000	.000	.000	.000	.000	.000	.000	.000	.000	.000	.000	.000
3	.059	.058	.057	.057	.056	.055	.055	.054	.054	.053	.053	.053	.053	.052	.052
6	.119	.117	.115	.113	.112	.111	.109	.108	.107	.107	.106	.106	.105	.105	.105
9	.179	.176	.173	.171	.169	.167	.165	.163	.162	.161	.160	.159	.159	.158	.158
12	.241	.236	.233	.229	.226	.223	.221	.219	.217	.216	.215	.214	.213	.213	.213
15	.303	.298	.293	.289	.285	.282	.279	.276	.274	.272	.271	.269	.269	.268	.268
18	.368	.362	.356	.350	.346	.342	.338	.335	.332	.330	.328	.327	.326	.325	.325
21	.435	.427	.420	.414	.408	.404	.399	.396	.392	.390	.388	.386	.385	.384	.384
24	.504	.495	.487	.480	.474	.468	.463	.459	.455	.452	.450	.448	.446	.446	.445
27	.577	.567	.558	.550	.542	.536	.530	.525	.521	.517	.515	.512	.511	.510	.510
30	.654	.642	.632	.623	.614	.607	.601	.595	.590	.586	.583	.581	.579	.578	.577
33	.735	.723	.711	.700	.691	.683	.676	.669	.664	.659	.656	.653	.651	.650	.649
36	.823	.808	.795	.784	.773	.764	.756	.749	.743	.738	.734	.731	.728	.727	.727
38	.885	.869	.855	.843	.831	.821	.813	.805	.799	.793	.789	.786	.783	.782	.781
40	.950	.934	.919	.905	.893	.882	.873	.865	.858	.852	.847	.844	.841	.840	.839
42	1.02	1.00	.986	.971	.958	.947	.937	.928	.921	.914	.909	.905	.903	.901	.900
44	1.09	1.07	1.06	1.04	1.03	1.02	1.00	.995	.987	.981	.975	.971	.968	.966	.966
46	1.17	1.15	1.13	1.12	1.10	1.09	1.08	1.07	1.06	1.05	1.05	1.04	1.04	1.04	1.04
48	1.26	1.24	1.22	1.20	1.18	1.17	1.16	1.14	1.14	1.13	1.12	1.12	1.11	1.11	1.11
50	1.35	1.33	1.30	1.29	1.27	1.25	1.24	1.23	1.22	1.21	1.20	1.20	1.19	1.19	1.19
52	1.45	1.42	1.40	1.38	1.36	1.35	1.33	1.32	1.31	1.30	1.29	1.29	1.28	1.28	1.28
54	1.56	1.53	1.51	1.48	1.46	1.45	1.43	1.42	1.41	1.40	1.39	1.38	1.38	1.38	1.38
56	1.68	1.65	1.62	1.60	1.58	1.56	1.54	1.53	1.52	1.51	1.50	1.49	1.49	1.48	1.48
58	1.81	1.78	1.75	1.73	1.70	1.68	1.66	1.65	1.64	1.63	1.62	1.61	1.60	1.60	1.60
60	1.96	1.93	1.90	1.87	1.84	1.82	1.80	1.79	1.77	1.76	1.75	1.74	1.74	1.73	1.73
62	2.13	2.09	2.06	2.03	2.00	1.98	1.96	1.94	1.92	1.91	1.90	1.89	1.89	1.88	1.88
LHA	118° 242°	116° 244°	114° 246°	112° 248°	110° 250°	108° 252°	106° 254°	104° 256°	102° 258°	100° 260°	98° 262°	96° 264°	94° 266°	92° 268°	90° 270°

ABC TABLES

C (correction) = A ± B

C	0.00	0.05	0.10	0.15	0.20	0.25	0.30	0.35	0.40	0.45	0.50	0.55	0.60	0.70
					Azimuth (°)									
0	90.0	87.1	84.3	81.5	78.7	76.0	73.3	70.7	68.2	65.8	63.4	61.2	59.0	55.0
10	90.0	87.2	84.4	81.6	78.9	76.2	73.5	71.0	68.5	66.1	63.8	61.6	59.4	55.4
20	90.0	87.3	84.6	82.0	79.4	76.8	74.3	71.8	69.4	67.1	64.8	62.7	60.6	56.7
24	90.0	87.4	84.8	82.2	79.6	77.1	74.7	72.3	69.9	67.7	65.5	63.3	61.3	57.4
28	90.0	87.5	85.0	82.5	80.0	77.6	75.2	72.8	70.5	68.3	66.2	64.1	62.1	58.3
30	90.0	87.5	85.1	82.6	80.2	77.8	75.4	73.1	70.9	68.7	66.6	64.5	62.5	58.8
32	90.0	87.6	85.2	82.8	80.4	78.0	75.7	73.5	71.3	69.1	67.0	65.0	63.0	59.3
34	90.0	87.6	85.3	82.9	80.6	78.3	76.0	73.8	71.7	69.5	67.5	65.5	63.6	59.9
36	90.0	87.7	85.4	83.1	80.8	78.6	76.4	74.2	72.1	70.0	68.0	66.0	64.1	60.5
38	90.0	87.7	85.5	83.3	81.0	78.9	76.7	74.6	72.5	70.5	68.5	66.6	64.7	61.1
40	90.0	87.8	85.6	83.4	81.3	79.2	77.1	75.0	73.0	71.0	69.0	67.2	65.3	61.8
42	90.0	87.9	85.7	83.6	81.5	79.5	77.4	75.4	73.4	71.5	69.6	67.8	66.0	62.5
44	90.0	87.9	85.9	83.8	81.8	79.8	77.8	75.9	73.9	72.1	70.2	68.4	66.7	63.3
46	90.0	88.0	86.0	84.1	82.1	80.1	78.2	76.3	74.5	72.6	70.8	69.1	67.4	64.1
48	90.0	88.1	86.2	84.3	82.4	80.5	78.6	76.8	75.0	73.2	71.5	69.8	68.2	64.9
50	90.0	88.2	86.3	84.5	82.7	80.9	79.1	77.3	75.6	73.9	72.2	70.5	68.9	65.8
52	90.0	88.2	86.5	84.7	83.0	81.2	79.5	77.8	76.2	74.5	72.9	71.3	69.7	66.7
54	90.0	88.3	86.6	84.9	83.3	81.6	80.0	78.4	76.8	75.2	73.6	72.1	70.6	67.6
56	90.0	88.4	86.8	85.2	83.6	82.0	80.5	78.9	77.4	75.9	74.4	72.9	71.5	68.6
58	90.0	88.5	87.0	85.5	84.0	82.5	81.0	79.5	78.0	76.6	75.2	73.8	72.4	69.6
60	90.0	88.6	87.1	85.7	84.3	82.9	81.5	80.1	78.7	77.3	76.0	74.6	73.3	70.7
62	90.0	88.6	87.3	85.9	84.6	83.3	82.0	80.7	79.4	78.1	76.8	75.5	74.3	71.8
64	90.0	88.7	87.5	86.2	85.0	83.7	82.5	81.3	80.1	78.8	77.6	76.4	75.3	72.9
66	90.0	88.8	87.7	86.5	85.3	84.2	83.0	81.9	80.8	79.6	78.5	77.4	76.3	74.1
C	0.00	0.05	0.10	0.15	0.20	0.25	0.30	0.35	0.40	0.45	0.50	0.55	0.60	0.70

C	0.80	0.90	1.00	1.10	1.20	1.40	1.60	1.80	2.00	2.20	2.40	2.60	2.80
					Azimuth (°)								
0	51.3	48.0	45.0	42.3	39.8	35.5	32.0	29.1	26.6	24.4	22.6	21.0	19.7
10	51.8	48.4	45.4	42.7	40.2	36.0	32.4	29.4	26.9	24.8	22.9	21.3	19.9
20	53.1	49.8	46.8	44.1	41.6	37.2	33.6	30.6	28.0	25.8	23.9	22.3	20.8
24	53.8	50.6	47.6	44.9	42.4	38.0	34.4	31.3	28.7	26.5	24.5	22.8	21.4
28	54.8	51.5	48.6	45.8	43.3	39.0	35.3	32.2	29.5	27.2	25.3	23.5	22.0
30	55.3	52.1	49.1	46.4	43.9	39.5	35.8	32.7	30.0	27.7	25.7	23.9	22.4
32	55.8	52.7	49.7	47.0	44.5	40.1	36.4	33.2	30.5	28.2	26.2	24.4	22.8
34	56.4	53.3	50.3	47.6	45.1	40.7	37.0	33.8	31.1	28.7	26.7	24.9	23.3
36	57.1	53.9	51.0	48.3	45.8	41.4	37.7	34.5	31.7	29.3	27.3	25.4	23.8
38	57.8	54.7	51.8	49.1	46.6	42.2	38.4	35.2	32.4	30.0	27.9	26.0	24.4
40	58.5	55.4	52.5	49.9	47.4	43.0	39.2	36.0	33.1	30.7	28.5	26.7	25.0
42	59.3	56.2	53.4	50.7	48.3	43.9	40.1	36.8	33.9	31.5	29.3	27.4	25.7
44	60.1	57.1	54.3	51.6	49.2	44.8	41.0	37.7	34.8	32.3	30.1	28.1	26.4
46	60.9	58.0	55.2	52.6	50.2	45.8	42.0	38.7	35.7	33.2	31.0	29.0	27.2
48	61.8	58.9	56.2	53.6	51.2	46.9	43.0	39.7	36.8	34.2	31.9	29.9	28.1
50	62.8	60.0	57.3	54.7	52.4	48.0	44.2	40.8	37.9	35.3	33.0	30.9	29.1
52	63.8	61.0	58.4	55.9	53.5	49.2	45.4	42.1	39.1	36.4	34.1	32.0	30.1
54	64.8	62.1	59.6	57.1	54.8	50.6	46.8	43.4	40.4	37.7	35.3	33.2	31.3
56	65.9	63.3	60.8	58.4	56.1	51.9	48.2	44.8	41.8	39.1	36.7	34.5	32.6
58	67.0	64.5	62.1	59.8	57.5	53.4	49.7	46.4	43.4	40.6	38.2	36.0	34.0
60	68.2	65.8	63.4	61.2	59.0	55.0	51.3	48.0	45.0	42.3	39.8	37.6	35.5
62	69.4	67.1	64.9	62.7	60.6	56.7	53.1	49.8	46.8	44.1	41.6	39.3	37.3
64	70.7	68.5	66.3	64.3	62.3	58.5	55.0	51.7	48.8	46.0	43.5	41.3	39.2
66	72.0	69.9	67.9	65.9	64.0	60.3	56.9	53.8	50.9	48.2	45.7	43.4	41.3
C	0.80	0.90	1.00	1.10	1.20	1.40	1.60	1.80	2.00	2.20	2.40	2.60	2.80

Naming the azimuth: If answer is +, azimuth is **South** in north latitudes and **North** in south latitudes; if answer is −, azimuth is **North** in north latitudes and **South** in south latitudes. If hour angle is **less than 180°**, azimuth is **West**; if **more than 180°**, azimuth is **East.**

ABC TABLES

C (correction) = A ± B

C	3.20	3.60	4.00	4.50	5.00	6.00	7.00	8.00	9.00	10.0	15.0	20.0	40.0
						Azimuth (°)							
0	17.4	15.5	14.0	12.5	11.3	9.5	8.1	7.1	6.3	5.7	3.8	2.9	1.4
10	17.6	15.8	14.2	12.7	11.5	9.6	8.3	7.2	6.4	5.8	3.9	2.9	1.5
20	18.4	16.5	14.9	13.3	12.0	10.1	8.6	7.6	6.7	6.1	4.1	3.0	1.5
24	18.9	16.9	15.3	13.7	12.3	10.3	8.9	7.8	6.9	6.2	4.2	3.1	1.6
28	19.5	17.5	15.8	14.1	12.8	10.7	9.2	8.1	7.2	6.5	4.3	3.2	1.6
30	19.8	17.8	16.1	14.4	13.0	10.9	9.4	8.2	7.3	6.6	4.4	3.3	1.7
32	20.2	18.1	16.4	14.7	13.3	11.1	9.6	8.4	7.5	6.7	4.5	3.4	1.7
34	20.7	18.5	16.8	15.0	13.6	11.4	9.8	8.6	7.6	6.9	4.6	3.5	1.7
36	21.1	19.0	17.2	15.4	13.9	11.6	10.0	8.8	7.8	7.0	4.7	3.5	1.8
38	21.6	19.4	17.6	15.8	14.2	11.9	10.3	9.0	8.0	7.2	4.8	3.6	1.8
40	22.2	19.9	18.1	16.2	14.6	12.3	10.6	9.3	8.3	7.4	5.0	3.7	1.9
42	22.8	20.5	18.6	16.7	15.1	12.6	10.9	9.5	8.5	7.7	5.1	3.8	1.9
44	23.5	21.1	19.2	17.2	15.5	13.0	11.2	9.9	8.8	7.9	5.3	4.0	2.0
46	24.2	21.8	19.8	17.8	16.1	13.5	11.6	10.2	9.1	8.2	5.5	4.1	2.1
48	25.0	22.5	20.5	18.4	16.6	14.0	12.1	10.6	9.4	8.5	5.7	4.3	2.1
50	25.9	23.4	21.3	19.1	17.3	14.5	12.5	11.0	9.8	8.8	5.9	4.4	2.2
52	26.9	24.3	22.1	19.9	18.0	15.1	13.1	11.5	10.2	9.2	6.2	4.6	2.3
54	28.0	25.3	23.1	20.7	18.8	15.8	13.7	12.0	10.7	9.7	6.5	4.9	2.4
56	29.2	26.4	24.1	21.7	19.7	16.6	14.3	12.6	11.2	10.1	6.8	5.1	2.6
58	30.5	27.7	25.3	22.8	20.7	17.5	15.1	13.3	11.8	10.7	7.2	5.4	2.7
60	32.0	29.1	26.6	24.0	21.8	18.4	15.9	14.0	12.5	11.3	7.6	5.7	2.9
62	33.6	30.6	28.0	25.3	23.1	19.6	16.9	14.9	13.3	12.0	8.1	6.1	3.0
64	35.5	32.4	29.7	26.9	24.5	20.8	18.1	15.9	14.2	12.9	8.6	6.5	3.3
66	37.6	34.3	31.6	28.7	26.2	22.3	19.4	17.1	15.3	13.8	9.3	7.0	3.5
C	3.20	3.60	4.00	4.50	5.00	6.00	7.00	8.00	9.00	10.0	15.0	20.0	40.0

Naming the azimuth: If answer is +, azimuth is **South** in north latitudes and **North** in south latitudes. If answer is −, azimuth is **North** in north latitudes and **South** in south latitudes. If hour angle is **less than 180°**, azimuth is **West**; if **more than 180°**, azimuth is **East.**

STAR CHARTS

NORTHERN HEMISPHERE

SOUTHERN HEMISPHERE

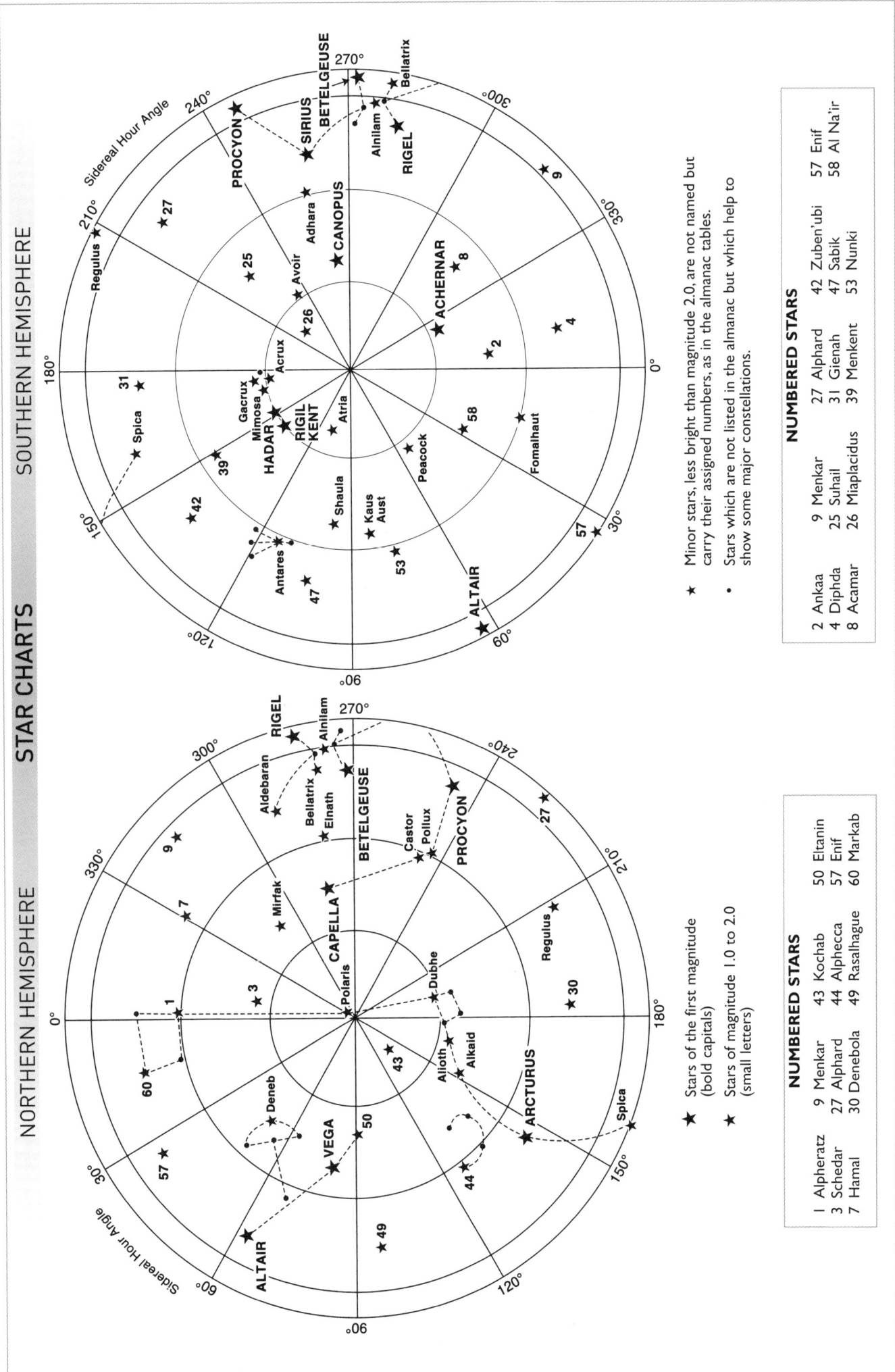

NUMBERED STARS

1 Alpheratz	9 Menkar	43 Kochab	50 Eltanin
3 Schedar	27 Alphard	44 Alphecca	57 Enif
7 Hamal	30 Denebola	49 Rasalhague	60 Markab

★ Stars of the first magnitude (bold capitals)

★ Stars of magnitude 1.0 to 2.0 (small letters)

★ Minor stars, less bright than magnitude 2.0, are not named but carry their assigned numbers, as in the almanac tables.

• Stars which are not listed in the almanac but which help to show some major constellations.

NUMBERED STARS

2 Ankaa	9 Menkar	27 Alphard	42 Zuben'ubi	57 Enif
4 Diphda	25 Suhail	31 Gienah	47 Sabik	58 Al Na'ir
8 Acamar	26 Miaplacidus	39 Menkent	53 Nunki	

68